MIX
Papier aus verantwortungsvollen Quellen
Paper from responsible sources
FSC® C105338

If you have any concerns about our products,
you can contact us on
ProductSafety@springernature.com

In case Publisher is established outside the EU,
the EU authorized representative is:
Springer Nature Customer Service Center GmbH
Europaplatz 3, 69115 Heidelberg, Germany

Printed by Libri Plureos GmbH
in Hamburg, Germany

Lecture Notes in Electrical Engineering

Volume 492

"Lecture Notes in Electrical Engineering (LNEE)" is a book series which reports the latest research and developments in Electrical Engineering, namely:

- Communication, Networks, and Information Theory
- Computer Engineering
- Signal, Image, Speech and Information Processing
- Circuits and Systems
- Bioengineering

LNEE publishes authored monographs and contributed volumes which present cutting edge research information as well as new perspectives on classical fields, while maintaining Springer's high standards of academic excellence. Also considered for publication are lecture materials, proceedings, and other related materials of exceptionally high quality and interest. The subject matter should be original and timely, reporting the latest research and developments in all areas of electrical engineering.

The audience for the books in LNEE consists of advanced level students, researchers, and industry professionals working at the forefront of their fields. Much like Springer's other Lecture Notes series, LNEE will be distributed through Springer's print and electronic publishing channels.

More information about this series at http://www.springer.com/series/7818

Daniel Thalmann · N Subhashini
K. Mohanaprasad · M S Bala Murugan
Editors

Intelligent Embedded Systems

Select Proceedings of ICNETS2, Volume II

Springer

Editors
Daniel Thalmann
Computer Graphics Lab
EPFL
Lausanne
Switzerland

K. Mohanaprasad
School of Electronics Engineering
VIT University
Chennai, Tamil Nadu
India

N Subhashini
School of Electronics Engineering
VIT University
Chennai, Tamil Nadu
India

M S Bala Murugan
School of Electronics Engineering
VIT University
Chennai, Tamil Nadu
India

ISSN 1876-1100 ISSN 1876-1119 (electronic)
Lecture Notes in Electrical Engineering
ISBN 978-981-10-8574-1 ISBN 978-981-10-8575-8 (eBook)
https://doi.org/10.1007/978-981-10-8575-8

Library of Congress Control Number: 2018933009

Printed on acid-free paper

This Springer imprint is published by the registered company Springer Nature Singapore Pte Ltd. part of Springer Nature
The registered company address is: 152 Beach Road, #21-01/04 Gateway East, Singapore 189721, Singapore

Preface

The first edition of the International Conference on NextGen Electronic Technologies: Silicon to Software (ICNETS2) was held at Vellore Institute of Technology, Chennai (VIT Chennai), during March 23–25, 2017. VIT being the highest ranked private engineering and research institution in India was the apt setting for the conference and its discussions on emerging technologies and its applications in both today's and tomorrow's society. The Intelligent Embedded Systems (Symposium-B) was one of the six symposia hosted by ICNETS2 and represented the upper layers of the computing stack culminating in system architectures as well as software design. This symposium was an avenue for researchers working in areas centered around Embedded Systems with more focus on Intelligence, to disseminate their research efforts to a wide audience including professionals and pioneers in the industry and academia. Intelligent Embedded Systems is the ability of a product or process to gain knowledge of the operational capabilities and limitations by itself and its ecosystem and utilize this knowledge in enhancing its functions and performance. This aspect was prominently at display during the symposium, which discussed both conceptual systems and also practical consumer-centric systems in a wide set of horizontals including smart health care, automotive systems focusing on safety, precision agriculture, smart grids, and consumer electronics. A total of 52 papers were presented during the 3 days of the symposium, which attracted researchers from around the globe both to present their work and to deliver invited talks in cutting-edge research areas in several aspects of Embedded Systems. All the participants, a significant majority of whom were students, had the opportunity to engage in many fruitful discussions that paved way to the success of the conference. Each day of the conference started with keynote addresses, and each session began with invited talks. The symposium was co-chaired by Dr. Daniel Thalmann (Director, Virtual Reality Lab, EPFL, Switzerland), who also gave a keynote address on the advances in Human–Computer Interaction. There were three invited talks covering the different system design and communication technologies: Eric Torres (Tata Communications) talked on the LoRaWAN standard for low-power wide-area networking in Internet of Things (IoT) applications, Adamu Murtala Zungeru (Botswana International University of Science and Technology) on underground sensor networks, and Elizabeth Chang (The University of New South Wales) on important aspects of trust and security on IoT. All in all, the very first ICNETS2 was very successful and set a milestone in conferences organized in India. The plenary lectures and other invited talks bridged the gap between aspects of Embedded Systems and also enthused participants to produce more engaging research work in the future. Also, since most of the participants were students, new perspectives were discussed on how to embed intelligence into systems and the practicality of such systems in various application domains. The next ICNETS2 will take place in VIT Chennai in 2019. Given the explosive growth of the Internet of Things and the associated analytics and machine learning components, Intelligence in Embedded Systems is bound to grow in leaps and bounds, and in the same vein, we expect that the future ICNETS2, and particularly the symposium on Intelligent Embedded Systems, will be as stimulating as the current one was, as indicated by the contributions presented in this proceedings volume.

Lausanne, Switzerland — Daniel Thalmann
Chennai, India — N Subhashini
Chennai, India — K. Mohanaprasad
Chennai, India — M S Bala Murugan

Acknowledgements

The organization of a conference, especially on the scale of ICNETS2, is as stressful as it is rewarding. Managing every detail, right down to the smallest issues, requires immense effort which would have been unlikely if not for the tremendous support rendered to the organizing committee. Organizing a conference covering all aspects of Electronics and Communication as part of six distinct symposia is not an easy task. On behalf of Symposium-B (Intelligent Embedded Systems), we would like to thank the management of VIT University, Chennai, for enabling the conference at every juncture. The conference would not have had its visibility is not for the partnership of Nadia Thalmann (Director, MIRALab, University of Geneva, Switzerland) and Daniel Thalmann (Director, VRlab, EPFL, Switzerland), who were co-chairs for the conference. We take this opportunity to thank SRS Prabaharan (Chair, ICNETS2) and Kanchana Bhaaskaran (Co-chair, ICNETS2), for their invaluable guidance in organizing our symposium. Thanks also go to the invited speakers, Elizabeth Chang, Eric Torres, and Adamu Murtala Zungeru, who shared their expertise to the eager participants of the symposium. We also thank the advisory committee, reviewers, and session chairs who took time off from their busy schedules to contribute to the symposium. The participants and presenters at the sessions only enhanced the quality of the symposium, and their willingness to publish their work as part of a fledgling symposium is highly appreciated. As the whole conference and our symposium was the result of well-coordinated teamwork, it is only fair that we acknowledge the efforts of the entire organizing team of more than 60 distinguished faculty members of the School of Electronics Engineering at VIT Chennai, who worked day and night for 8 months, way beyond their academic schedules. The contributions of the laboratory technicians and other maintenance staff at VIT Chennai should also be noted. Tata Communications, our corporate legend supporter, and Tenet Technetronics, our premium partner, provided the necessary technical and financial impetus in moving the conference forward. We also acknowledge the endorsements of ISRO, CSIR, INSA, and the Department of

State (USA) in increasing the conference visibility. Lastly, the acceptance of Springer to publish the proceedings of the conference and all its symposia established the credibility of the event, which aims at setting the standard for future conferences organized in the country.

Contents

About the Editors

Prof. Daniel Thalmann is a Swiss and Canadian Computer Scientist. He is one of the most highly cited scientists in Computer Graphics. He is currently Honorary Professor at EPFL, Switzerland, and was since 2009 with the Institute for Media Innovation, Nanyang Technological University, Singapore. Pioneer in research on Virtual Humans, his current research interests also include Social Robots, Crowd Simulation, and Virtual Reality. He has been the Founder of Virtual Reality Lab (VRlab) at EPFL, Switzerland; Professor at University of Montreal; and Visiting Professor/Researcher at CERN, University of Nebraska, University of Tokyo, and National University of Singapore. He is Co-Editor-in-Chief of the Journal of Computer Animation and Virtual Worlds and member of the editorial board of 12 other journals. He was Program Chair and Co-Chair of several conferences including IEEE-VR, ACM-VRST, and ACM-VRCAI. He has published more than 600 papers in Graphics, Animation, and Virtual Reality. He is co-editor of 30 books and co-author of several books including 'Crowd Simulation' (second edition 2012) and 'Stepping into Virtual Reality' (2007), published by Springer. He received his Ph.D. in Computer Science in 1977 from the University of Geneva and an Honorary Doctorate from Paul Sabatier University, Toulouse, France, in 2003. He also received the Eurographics Distinguished Career Award in 2010, the 2012 Canadian Human Computer Communications Society Achievement Award, and the CGI 2015 Career Achievement.

N Subhashini received her B.E. in Electronics and Communication Engineering from the University of Madras, Chennai. She obtained her master's degree in Systems Engineering and Operations Research from the College of Engineering, Guindy, Chennai. She was awarded a gold medal in PG for securing the highest rank in the University and also awarded a gold medal for being the Best Outgoing Student in the year 2006. With over 12 years of experience in teaching UG and PG students, she is currently working as Assistant Professor in the School of Electronics Engineering, Vellore Institute of Technology (VIT), Chennai. She has guided a number of UG and PG students in many projects. Her research interests include Optical Metro/Access Networks, Fiber-to-the-X (FTTX) Technologies,

Next-Generation Architectures and Services, Optical Fiber Technology and Wavelength-Division Multiplexing (WDM) Systems. She has presented and published a number of papers in peer-reviewed journals and conferences. She is currently working in optical networks and her research focuses on the design and performance evaluation of optical networks, especially passive optical networks. She is an active member of the Optical and Microwave Research Group in VIT, Chennai, and has organized several workshops and seminars in the field of optical networks.

Dr. K. Mohanaprasad graduated from VIT University, Vellore, Tamil Nadu, India, in the field of Signal Processing. He is currently associated with the School of Electronics Engineering, VIT University, Chennai, Tamil Nadu, India, as Associate Professor. His major research interests are in the areas of Speech Processing, Signal Processing, Wavelet Transform, Image Processing, and Biomedical Signal Processing. He has co-authored a chapter and published over 20 reputed international journal/conference papers, several of them winning best paper awards. He is a regular reviewer of several top signal processing journals.

M S Bala Murugan has 10 years of experience in teaching industry, with a year of research experience at Centre for Electronics Design and Technology (CEDT), IISc, Bangalore, India. He is currently working as Assistant Professor at VIT University, Chennai, and also chairs the IoT working committee in the School of Electronics Engineering. His areas of interest include Embedded Systems and IoT. He has published ten papers in international conferences and journals. His recent research includes deploying real-time operating system (RTOS) in heterogeneous multicore architectures and multistandard protocol gateway for IoT.

Design and Implementation of Dialysate Temperature Control System for Hemodialysis: A Pilot Study

Mohamed Haroon Abdul Jabbar, S. Anandan Shanmugam and Poi Sim Khiew

Abstract In hemodialysis, the control of body temperature by altering the dialysate temperature would reduce the intradialytic complications. Several studies show that the constant dialysate temperature affects the patient's quality of life due to their different temperature threshold. Thus, these factors serve as a motivation factor to design an individualized dialysate temperature control for hemodialysis patients, which can actively control the body temperature even in the case of any external disturbances. In this paper, a novel dialysate proportioning model has been proposed. Then, initial implementation of proposed model was designed with fuzzy logic control and implemented on a low-cost microcontroller—Raspberry Pi 3. A Simulink model was also designed by incorporating fuzzy logic control to implement in real time. The pumps' flow rates are varied using Pulse Width Modulation (PWM) according to the controller signal, while the temperature sensors are placed to acquire actual temperature in this model. Subsequently, it has been tested and verified by comparing the simulation and experimental results. Furthermore, the dialysate temperature trend was studied for various input conditions to analyze its controller behavior in real-time implementation. The results showed the potential to develop robust control by optimizing the fuzzy rules and membership functions. The system response time is found to be minimal (less than 300 ms), and the performance error is acceptable (less than 0.55%). Further work is ongoing to implement the dialysate temperature controller incorporating in vitro studies.

Keywords Hemodialysis · Temperature controller · Fuzzy control system Raspberry Pi · MATLAB/Simulink

---

M. H. A. Jabbar (✉) · S. Anandan Shanmugam
Department of Electrical and Electronic Engineering, University of Nottingham Malaysia Campus, Semenyih, Malaysia
e-mail: kecx4mhr@nottingham.edu.my

P. S. Khiew
Division of Materials, Mechanics and Structures, University of Nottingham Malaysia Campus, Semenyih, Malaysia

D. Thalmann et al. (eds.), *Intelligent Embedded Systems*, Lecture Notes in Electrical Engineering 492, https://doi.org/10.1007/978-981-10-8575-8_1

1 Introduction

Over many years, hemodialysis (HD) had been recognized as most effective treatment for patients suffering from kidney failure. Yet, it has been associated with frequent intradialytic complications, while intradialytic hypotension (IDH) remains the most common complication in HD [1]. In addition to toxin clearance, there is also a heat transfer taking place in dialyzer other than the heat loss from the blood tubing to the environment. This tends to fluctuate body temperature during HD, subsequently interrupting the patient's thermal equilibrium. However, if the core temperature changes beyond a critical threshold, the increase in the thermoregulatory mechanisms will lead to IDH and an increased risk of morbidity. Likewise, the temperature threshold differs in individual patients. A long-term study showed that the highest mortality was observed in patients whose post-dialysis body temperature increased or decreased, irrespective of baseline body temperature [2]. This serves as strong evidence in mortality due to fluctuation in body temperature during HD. Therefore, the control of body temperature plays a vital role in the onset of hypotension.

The most common practice to control body temperature is to alter the dialysate temperature in extracorporeal circuit, which was first described by Maggiore et al. in the 1980s [3]. Moreover, maintaining the dialysate temperature within the physiological range is vital for patients' safety. Until recently, a dialysate temperature of 37 °C was considered as standard temperature, which is somewhat higher than the average physiological body temperature. Hence, warm dialysate (37–37.5 °C) frequently causes an increase in body temperature of approximately 0.3–0.5 °C [4]. In contrary, studies have confirmed that mild cool dialysate improves the hemodynamic stability compared to warm dialysate [5, 6]. But the use of cool dialysate in the range of 35–35.5 °C showed unpleasant effects in some patients such as shivering and cold sensation [7]. These tolerability can be optimized by an individualized approach to dialysate temperature prescription.

Until now, there has been only one commercially available system in hemodialysis machine that was able to measure and control body temperature—Blood Temperature Monitor (BTM) by Fresenius, Germany. The control of BTM regulates the temperature of the dialysate to compensate for increase or decrease in body temperature. Even though, it shows impressive improvements in hemodynamic stability during the treatment, there is a possibility for further improvement. In the recent past, studies have been published on dialysate temperature control system with the intention to minimize complications during the treatment [8, 9]. Based on these literatures, it is probably best to prevent a fluctuation in body temperature during HD, when concerned about optimal dialysate temperature. It can also be seen that the idea of active regulation of dialysate temperature according to the patients' body temperature is much needed for our current society. The control which does not require the predialysis body temperature prescription to be inputted unlike BTM would be one of the major developments in hemodialysis machine technology. The fuzzy logic control (FLC) system is found to be the most suitable adaptive controller for this application due to its decision-making capability.

In this paper, a novel dialysate proportioning model has been proposed for the effective temperature control. So, the desired dialysate temperature can be achieved by a suitable controller. Later, a preliminary controller model was designed with fuzzy logic control and then implemented on a low-cost microcontroller—Raspberry Pi 3. The main aim of this pilot study was to analyze the performance and verify the controller behavior in real-time environment. Thereby, the same strategy can be further applied to implement in full-fledged dialysate model.

2 Proposed Model

The model which consists of two dialysate tanks at constant temperature of 35 and 37 °C, respectively, has been proposed for the benefit of temperature controller as shown in Fig. 1. As the dialysate temperature range is narrow (35–37 °C), the efficient way to control the temperature is by varying the flow rates using peristaltic pumps. Contrary, the control of temperature can also be made possible by implementing heating elements through the tubing, which would be ineffective in active regulation. Moreover, the response time and accuracy of this narrow temperature range would be challenging to control using heating element. Hence, the control of

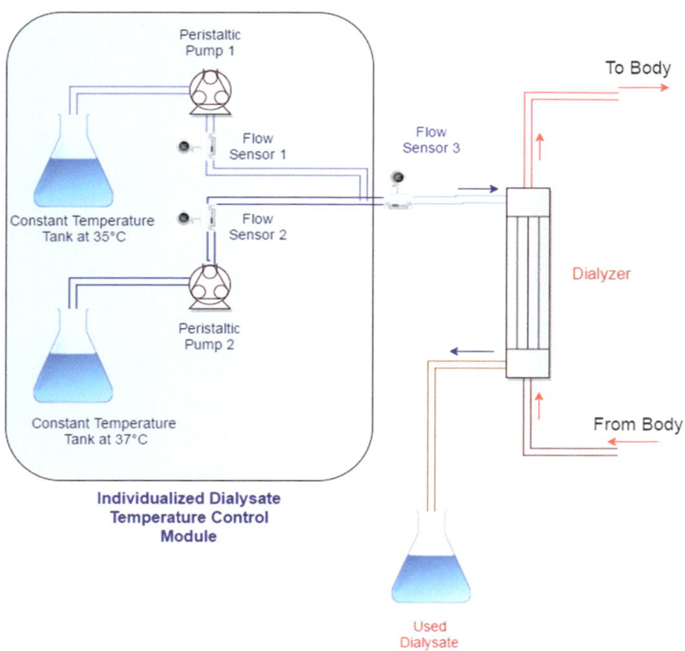

Fig. 1 Proposed model for individualized dialysate temperature control module

peristaltic pumps by varying flow rates would be superior to conventional heaters for this application.

This dialysate proportioning model was designed as an external module that can interface in existing HD machines. This module consists of three temperature sensors, three flow sensors, and two peristaltic pumps to ensure the dialysate flow to dialyzer in an efficient way. The temperature sensors were placed in each temperature tanks and at the inlet of dialyzer, while the flow sensors were positioned to monitor the flow rates of both tanks and final flow rate to dialyzer. Since the efficiency of dialysis depends on dialysate flow rate [10], extra care needs to be taken to control the final dialysate flow rate. Therefore, the required dialysate temperature can be achieved using a robust controller, which controls body temperature in HD by regulating the two dialysate flow rates effectively.

3 Implementation Design

As this is an ongoing project, this paper presents a pilot model of the proposed design, consisting of two tanks with temperature sensor, a peristaltic pump with encoder and Raspberry Pi as shown in Fig. 2. Raspberry Pi 3 model B was selected as the microcontroller in this design due to its high processing power and peripheral interface. The other components such as sensors and actuator were interfaced to Raspberry Pi using the predefined functions provided by MATLAB/Simulink. A 12-V peristaltic pump was chosen, which can drive the fluid up to 400 mL/min suitable for this application. However, a switching circuit was constructed to create the interface between the pump and Raspberry Pi. To measure the flow rate non-invasively, a 3D-printed encoder wheel with 16 evenly spaced holes and

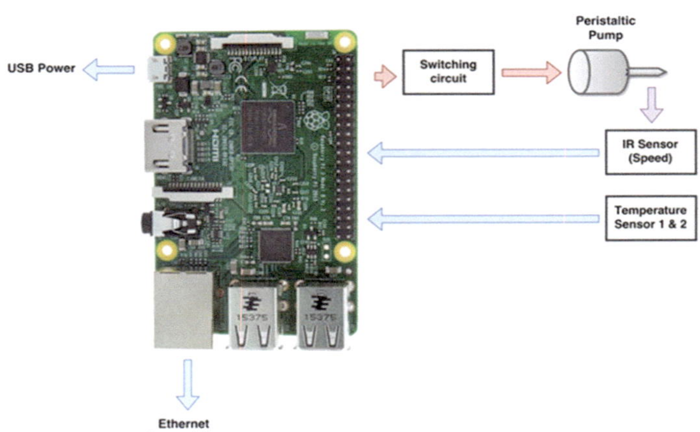

Fig. 2 Block diagram of the implementation design

infrared sensor are attached to the shaft of the DC motor. An additional limitation is that the Raspberry Pi does not have an in-built analog-to-digital converter (ADC). Hence, two DS18B20 digital temperature sensors were selected, which are accurate and waterproof for this model. Meanwhile, water was used to mimic dialysate fluid in this study as it is the major component in dialysate.

3.1 Fuzzy Logic Control Design

In this study, a multiple-input and single-output (MISO) fuzzy logic control was designed using fuzzy logic toolbox in MATLAB. In hemodialysis, the input parameters should be capable of reflecting the hemodynamic status of patient and should be measurable by using non-invasive sensors, while the output parameters should be adjustable. Our earlier simulation study showed the comparison of Takagi–Sugeno and Mamdani fuzzy inference techniques. Accordingly, this paper focuses on Mamdani model due to its high reliability. The inputs 'TS1' and 'TS2' denote the two temperature sensors, while the output 'Pump' denotes the PWM value to vary the speed of pump. By considering these properties, membership functions were defined for each of the input and output variables as shown in Fig. 3.

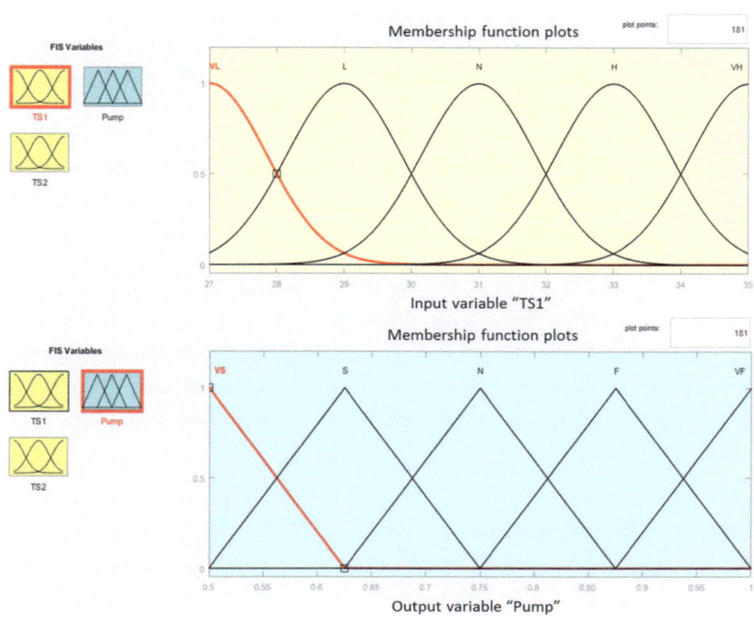

Fig. 3 Membership function for input (above) and output (below)

Table 1 Rule base of the implementation design

TS1	TS2				
	Very low (VL)	Low (L)	Normal (N)	High (H)	Very high (VH)
VL	VS	VS	F	VF	VF
L	S	S	F	F	VF
N	F	N	N	N	F
H	VF	F	S	S	S
VH	VF	VF	S	VS	VS

One of the most important factors that depend on the efficiency of controller is based on fuzzy rule base. In this system, rule base was defined randomly with the purpose of analyzing its behavior in the hardware. Hence, a total of 25 rules were created for this initial implementation design as shown in Table 1. In future, by analyzing its performance, the rules can be continuously added to improve the accuracy of this system.

3.2 *Simulink Model*

Simulink provides the environment to deploy it to hardware directly and also able to run it in external mode. This makes the fuzzy logic control implementation effective using fuzzy logic control Simulink block. The MATLAB functions were used to read the temperature from the sensors using 1-wire communication bus, whereas the PWM signals were generated using SFunction block along with WiringPi libraries for the pump output. Since infrared sensor is a digital sensor, a predefined Raspberry Pi block was used to read the signals based on the speed. This speed was then calibrated with known tachometer and flow meter to convert the rpm to flow rate. The overall Simulink model is shown in Fig. 4.

Then, the temperature trend was studied for various input conditions to analyze the fuzzy logic controller performance. Further tests were carried out to verify the results between the simulation and experimental PWM values for random inputs with the help of oscilloscope. Moreover, the time response was also analyzed by initializing trigger blocks at random inputs and corresponding outputs.

4 Results and Discussion

The Simulink model was simulated and deployed to Raspberry Pi by defining its unique IP address. The fuzzy rule viewer and other scope were able to monitor the corresponding readings while running the controller. This helps to analyze the overall behavior of the model. In external mode, the temperature and flow rate

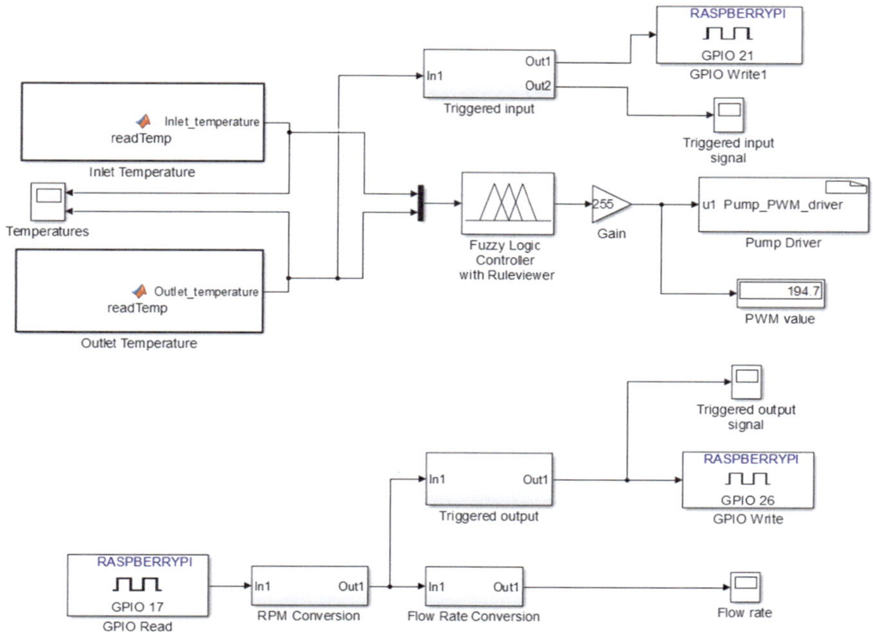

Fig. 4 Simulink model of the implementation design

reading were analyzed, and it can be seen that the flow rate varies with the change in temperature in agreement with fuzzy rule as shown in Fig. 5. Additionally, it shows that the output has the potential to adapt to various operating conditions and disturbances. Thus, the fuzzy logic control allows more flexibility to the changing environment.

To further strengthen the effective implementation, the simulated results of PWM must show a strong agreement with its experimental results. As the output from fuzzy logic is PWM signal, it is best practice to compare the raw data with the intention to verify the fuzzy logic implementation. The comparison of these results was shown in Table 2. It was confirmed that the error between the actual and simulation results is quite negligible. However, the maximum error was found to be 0.55%, which can be regarded as acceptable.

The response time is considered as one of the vital parameter in hemodialysis machine technology. This is to ensure that the output changes when the input triggers within a fraction of second. In this design, the response time was analyzed by increasing and decreasing the defined trigger input values. The change in state and time taken to reach 235 rpm when increasing and decreasing the temperature is shown in Fig. 6. Overall, the response time for random inputs was shown in Table 3. The maximum time taken to alter the speed of pump with increased input is 300 ms, while the maximum when decreased is 270 ms. It was also noticed that

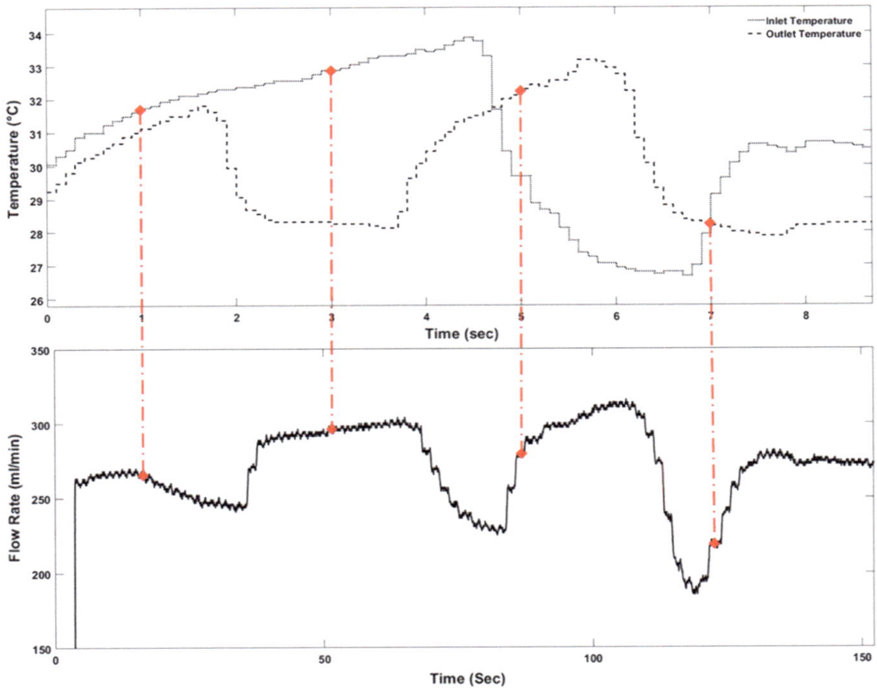

Fig. 5 Output of temperatures (above) and flow rate (below) from Simulink

Table 2 Validation of output with simulation and experimental results

TS1 (°C)	TS2 (°C)	Simulation	Experiment		Relative error (%)
		PWM value	Duty cycle	PWM value	
27	34.5	240.5	0.94	239.2	0.55
28	31.1	216.2	0.84	215.2	0.48
30	31.1	204.9	0.80	204.0	0.44
30.6	30.4	191.3	0.75	191.0	0.16
29.8	28.6	179.6	0.70	179.0	0.33
34.5	32.2	152.8	0.60	152.0	0.52
34.5	34.5	148	0.58	148.0	0.00

the rapid change in the input results in faster response time to trigger the output value.

The real-time implementation of fuzzy logic on Raspberry Pi showed us the possibility of controlling the hemodynamic parameters during hemodialysis. These results motivate us to develop the proposed model of dialysate temperature control. The improvements in fuzzy control design have high potential to make it a robust control by optimizing the fuzzy rules and membership functions. Therefore, our

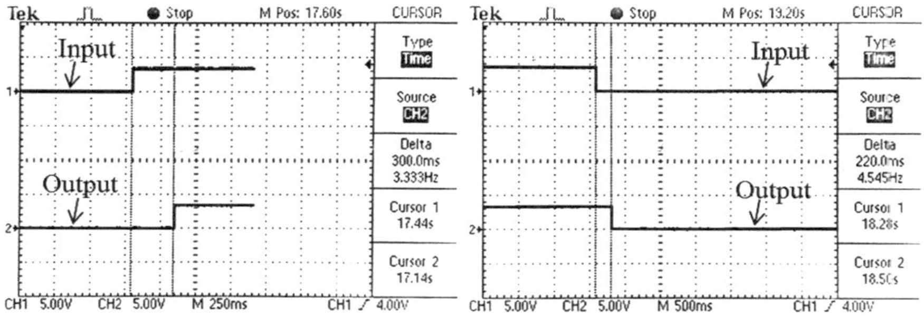

Fig. 6 Response time for the output at 235 rpm when input increasing (left) and decreasing (right)

Table 3 Time response for various increasing and decreasing inputs

TS1 (°C)	TS2 (°C)	Triggered RPM	Temperature (increase/ decrease)	Time response (in ms)
29.1	27.3	215	Increase	300
			Decrease	270
27.3	30.4	235	Increase	300
			Decrease	220
27.6	32.5	255	Increase	280
			Decrease	160

future study shall implement fuzzy logic control system on dialysate temperature biofeedback system to maintain stable body temperature during hemodialysis.

5 Conclusion

The body temperature is one of the vital parameters to be controlled during hemodialysis. Thus, an innovative design for the active regulation of dialysate temperature by varying flow rates has been proposed. Moreover, it is evident that the fuzzy logic control system has been successfully implemented in a low-cost microcontroller—Raspberry Pi 3. It also showed that the performance is encouraging with faster response time (less than 300 ms) and minimum error (less than 0.55%). Further work is ongoing to develop the proposed dialysate model and its in vitro studies.

Acknowledgements This work was supported by Faculty of Engineering, University of Nottingham, Malaysia Campus.

References

1. Bradshaw W, Bennett PN (2015) Asymptomatic intradialytic hypotension: the need for pre-emptive intervention. Nephrol Nurs J 42(5):479–485 (quiz 486)
2. Usvyat LA et al (2012) Relation between trends in body temperature and outcome in incident hemodialysis patients. Nephrol Dial Transplant 27(8):3255–3263
3. Maggiore Q et al (1982) Blood temperature and vascular stability during hemodialysis and hemofiltration. Trans Am Soc Artif Intern Organs 28(1):523–527
4. Pergola PE, Habiba NM, Johnson JM (2004) Body temperature regulation during hemodialysis in long-term patients: Is it time to change dialysate temperature prescription? Am J Kidney Dis 44(1):155–165
5. Korkor AB, Bretzmann CM, Eastwood D (2010) Effect of dialysate temperature on intradialytic hypotension. Dial Transplant 39(9):377–385
6. Chesterton LJ, Selby NM, Burton JO, McIntyre CW (2009) Cool dialysate reduces asymptomatic intradialytic hypotension and increases baroreflex variability. Hemodial Int 13 (2):189–196
7. Van Der Sande FM et al (2009) Control of core temperature and blood pressure stability during hemodialysis. Clin J Am Soc Nephrol 4(1):93–98
8. Busono Ario PF, Handoyo T, Barkah A, Suryana Y, Riyanto R, Febryarto R (2015) Development of fuzzy logic based temperature controller for dialysate preparation system. In: Proceeding of the electrical engineering computer science and informatics, 2015, vol 2. EECSI, pp 264–268
9. De Capua C, Fabbiano L, Morello R, Vacca G (2014) Optimized procedure to evaluate the thermal energy transfer in hemodialysis treatment. Instrum Sci Technol 42(4):458–468
10. Albalate M et al (2015) Is it useful to increase dialysate flow rate to improve the delivered Kt? BMC Nephrol 16(20):1–6

Raspberry Pi in Computer Science and Engineering Education

S. Alex David, S. Ravikumar and A. Rizwana Parveen

Abstract Sustainable Environmental Development is one of the hot topics nowadays. Industries were advised to reduce and control pollution by using pollution control equipment. Computers are also contributing some amount in the pollution and power consumption. On the other hand, miniaturization, less power consumption, and environment-friendly devices are invented by the researchers every day. One such device is Raspberry Pi. Size of the Raspberry Pi is not bigger than a credit card with high computing capacity and low power consumption. Pi can run Linux as well as Windows 10 in its higher versions. This paper aims to suggest the Pi can be used for practicing most of the laboratory courses in the computer science engineering curriculum. Most laboratory courses are practiced in C, C++, and Java languages. Apart from above-mentioned languages, some laboratory courses make use of front end and back end tools. Above-mentioned languages can be executed in Raspberry Pi. Following points are analyzed between normal computer and Pi, execution time, power consumption, and environmental effect. In all the comparison, Pi gives much higher advantage over existing system.

Keywords Raspberry Pi · Engineering education · Eco-friendly computer

S. Alex David (✉) · S. Ravikumar · A. Rizwana Parveen
Department of Computer Science and Engineering, Vel Tech University,
Avadi, Chennai, India
e-mail: alex_art2002@yahoo.co.uk

S. Ravikumar
e-mail: ravikumar.086@gmail.com

A. Rizwana Parveen
e-mail: rizwanaparveen64@gmail.com

D. Thalmann et al. (eds.), *Intelligent Embedded Systems*, Lecture Notes
in Electrical Engineering 492, https://doi.org/10.1007/978-981-10-8575-8_2

1 Introduction

Computers are much useful in the education. Most of the complex problems can be expressed in simple manner with a help of computers. For example, in childhood everybody learned the poem by seeing the book with some pictures. Reading and understanding was made easy with the help of those pictures. Later when animation was introduced in the education, it takes the understanding one step higher. In school days, many of us who studied during the 1990s used the picture to understand the working heart. Now, the same has been available as animated video shows the flow of blood, arrangements of nerves clearly; hence, the complex structures become very easy to understand. Nowadays, the miniaturization, cost-effective, pollution controlled, or environment-friendly device attracted the researchers over past one decade. The size of the first computer occupies a room, and then the changes comes in the hardware technologies reduced the size of the computer to hold with in the palm. A team of students from University of Cambridge invented a credit card size computer in the year 2012. It competes in the computer world to make a place permanently in this field. It is cheap, compact and more computing power compare to other computers. The name of this credit card size computer is "Raspberry Pi" [1].

This paper aims to give some idea about Raspberry Pi versions. What role Raspberry Pi can play in the educational institute in India. The following chapters discussed the laboratory courses in the curriculum in major intuitions in India, the software which is used in each laboratory and what laboratory courses can make use of Raspberry Pi [2]. The advantages of Raspberry Pi over the other computers are discussed in the conclusion.

2 Raspberry Pi an Intro

Raspberry Pi has an ARM processor as a computing component in single board, which makes use of low power to boot and run and occupies the size not more than the size of a credit card. Raspberry Pi comes in many versions: Raspberry Pi A, A+, B, B+, and Pi 2. At present, A+, B+, and Pi 2 are available in the market. Remaining were versions stopped from the production. Raspberry Pi can be connected to TV through HDMI or RCA. It has a computing power equal to few years back UNIX workstation. Many operating systems come based on Linux kernel and in last version; i.e., Raspberry Pi 2 can run Windows 10 operating system. Other hardware can be interfaced with Raspberry Pi with I/O pins which makes the Raspberry Pi as server. It can be used as a super computer by clustering number of Raspberry Pi's. Graphics capabilities are more or less equal to some Xbox which comes during 2000–2001. Model A versions require very low power and it can be used in school level embedded project. Model B version can be used in many areas like development, high computing environment [3].

Fig. 1 Raspberry Pi 2

All versions (except Raspberry Pi 2) of the RPi are based on the SoC Broadcom BCM2835; it has an ARM CPU together with a VideoCore IV GPU. The RAM is shared between the CPU and the GPU. The model A has 256 Mbytes of RAM, and model B/B+ has 512 Mbytes [4].

The Raspberry Pi 2 is based on Broadcom BCM2836 system on chip (SoC) that has 1 GB RAM (Fig. 1).

Raspberry Pi 2 has the following features 4 USB ports, 40 GPIO pins, full HDMI port, Ethernet port, combined 3.5 mm audio jack and composite video, camera interface (CSI), display interface (DSI), micro SD card slot, VideoCore IV 3D graphics core. Comparison between all B models is listed below.

SoC

Model B: Broadcom BCM2835 SoC
Model B+: Broadcom BCM2835 SoC Model Pi 2: BroadcomBCM2836 SoC

CPU

Model B: 700 MHz single-core ARM1176JZF-S
Model B+: 700 MHz single-coreARM1176JZF-S Model Pi 2: 900 MHz quad-core ARMCortex-A

GPU

Model B: Broadcom VideoCore IV @250 MHz, OpenGL ES 2.0
Model B: Broadcom VideoCore IV @250 MHz, OpenGL ES 2.0
Model Pi 2: Broadcom VideoCore IV @250 MHz, OpenGL ES 2.0

Primary Memory (SDRAM)

Model B: 256 MB in first version of model B, 512 MB SDRAM@ 400 MHz in the second version of model B
Model B+: 512 MB SDRAM@ 400 MHz
Model Pi 2: 1 GB SDRAM@ 400 MHz

Storage
Model B+: SD/MMC/SDIO card slot
Model B+: Micro SD
Model Pi 2: Micro SD

USB Ports
Model B: 2
Model B+: 2
Model Pi 2: 4

GPIO
Model B: 26 pin
Model B+: 40 pin
Model Pi 2: 40 pin

Power
Model B: 700 mA
Model B+: 600 mA
Model Pi 2: 800 mA

Operating system
Model B: Linux, OpenELEC, XBMC, RetroPie, RISC OS, Firefox OS, Plan 9, Android

Model B+: Linux, OpenELEC, XBMC, RetroPie, RISC OS, Firefox OS, Plan 9, Android

Model Pi 2: Linux, OpenELEC, XBMC, RetroPie, RISC OS, Firefox OS, Plan 9, Android, Windows 10.

Among all the B models, only Pi 2 has additional support that it can run Windows 10 operating system.

3 Laboratory Courses in Computer Engineering

Bench marking of computer laboratory courses in India is shown in Table 1.

From the above benchmarking among the universities and top engineering college, it is clearly showing most of the practical executed in C or C++ environment. Most of the programs have been executed in Pi, and the average execution time has been calculated in order to compare with other existing system [5].

Computer Practice Laboratory: In all the engineering degree programs, this course is mandatory. Every freshman will undergo this course. Basic concepts in the C programming will be taught and practiced in the laboratory sessions [6].

Data Structures Laboratory: In this course, learners will practice all the experiment in C or C++.

Table 1 Computer laboratory courses in engineering

S. No.	Practical name	Anna University, Chennai, Tamil Nadu	JNTU, Hydrabed, Telungana/AP	NIT Silchar, Assam	SRM University, Chennai, Tamil Nadu/ Delhi	Jadavpur University. Kolkata, West Bengal
1	Computer Practices Laboratory	C/C++ language	C/C++ language	C/C++ language	C/C++ language	C/C++ language
2	Data Structures Lab	”	”	”	”	”
3	Operating Systems Lab	”	”	”	”	”
4	Computer Networks Lab	”	”	”	”	”
5	Compiler Llaboratory	C, YACC and Lex	C, YACC and Lex	C, YACC and Lex	C, YACC and Lex	C, YACC and Lex
6	Computer Graphics Laboratory	C/C++ language	C/C++ language	C/C++ language	C/C++ language	C/C++ language
7	Java Programming	JDK	JDK	JDK	JDK	JDK
8	Data Base Management System	MySQL/Oracle	MySQL/Oracle	MySQL/Oracle	MySQL/Oracle	MySQL/Oracle
9	Web Technology Laboratory	HTML, Applet, Java Script	HTML, Applet, Java Script	HTML, Applet, Java Script	HTML, Applet, Java Script	HTML, Applet, Java Script
10	CASE Tools Lab	Rational suite open source alternatives: ArgoUML	Rational rose	Rational rose	Rational rose	Rational rose
11	Visual Programming	Microsoft visual studio	Microsoft visual studio	Microsoft visual studio	Microsoft visual studio	Microsoft visual studio
12	Mobile Applications Development Lab	ADK, iOS development kit, JAVA	ADK, iOS development kit, JAVA	ADK, iOS development kit, JAVA	ADK, iOS development kit, JAVA	ADK, iOS development kit, JAVA

Operating Systems Laboratory: Shell programming and C programming will be executed in the Linux environment.

Compiler Laboratory: Various phases of the compiler will be executed in the Linux C environment.

Computer Graphics Laboratory: Graphical library had been used to experiment the basic graphics concepts such as 2D primitives and 2D, 3D transformation.

Java Programming Laboratory: Basic Java concepts will be executed in the Java Development Kit (JDK).

Web Technology Laboratory: Basic concepts of HTML and Java Script will be practiced in this course.

All the above courses can be executed in Raspberry Pi B model. Some of the programs were executed in high-end computer and Raspberry Pi in order to compare the execution time. Most of the programs had the execution time more or less equal to the high-end system.

A. *Advantages of using Raspberry Pi*:

- Power consumption very less than high-end system
- Cost: Very less than high-end system
- Environment: Raspberry Pi produces less heat than high-end system.

4 Conclusion

Raspberry Pi has a computing capacity equal to the high-end systems. It can be used for most of the laboratory courses in the engineering studies. Comparison on execution time has been made between high-end system and Raspberry Pi. Results clearly show that Raspberry Pi can be used for laboratory courses. In many ways, Raspberry Pi shows advantage over existing laboratory system.

References

1. Alex David S, Grace Priyanka J (2014) Study on Raspberry Pi. IJMEIT 2(7) July 2014
2. Paramanathan A, Pahlevani P, Thorsteinsson S, Hundebøll M, Lucani DE, Fitzek FH Sharing the Pi: testbed description and performance evaluation of network coding on the Raspberry Pi
3. Bruce RF, Brock JD, Reiser SL (2015) Make space for the pi. IEEE SoutheastCon, 9–12 Apr 2015, Fort Lauderdale, Florida
4. The Making of Pi [Online]. Available: http://www.raspberrypi.org/about/. Accessed 22 Jan 2015
5. Sharma A, Williams B (2014) Build your own supercomputer at home with Raspberry Pi computers. In Proceedings of the southern association for information systems conference, Macon, GA, 21–22 Mar 2014
6. Cunningham J (2014, February 28) Tech in the classroom: Raspberry Pi, (Education World), [Online]. Available: http://www.educationworld.com/a_tech/tech-inthe-classroom/raspberry-pi.shtml. Accessed 23 Jan 2015

Advanced Tele-surgery with IoT Approach

N. Shabana and G. Velmathi

Abstract Availability of special doctors during emergency surgeries is decreasing. It is very vital to assist doctors in their surgeries of surgical tools. As a solution to this problem, tele-surgery concept prevails and many simulators have been developed for education, planning rehearsal, and so on. Some of the defects in the existing systems are interaction between the surgeon and the remote surgical room, display of the image, camera motion, data transfer quality, and accuracy in the motion capturing. These issues are overcome in this approach with an enhanced vision-oriented VR goggle and hand motion acquisition system (a control system) accessed by the surgeon. Data transfer is done using IoT which makes this system stand superior to the other approaches of tele-surgery.

Keywords 3D robotic surgery · VR goggle · Data transfer quality
Sensor UDP · IoT · Hand acquisition system

1 Introduction

Tele-surgery or surgery paves way for revolutionary in the field of health care today. This is made possible by the assistance of the robots which are programmed to mimic the work done by the surgeons during the surgery [1]. The surgeon's presence is made unnecessary because of this method. There are multiple advantages and benefits because of this remote robot-assisted surgery [2, 3]. Geographical distance will no longer be a major constraint to perform surgeries in the absence of surgeons. There are some major problems such as inaccuracy, delay, bulky structure, freedom for robotic arm, data transmission from the surgeon's section, and other physical issues. The enumerated mechanical advantages of robotic surgery

N. Shabana (✉) · G. Velmathi
VIT University, Chennai Campus, Chennai, Tamil Nadu, India
e-mail: Shabananizam3@gmail.com

G. Velmathi
e-mail: velmathi.g@vit.ac.in

D. Thalmann et al. (eds.), *Intelligent Embedded Systems*, Lecture Notes
in Electrical Engineering 492, https://doi.org/10.1007/978-981-10-8575-8_3

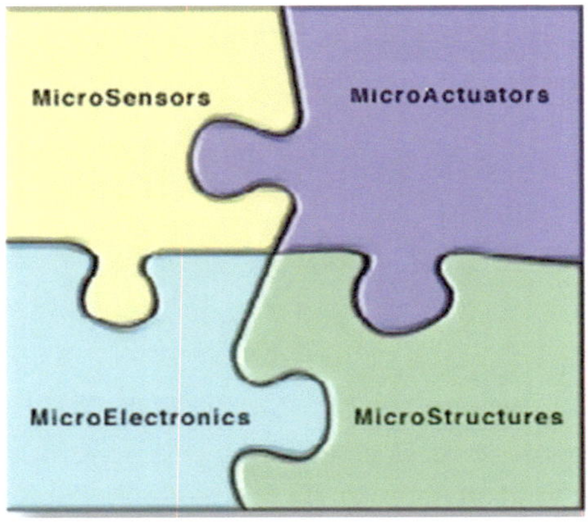

Fig. 1 MEMS components

[4, 5] are—removing tremor, elevated fidelity of hand gesticulation dropsical scrutiny, revived dexterity, and remote surgery [5–7]. MEMS is the amalgamation of sensors, mechanical elements, actuators, sensors, and electronics stowed on a common silicon substrate by the virtue of microfabrication technology. Computer chip fabrication [8] technique is originally used. MEMS exists as a conventional technology in today's world, by forming the best platform for miniaturization of devices and production of cost-efficient products (Fig. 1).

The MEMS is well developed and established in field of the miniaturization. The field of MEMS is an extensional block of the strongly laid foundation of sensors and actuators. MEMS is a valid outcome of a long-term prominence, given to the concept of miniaturization. MEMS technologies are becoming dependent on nanotechnologies for successful new products, and their application extends to different fields with more advantages compared to the other upcoming miniaturization techniques (Fig. 2).

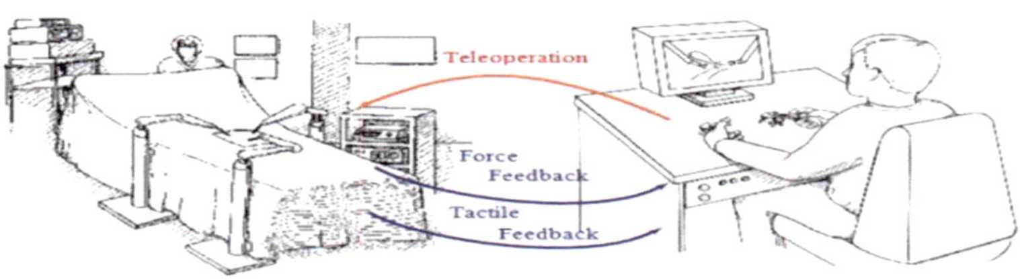

Fig. 2 Basic tele-surgery system

2 Implementation of System Design

There are two systems, namely (1) surgeon Section (2) surgical room section. The movement of the manipulator holding the surgical tool will be controlled by the surgeon wearing a hand acquisition system programmed to a microcontroller which is connected to a serial port communication from which the motion and inclination of sensor will be captured and sent to the surgical room. In this system, the surgeon available in long distance will have a virtual environment of the surgical room through the VR goggle worn by him. These procedures are achieved by experimenting various possible ways for reading the sensor values from hand acquisition system, and a unique hardware is set up with low-power microcontroller connected to two Zigbees at the surgeon's section and with a repeated hardware section at the surgical room, which is additionally connected to servomotors which control the motion of the manipulator, holding the surgical tool in the surgery section. The two-way data transfer is done wirelessly without delay and data loss. The advanced tele-surgery system consists of two parts, hardware design and software design (Fig. 3).

2.1 *Hardware Design*

Atmega32 is a high-performance, low-power, and 8-bit microcontroller combines 32 KB ISP flash memory. Zigbee is the next important hardware component used. Due to new inventions, new protocols were introduced and interfacing becomes a difficult. So, there was a standard protocol, namely Zigbee. MEMS sensors occupy the major part in the hand acquisition system. The ADXL335 is a complete three-axis accelerometer, with features of being small and thin. It also provides signal-conditioned voltage outputs. Servomotors of 12 V/300 rpm are fitted to the manipulator and take control over the motion of the manipulator holding the surgical tools. LCD, UART, camera module for tilt and pan setup, end manipulator, robot case, smartphones with Bluetooth, and gyroscopes are other hardware components that enhance the system.

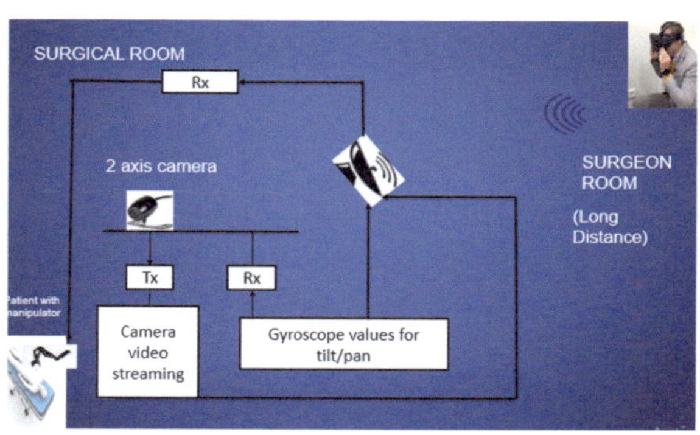

Fig. 3 Block diagram of advanced tele-surgery for entire setup

2.2 *Software Design*

Embedded C programming, MATLAB R2013a, Keil software, and HyperTerminal are the major software used, and many other IoT supporting software will be included. Embedded C programming is used to program the microcontroller. It is used to set the frequency of the controller, to send the string through UART, to get the values at the LCD display, to display the *XYZ* values of the MEMS sensors, to display *X* value on the LCD, to send the *X* value via UART with delay of 10 MS, and for the other two *Y*-axis and *Z*-axis, and also to send the characters to UART.

3 Hand Acquisition System

The inclination of MEMS sensor is tested by connecting it to Atmega32 micro-controller, and a LCD display is connected to get the corresponding sensor values. The data from the sensor with respect to the motion of the surgeon is fed inside the controller. They are analog accelerometers, and their output is voltage. Its sensitivity is 300 m V/g, and the range is −3 to +3 g. At 0 g, it gives output of 1.65 V at three axes and the corresponding output is calculated as (1023/5) * 1.65 = 338. So, above 338 the acceleration error (g) is positive, and below, it is negative *X*-axis and *Y*-axis: It reaches maximum at 395, minimum at 265, and zero at 338. *Z*-axis: It reaches minimum at 275 and maximum at 405, and this sensor only can sense 180°. The sensor values fed to microcontroller are transmitted to first Zigbee used in the system. This one acts as the transmitter. It is used to receive the data as single serial bit streams. These serial bit streams are converted to bytes by UART, so that it will be convenient for the further applications in this system. The data received by this Zigbee is transmitted to the second Zigbee which acts as the receiver in this system. USB 232 converter is used to feed this data to the personal computer in the surgeon's section which is followed by IoT data transfer (Fig. 4).

Fig. 4 Hardware of the tele-surgery system with hand acquisition system

4 IoT Data Transfer

IoT in this system exists as a network of the machines which utilizes the data received from the MEMS sensors and the networks which help in collecting and exchanging the processed data. The data received by the system from the device is retrieved back by the server using php Wamp server, and the data is fed to the receiver section using the .net service available in the system. This data is used to control the motion of the motors fitted to the manipulator holding the surgical tools (Fig. 5).

4.1 IoT Security

IoT products are now mainly designed with idea of networking and other devices connected to it. Security is least considered in the product design. But, it is very important to safeguard the connected devices and networks in IoT. The IoT device which is directly accessible using the Internet should be partitioned as own network and for others, and the network access must be restricted. It is very essential to take serious actions, in notice of anomalous traffic. It is important to maintain guest networks, so that other users will not gain access to our shared files and data. It is also necessary to update the system firmwares regularly. This practice will help in fixing the vulnerabilities and exploits as they emerge. The most important practice for the security is to track and assess the devices based on their traffic, level of access, and alert flags in case of intervention of unknown devices in the network (Fig. 6).

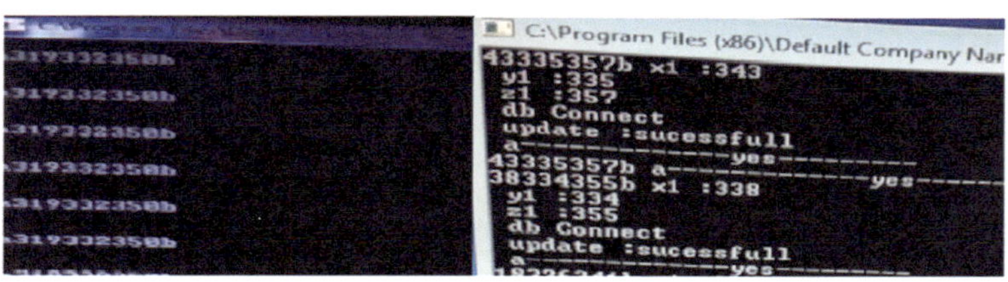

Fig. 5 Sensor values received at machines in transmitter and receiver screens through internet

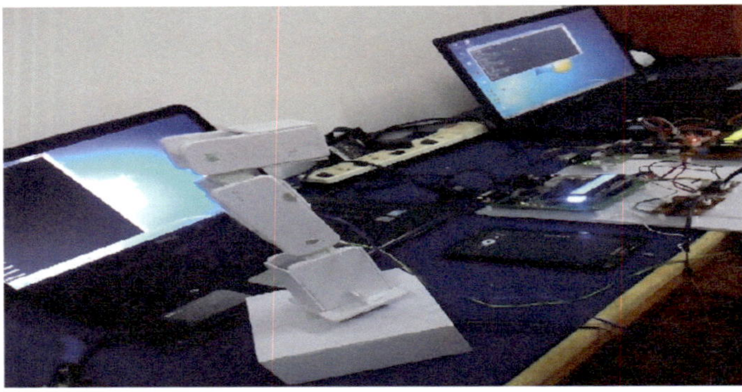

Fig. 6 Prototype of IoT-based robotic tele-surgery

5 Surgical Room Camera Mechanism

The surgeon at a long distance should have a proper view of the patient in the surgical room. This need is satisfied by controlling the inclination of two-axis nylon camera setup, which has focus on the patient. The VR goggle worn by the surgeon is fitted with an android phone, and the sensor values of the android phone are read by the MATLAB software and transmitted wirelessly using additional hardware components. Sensor UDP program is developed by the Takashi Sasaki. This program is used to send the sensor data from the android phone fitted to the surgeon's VR goggle, which transfer to a MATLAB session through local are network's UDP. MATLAB has a file exchange which allows to read the upcoming UDP that was sent by the Sensor UDP program and reads the sensor data into graph. LAN connection is set up between the phone and the PC. After the establishment of LAN connection, the Sensor UDP is set up. The sensor data should be read by the MATLAB, processed, and visualized (Fig. 7).

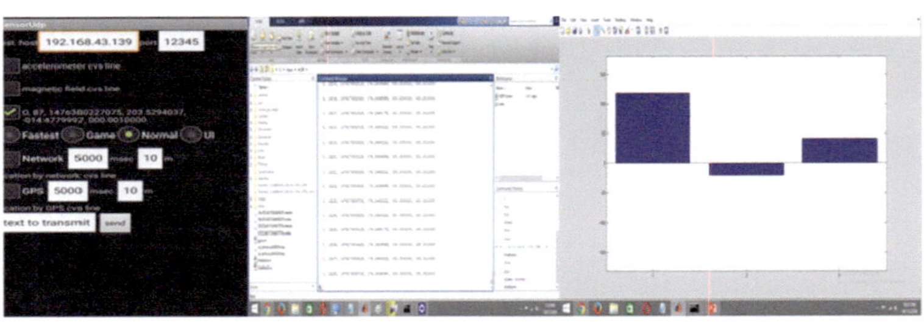

Fig. 7 Reading android phone sensor to MATLAB using sensor UDP

6 Problems Occurred

- Security of the data must be considered as important while using IoT and wireless data transfer.
- System must be reliable to work in present infrastructure considering all the environmental effects such as shocks, tremors, and vibrations.
- Processing speed should be increased while data transmission and reception.
- Network attacks and penetration must be handled.
- Memory size must be increased to store and handle the data.
- As the data transmission is entirely wireless, it is necessary to ensure the availability of network coverage, throughout the process.

7 Result and Conclusion

This advanced tele-surgery system overcomes the drawback of distance constraint and makes the system compact by adhering to wireless data transmission. This system has an improvised data transmission with accuracy in motion of the end manipulator. The server adopts IoT for data transfer, which makes this system stand superior to other existing approaches toward tele-surgery.

8 Future Work

The hardware of the entire system must be miniaturized in both surgeon section and in surgical room. The distance constraint must be minimized. The delay prevailing in the end manipulator must be eliminated with more fine mechanism applied to the motor section. The live video streaming of the surgical room must be improved with fine quality and good data transfer speed providing the most reliable and flexible tele-surgery system. The incision of the surgical instrument and the positioning of the patients must be handled very carefully with respect to the limitations of the working model. It is mandatory to have a clear view on cyber threats. The security should be considered in every stage of the product design, which includes coding, testing, and final evaluation. It is necessary to develop a standard model to handle the unstructured data involved in the process. The algorithm limitations of the data caused by the false positives and false negatives must be quantized more accurately in the future system. It must be improvised with integration of IoT with control systems, thus providing feedback for the system. The future system must be a dedicated on-site end-to-end control system, so that it will be the most prominent application of IoT in healthcare sector.

References

1. Smith R, Day A, Rockall T et al (2012) Advanced stereoscopic projection technology significantly improves performance of minimally invasive surgical skills. Surg Endosc 26 (6):1522–1527
2. Janetschek G, Reissial A, Peschel R, Drax H, Bartsch G (1993) Chipon a stick technology: first clinical experience with this new video laparoscope. J Endurol 7:S195
3. Bahayan RK, Chiu AW, Este-McDonald J, Birkett DH (1993) The comparison between 2-dimensional and 3-dimensional laparoscopic video systems in a pelvic trainer. J. Endourol 7: S195
4. Marescaux J, Leroy J, Ganger M et al (2001) Translantic robot-assisted telesurgery. Nature 413 (6854):379–380
5. Strebis JR, Hanly EJ, Herman BC, Marohn MR, Broderick TJ, Shih SP, Harnett B, Doarn C, Schenkman NS (2008) Transcontinental telesurgical nephrectomy, using the da Vinci robot in a porcine model. Urology 71(5):971–973
6. Cosman PC, Davidson HC, Bergin CJ et al (1994) Thoracic CT images: effect of lossy image compression on diagnostic accuracy. Radiology 190(2):517–524
7. Choi JG, Bahk S (2007) Cell-throughput analysis of the proportional fair scheduler in the single-cell environment. 766–778
8. Ameigeiras P, Wigard J, Mogensen P (2004) Performance of the MLWDF scheduling algorithm for streaming services in HSDPA. In: IEEE 60th vehicular technology conference, vol 2. Los Angeles, USA, September 2004, pp 999–1003

Xilinx System Generator-Based FPGA Control of Power Flow for DC/DC Converter

Anurag Sharma, Rajesh Gupta and Muskan Gupta

Abstract In recent years, Xilinx system generator platform (XSG) is becoming more popular in real-time simulations. This is especially true for applications involving FPGA as controllers such as power electronic converter control. Since XSG is integrated with MATLAB/Simulink, it eliminates the need of special skills in VHDL or HDL and enables the user to design control techniques with the help of Xilinx block sets and to test these control designs simultaneously. This paper aims to design the power flow control method in dual active bridge (DAB) bidirectional DC/DC converter using XSG for real-time simulation and to verify the control strategy. The result obtained through XSG-based hardware-in-the-loop (HIL) simulation is helpful to know the controller performance a priori in real-time implementation.

Keywords Dual active bridge (DAB) · Field-programmable gate array (FPGA)
Hardware co-simulation · Hardware-in-the-loop (HIL)
Xilinx system generator (XSG)

This work is supported by the Council of Scientific & Industrial Research (CSIR), New Delhi, Sanction No. 22(0691)/15/EMR-II.

A. Sharma (✉) · R. Gupta
Department of Electrical Engineering, M. N. National Institute of Technology Allahabad, Allahabad, India
e-mail: anurag.mnnit2015@gmail.com

R. Gupta
e-mail: rajgupta310@gmail.com

M. Gupta
School of Electronics Engineering, VIT University, Chennai, India
e-mail: muskangupta175@gmail.com

D. Thalmann et al. (eds.), *Intelligent Embedded Systems*, Lecture Notes in Electrical Engineering 492, https://doi.org/10.1007/978-981-10-8575-8_4

1 Introduction

Power electronic converters are widely used in the electrical power circuits for conversion from AC/DC to DC/AC. The power electronic devices such as power MOSFETs and IGBTs are commonly used for this purpose in a power electronic converter. These switching devices are turned ON and OFF by the pulse signal to the gate terminal of the switching devices. Thus, there is a need of control of the power electronic switches. This control is generally employed by the use of digital control techniques. Earlier, the switching pulse or the pulse width modulation (PWM) signal was generated by the use of microprocessors, microcontrollers, or digital signal processors (DSPs), but now field-programmable gate arrays (FPGAs) have proved to be better option.

A FPGA combines a large number of logic gates onto a single chip. Some of these logic gates can be grouped together to form a block. These blocks can perform specific functions and thus can simplify the higher-level circuit design. Because of easy programmability, configurability, and parallelism, FPGAs prove to be better. However, the design of control algorithm in FPGA for actual implementation in hardware requires a high level of expertise in the hardware description language (HDL). However, writing codes in HDL is a tedious process as it requires the programmer to write many lines of code for simple tasks.

To reduce the task of researchers in order to implement control logic using HDL, the Xilinx system generator (XSG) platform can be used. The FPGA-based Xilinx system generator hardware-in-the-loop (HIL) simulation is a useful tool for developing, designing, and testing hardware prototypes without any prior knowledge of HDL coding [1, 2]. Using this platform, the VHDL code can be automatically generated in XSG using HDL co-simulation, enabling quick prototyping of power electronic circuit based on FPGA controller [3]. Following are the advantages of using XSG over other platforms such as Opal-RT, dSPACE for simulating HIL applications of power electronic circuits.

i. No need of hardware implementation for testing of control design of FPGA.
ii. Easier to find delays associated with the control logic.
iii. Time required to develop control logic is reduced by using HDL co-simulation.

The DC/DC converters are mostly employed in renewable energy and energy storage systems for supplying power both to and from the storage device. Among many DC/DC converter topologies, the dual active bridge bidirectional converter has advantages of higher power density and compact size [4, 5]. The DAB converter consists of two MOSFET-based H-bridges, both having DC voltages as input. The input DC voltages to the H-bridges are converted to high-frequency AC output by switching signals in the H-bridge by high-frequency pulses. The circuit schematic of DAB is shown in Fig. 1. The input DC voltages are defined as V_1 and V_2. The turns ratio of the high-frequency transformer from HV side to LV side is kept as n:1. The transfer of power is done by inductance L. The AC output voltages of converters are V_{HV} and V_{LV}. Voltage and current through the inductor L are V_L and I_L, respectively. The switches S_1 to S_8 in Fig. 1 are MOSFETs, and D_1 to D_8 are the inbuilt antiparallel diodes. From Eq. (1), we can say that by changing the phase shift we can vary the power flow. Also, the power flows from the leading output bridge voltage toward the bridge with lagging output voltage.

$$P = \frac{nV_1 V_2}{2\pi L f_{sw}} \frac{\phi(1-\phi)}{\pi} \tag{1}$$

where ϕ is the phase shift and f_{sw} is the switching frequency.

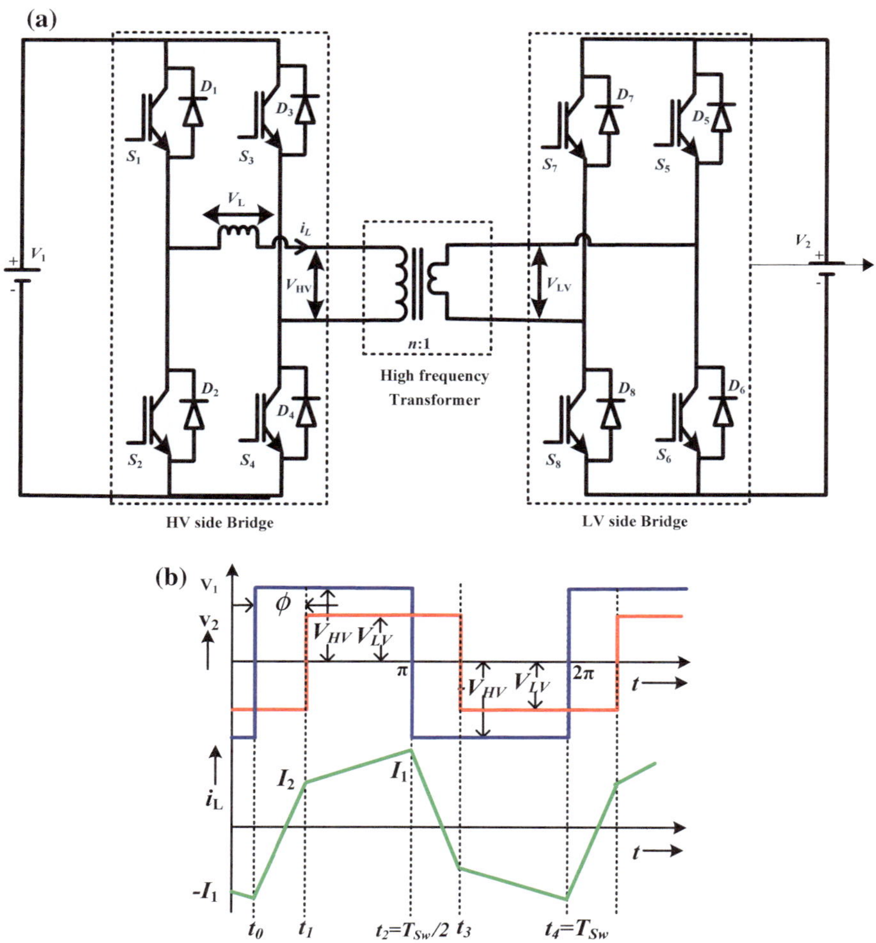

Fig. 1 **a** Circuit schematic of DAB DC/DC converter, and **b** HV bridge, LV bridge voltages and inductor current

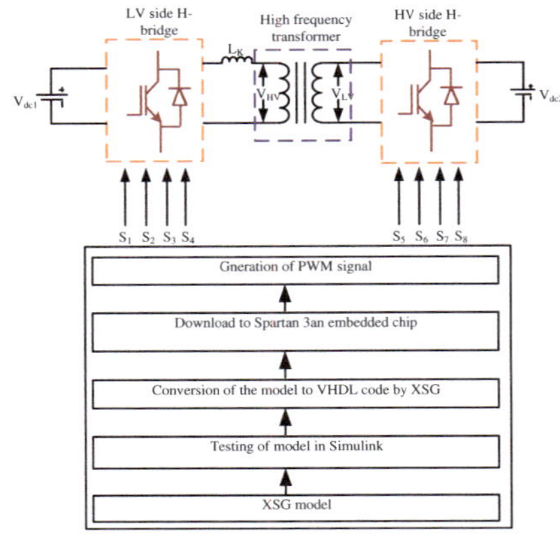

Fig. 2 Control scheme for power flow control in dual active bridge converter

In this paper, pulses for the PWM of DAB are generated using Xilinx system generator and then interfaced with the MATLAB/Simulink model of DAB converter to check the functioning of the PWM logic as shown in Fig. 2. Also, a comparison of modulation scheme is also performed.

2 PWM Modulation of Dual Active Bridge

There are many modulation techniques for power flow control in DAB converter such as single-phase shift (SPS), dual-phase shift (DPS), triple-phase shift (TPS), and extended-phase shift (EPS) modulation [4, 5]. This paper presents the implementation of SPS and DPS modulation techniques using XSG.

2.1 Single-Phase Shift Modulation

In SPS modulation, the HV and LV bridges are excited by the two-level phase-shifted pulses, where the phase shift is equal to one-fourth of the switching

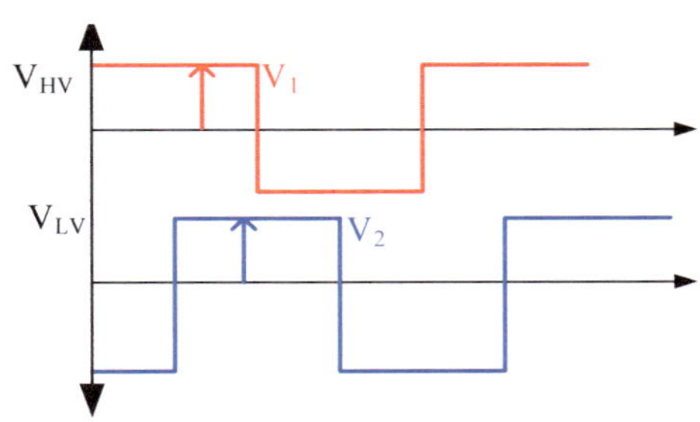

Fig. 3 HV and LV side voltages with SPS modulation

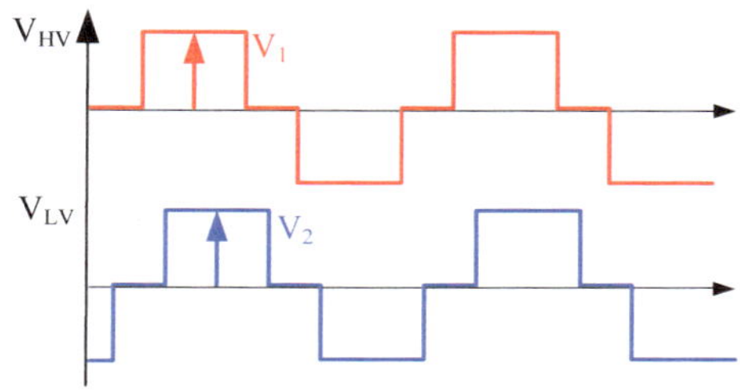

Fig. 4 LV and HV side voltages of transformer for DPS modulation

time period T_{sw}, as shown in Fig. 3. The cross switches (S_1, S_4) and (S_2, S_3) of the HV bridge are given same switching pulse generating two-level voltage of magnitude (V_1 to $-V_1$), which appears across HV side of the transformer. Similar switching is done on LV side, but the switching pulse is phase-shifted; thus, the LV voltage is also of two level of magnitude (V_2 to $-V_2$), but with a phase shift, the power flows from the HV side to LV side of the transformer.

2.2 Dual-Phase Shift Modulation

In DPS modulation, the switching pulses for cross-connected switches (S_1, S_4) and (S_2, S_3) are also phase-shifted; thus, there are two phase shifts for DPS modulation: One is the inner phase shift and the other is the outer phase shift. Due to the inner phase shift, the HV and LV voltages of the transformer are of three levels as shown in Fig. 4.

3 PWM Generation

To operate the switches in the H-bridges, pulses are required. These pulses are generated in the XSG by comparing the sawtooth waveform with a constant value to produce 50% duty cycle pulse; thus, the generation process of PWM can be divided into two parts: (a) generation of sawtooth waveform and (b) generation of pulse.

3.1 Sawtooth Generation

The counter block in XSG generates the sawtooth as shown in Fig. 5. The counter counts up to value set in the block, and the frequency of the sawtooth is decided by

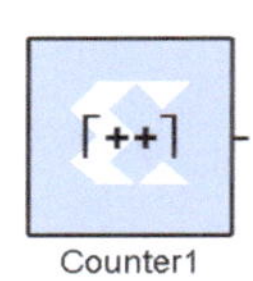

Fig. 5 Counter block in XSG for sawtooth waveform generation

both the count value and the sampling period of the counter. The count value, frequency of sawtooth, and the sampling period can be related by the following equation:

$$Nf_{sw} = \frac{1}{T_s} \tag{2}$$

where T_s = sampling frequency; N = count value;
f_{sw} = switching frequency or sawtooth waveform frequency.

3.2 Generation of Pulses

To generate pulses of 50% duty cycle, the counter output is compared with constant value equal to the half of the maximum count value. Figure 6 shows the logic for PWM generation.

4 Logic for SPS and DPS Modulation and HDL Code Generation

For SPS modulation, to provide the phase shift between the output voltages of the two bridges, the pulses are generated with a phase delay as shown in Fig. 7.

A similar method can be adopted for implementing DPS modulation by providing a delay between the pulses exciting the cross switches of a bridge. The logic for generating pulses in DPS is shown in Fig. 8. The pulses S_1 and S_4 of the same bridges are phase-shifted, and S_1 and S_5 are also phase-shifted. Thus, there are two kinds of phase shift: inner phase shift and outer phase shift.

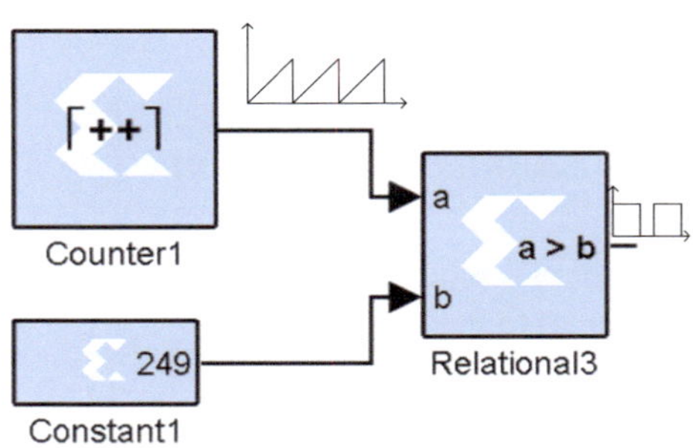

Fig. 6 Pulse generation

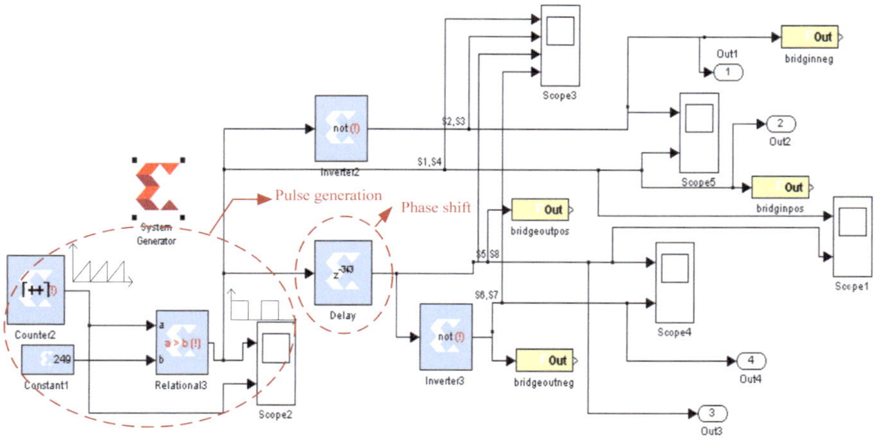

Fig. 7 PWM generation for SPS modulation

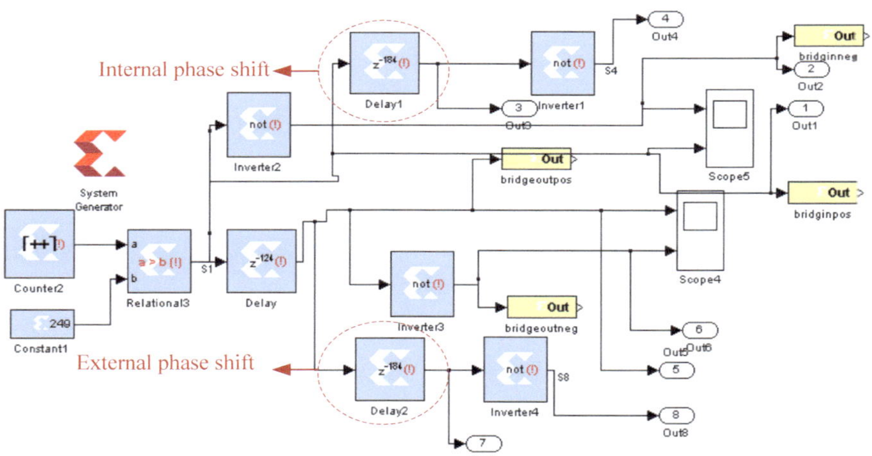

Fig. 8 PWM generation for DPS modulation

5 Simulations Results for SPS and DPS Modulation

The developed logic for SPS operation and DPS operation of DAB is tested by using the pulses generated in XSG with MATLAB/Simulink model of the DAB [6]. The two modes are separately performed by open-loop simulations. Figures 9 and 10 show the waveforms for the voltages for the HV side, LV side, inductor current under the two cases: (a) V_{LV} is leading V_{HV}, and (b) V_{HV} is leading V_{LV}.

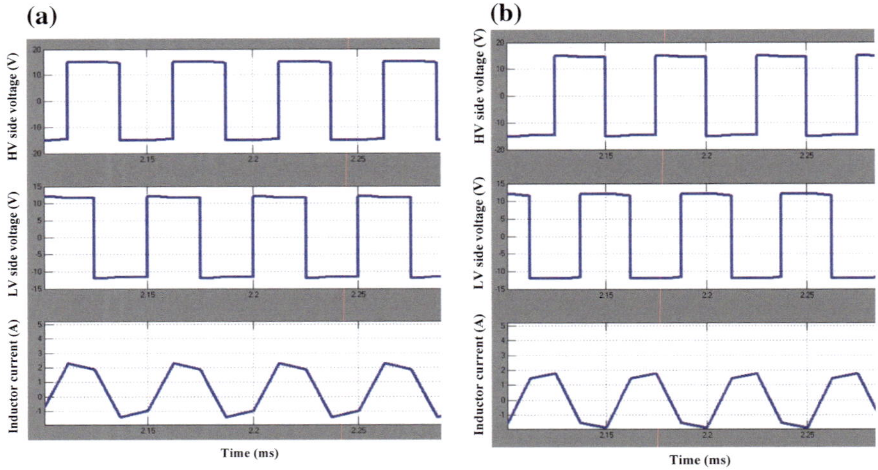

Fig. 9 Waveform for SPS modulation for HV side voltage, LV side voltage, and inductor current for **a** V_{LV} voltage is leading, and **b** V_{HV} voltage is leading

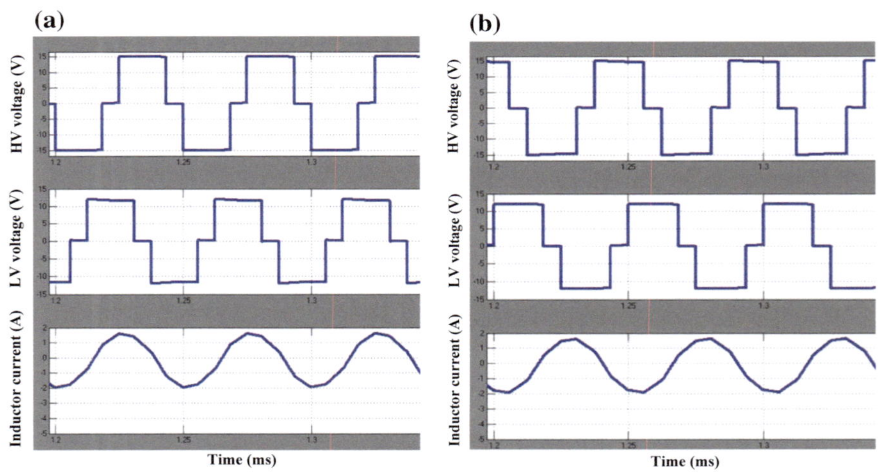

Fig. 10 Waveforms for DPS modulation of DAB for HV side voltage, LV side voltage, and inductor current for **a** V_{LV} voltage is leading, and **b** V_{HV} voltage is leading

6 XSG Logic to VHDL Code Conversion and Experimental Results

After successfully testing, the logic is developed in XSG with the help of MATLAB/Simulink model and simulation. The logic is converted into the HDL code by opening the system generator block shown in Fig. 11 and selecting VHDL option under the hardware description language. It is also required to mention the FPGA board used for the HIL system. For this paper, SPARTAN 3AN FPGA processor has been used.

After selecting the processor, the destination to save the program is mentioned under the target directory. After all the above procedures, the VHDL code can be generated. Figure 12 shows the experimental result obtained after implementation of the logic developed in XSG to the hardware. Figure 12a shows the SPS modulation of DAB with V_{HV} as leading, and Fig. 12b shows the SPS modulation of DAB with V_{LV} as leading. Similar results are shown for DPS modulation in Fig. 12c and d. The experimental results verify the proposed XSG generation of PWM pulses and interface with the MATLAB/Simulink model of DAB converter. (Figure 13)

Fig. 11 System generator block parameters

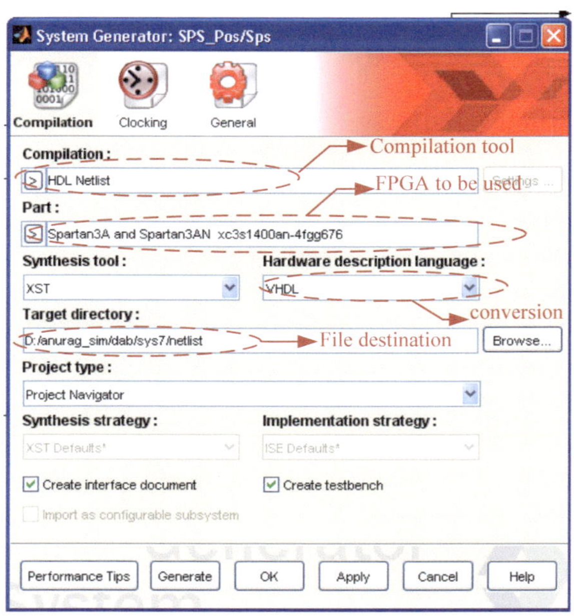

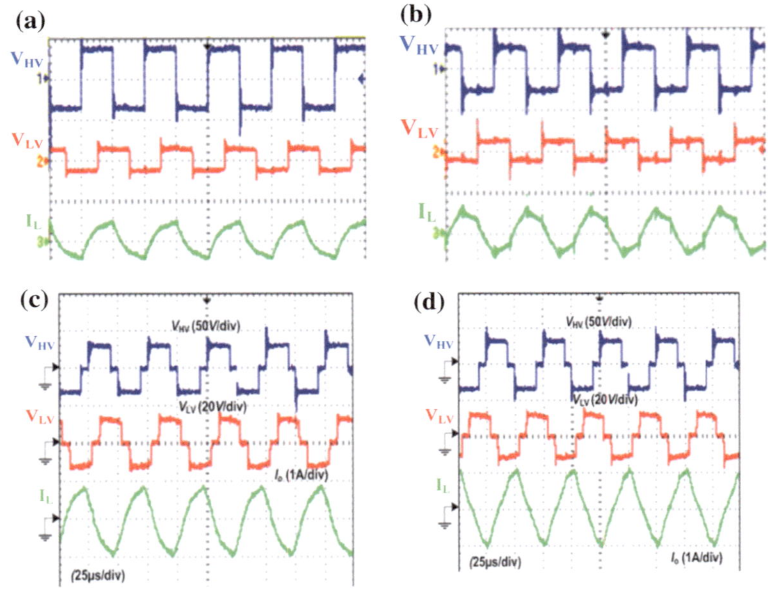

Fig. 12 Experimental results for **a** SPS modulation of DAB with V_{HV} as leading, **b** SPS modulation of DAB with V_{LV} as leading, **c** DPS modulation of DAB with V_{HV} as leading, and **d** DPS modulation of DAB with V_{LV} as leading

Fig. 13 Experimental setup

7 Conclusion

The hardware-in-the-loop simulation through XSG platform can help to identify switching losses, circulating power flow, and net power transfer, using different modulation methods adopted in the DAB converter. The PWM generation with SPS and DPS modulation has been verified using XSG-based HIL simulation of the DAB converter with FPGA controller. The power flows from the bridge with leading voltage toward the bridge with lagging voltage. To reverse the power flow, the switching pulses of the bridges need to be interchanged. As compared to SPS modulation, the DPS modulation has increased range of controllability and better quality of power flow.

References

1. Selvamuthukumaran R, Gupta R (2014) Rapid prototyping of power electronics converters for photovoltaic system application using Xilinx System Generator. IET Power Electronics 7 (9):2269–2278
2. Rajesh P, Rajasekar S, Gupta R, Samuel P (2014) Solar array system simulation using FPGA with hardware co-simulation. In: 2014 IEEE 23rd international symposium on industrial electronics (ISIE), Istanbul, 2014, pp 2291–2296
3. Xilinx Inc. System Generator for DSP User Guide, December 2010
4. De Doncker RWAA, Divan DM, Kheraluwala MH (1991) A three-phase soft-switched high-power-density DC/DC converter for high-power applications. IEEE Trans Ind Appl 27 (1):63–73
5. Zhao B, Song Q, Liu W, Sun Y (2014) Overview of dual-active-bridge isolated bidirectional DC–DC converter for high-frequency-link power-conversion system. IEEE Trans Power Electron 29(8):4091–4106
6. https://in.mathworks.com/solutions/fpga-design/simulink-with-xilinxsystemgenerator-fordsp.html

An Improved Algorithm for the Estimation of Multibody Motion

K. Raghavan and R. Prithiviraj

Abstract The problem of estimating ego motion and eoru motions through a vehicle mounted with a camera is related in this paper. Localization of multiple moving objects and estimating their motion are crucial for autonomous vehicles, but it is not that much successful in estimating the motion of the moving vehicles and objects. Ego motion can be calculated only by conventional localization and mapping techniques. The framework for estimation of multiple motions in addition to the camera ego motion is presented. The video is processed through MATLAB for pre-processing. The video is then segmented into frames, and then the framework is done to estimate the multibody motion through different algorithms. The algorithms like block matching algorithms, corner detection, and background subtraction algorithm are used to estimate the multibody motions of the moving objects. From this, we can detect and estimate the motion and speed of the object in the frame. Then, it is processed in hardware (raspberry pi3) using the same algorithms, so that we can effectively use it in any autonomous cars.

Keywords MATLAB · Block matching algorithm · Harris corner detection Background subtraction

1 Introduction

Visual odometry is the process of locating a vehicle from its mounted camera. In the current system, the motion or speed of the moving vehicle is calculated only by assumption; moving objects are normally treated as outliers. Estimating the moving objects and its problem is called as multibody structure from motion. In this paper,

K. Raghavan (✉) · R. Prithiviraj
Department of Electronics and Communication, SRM University, Chennai 603203, India
e-mail: mailtoraghavkrish@gmail.com

R. Prithiviraj
e-mail: prithiviraj.r@ktr.srmuniv.ac.in

D. Thalmann et al. (eds.), *Intelligent Embedded Systems*, Lecture Notes in Electrical Engineering 492, https://doi.org/10.1007/978-981-10-8575-8_5

theoretical framework for estimating the vehicles in motion with a car mounted camera. The proposed method is inspired from multibody structure from motion. Basically, there are two types of motion, eoru motion and ego motion. The eoru motion is used to find the objects in motion, and ego motion is used to locate the objects which are stationary. Our work mainly targets on estimating the speed of the objects in motion using three algorithms to improve the accuracy and efficiency of the system. For this, we are using cameras mounted over the car to record the videos; using monocular cameras will give effective results, and it is also cost effective in terms of price and power consumption. This can be used in autonomous cars as driver-assistance system and for robotics application. This process is implemented using MATLAB and raspberry pi3.

2 System Overview

2.1 *Need for Estimation of Motion*

There are several papers that are proposed to detect the ego motion and its structure, but not much paper had been proposed to detect the speed of the objects in motion. So in this paper, the estimation of speed of the vehicles in motion is estimated through camera mounted over the car by three various algorithms; these are processed through MATLAB. First, segment the video into frames; then by applying the algorithms, we can estimate the motion of the multiple moving objects.

2.2 *Existing Scenario and Drawbacks*

Scaramuzza [1] proposed an algorithm called as simultaneous localization and mapping (SLAM) to detect the object in motion in which they said only about what are the problems they had faced during the observation of the objects in motion. Visual odometry [2] is there to tract the objects through on-board cameras. No projects have been proposed to detect the speed of the multiple moving objects [4, 5] in motion.

2.3 *Objective*

Localization of multiple moving objects and estimation of their motion are crucial for autonomous vehicles. Estimating the speed of the multiple moving objects is the main objective. Estimating the speed of the multiple moving objects is very helpful for various applications such as driver-assistance system, android humanoid robots.

3 System Architecture

3.1 Proposed System

The estimation of multibody motion and its speed can be calculated using three various algorithms such as block matching algorithm, background subtraction and Harris corner detection. By processing the videos through these algorithms, the speed of the moving objects in the frames can be estimated. The implementation process is shown in Fig. 1, and the detailed description of the algorithms is given in Fig. 1.

3.2 Implementation Process

The input video is taken from the monocular camera which is fixed over the car to capture the video and is shown in Fig. 2. The captured video is then fed into the MATLAB for processing. Some of the videos are inbuilt in MATLAB itself. The monocular cameras are cost effective, and the power consumption is very less. The input videos can be of avi format or mp4. The videos are then converted into frames. Number of frames segmented in MATLAB depends on the given input video and also depends on the type of processor.

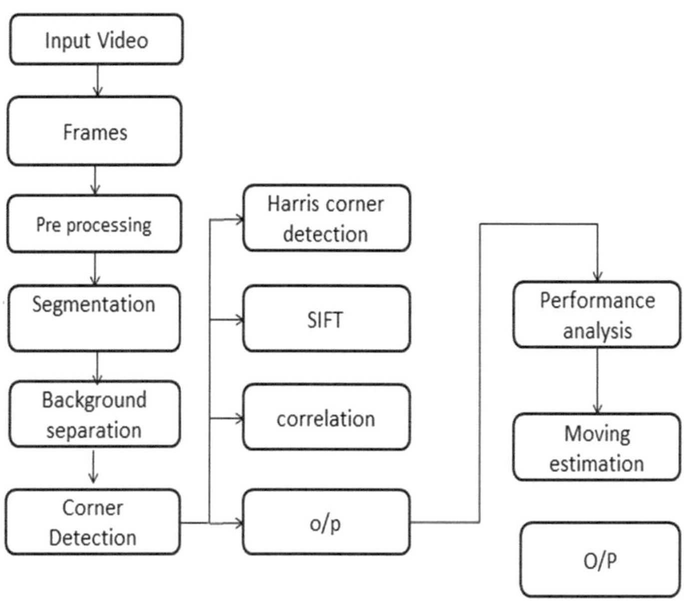

Fig. 1 Implementation process

Fig. 2 Input video

A very high-quality video cannot run fast on a system that has a very low processor. It takes much time to convert into frames. The pre-processing suppresses unwanted distortions in image, and it helps in further processing. There are four types of image pre-processing methods used for intensity transformation. Pre-processing which uses a local neighborhood image of the processed pixel and image restoration requires knowledge about the entire image. Using pre-processing technique, we can change the image based upon our convenience.

Image segmentation is the process of partitioning an image into multiple segments. The segmentation processes the image into meaningful and easy way to analyze the given input image. Image segmentation assigns a number to every pixel in image, and each number has certain characteristics. The segmented frames are then moved to background subtraction. Marking moving objects from a video is a fundamental and major task in many computer vision-based applications. The framework is done for original and segmented frame. Then, it will compare both the frames and it will start to subtract both the frames from foreground to foreground and then to background to background. After subtraction, we will get the values for only foreground frames (Fig. 3).

Harris corner detection is used to detect the structure of the objects in motion, it may be stationary or in motion which is shown in Fig. 4. An interest part of an image has a defined position and can be detected robustly. This shows an interest point can be a cornered; a detected point can have a intensity maximum or minimum. It is simply defined as the dramatic change in the area of data. All the corners in the frame are pointed. A lower number shows that it has more similarity. When the pixels have the same intensity, then the nearby portions will look similar. The pixel defined on the edge will be perpendicular to another patch that seems different, but nearby patches parallel will result only in a small change.

Fig. 3 Background subtracted image

Fig. 4 Harris corner detection image

In block matching algorithm, the frames are divided into blocks and it will check the block one by one. It checks for any movement of object in the frame. The reason for block matching is to segment the frames into a matrix of 'macroblocks'; then, they compare with corresponding blocks and its nearby frame will create a vector that makes motion of a macroblock from one location to another in the previous frame. This movement makes connection with all other macroblocks of the frames frame, constitutes the motion estimated in the current frame. The portion to search for good blocks match depends on pixels on all the other sides of the block in previous frame [3]. Estimation process determines the motion vectors for transformation from 2D image to another; usually, they transform [6–8] adjacent frames from the video sequence. It is the motion in three dimensions, but the images are a projection of the 3D scene onto a 2D plane. The estimated motion vectors connect the full image or some specific parts of the image, such as rectangular blocks, arbitrary-shaped patches and even in every pixel. After all the processes, we can detect the motion of vehicle in the frame and then performance analysis is done to get an effective output.

4 Results

The speed is estimated from the three algorithms, and it is represented in Fig. 6. The figure shows the speed estimated for 120 frames that are separated from the video. The speed is estimated for all the frames that are separated; the *Y*-axis represents the number of frames. Figure 5 shows the peak signal to noise ratio (PSNR), and it can be easily defined from mean square error, and it helps to know about the corruption rate of the frames, lesser value of PSNR gives accurate and effective speed estimation.

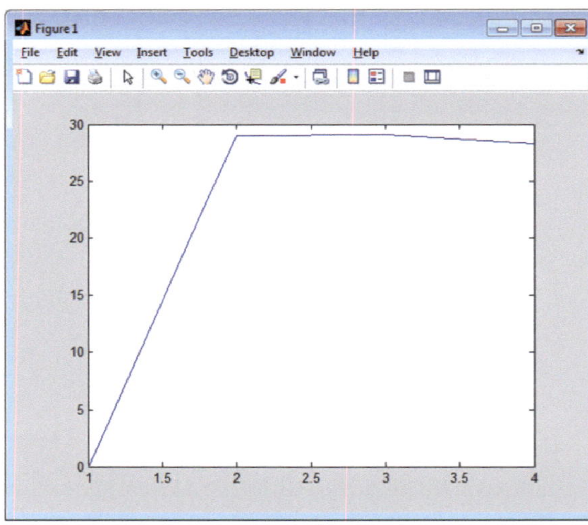

Fig. 5 Peak signal noise ratio

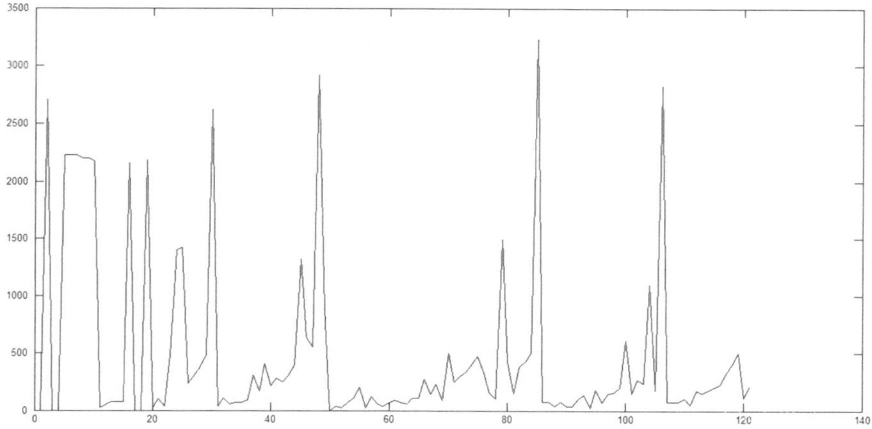

Fig. 6 Speed estimation for the input video

5 Conclusion

Thus, estimating the motion of multiple moving objects is done using three different algorithms to improve the efficiency and accuracy. The speed of the object is calculated in each frame; this is very important for an autonomous vehicle. It helps to prevent autonomous vehicle from accidents, and it also helps as a relief buster to humans. This can be used in driver-assistance system and in android humanoids robots. This process is done using MATLAB. The drawback in the process is that if the input video is too high then the processor speed will be too slow. In the future, this can be rectified by implementing the same process in hardware of higher processing power which can be more effective and more reliable for many applications.

Acknowledgements The author would like to thank the SRM University, Head of the Department, Dr. T. Rama Rao, Project Coordinator, Dr. A. Ruhanbevi for the support during the course of the project.

References

1. Sabzevari R, Scaramuzza D (2014) Monocular simultaneous multi bodymotion segmentation and reconstruction from perspective views. In: Proceedings of IEEE international conference on robotics and automation (ICRA), 2014, pp 23–30
2. Scaramuzza D, Fraundorfer F (2011) Visual odometry: Part I: the first 30 years and fundamentals. IEEE Robot Autom Mag 18(4):80–92
3. Vogel C, Roth S, Schindler K (2014) View-consistent 3D scene flow estimation over multiple frames. In: Proceedings of European conference on computer vision, 2014, pp 263–278

4. Zappella L, Del Bue A, Lladó X, Salvi J (2013) Joint estimation of segmentation and structure from motion. Comput Vis Image Underst 117(2):113–129
5. Kundu A, Krishna KM, Jawahar C (2011) Realtime multibody visual SLAM with a smoothly moving monocular camera. In: Proceedings of IEEE international conference on computer vision, 2011, pp 2080–2087
6. Enzweiler M, Gavrila DM (2011) A multilevel mixture-of-experts framework for pedestrian classification. IEEE Trans Image Process 20(10):2967–2979
7. Zhang T, Szlam A, Wang Y, Lerman G (2012) Hybrid linear modelling via local best-fit flats. Int J Comput Vis 100(3):217–240
8. Forster C, Lynen S, Kneip L, Scaramuzza D (2013) Collaborative monocular SLAM with multiple micro aerial vehicles. In: Proceedings of IEEE/RSJ international conference on intelligent robots and systems (IROS), 2013, pp 3962–3970

An IoT-Based Smart Shopping Cart Using the Contemporary Barcode Scanner

V. N. Prithvish, Shraddha Agrawal and John Sahaya Rani Alex

Abstract An efficacious prototyping of an intelligent Internet of Things (IoT)-based smart cart that primarily enhances the shopping experience of the customers as well as owners. The assembly of the smart cart consists of a contemporary barcode scanner from the present-day shopping scenario, feedback systems providing product weight and product imaging to avoid discrepancies, a cloud-based database and an embedded hardware to connect the above. The mechanism facilitates the customer to add and remove products throughout the shopping, updating the bill instantaneously, and further reflecting changes in the inventory. The framework incorporates a Parse cloud-based inventory to ease the managerial task at the owner end. Further focus has been also given on the proper placement of all the components on a cart for the efficient working of the shop, keeping user's ease of usage in mind.

Keywords Smart cart · IoT · Pars · Beagle Bone Black · TVS electronics BSC-101 star · RTL-8188EU · SIFT algorithm · Load cell · Barcode scanner

1 Introduction

Retail market is growing at an alarming rate, and the modern customer needs better than the best experience. Technology will play a very important role in improving and developing the experience which sets the modern retail apart from the brick and

V. N. Prithvish (✉) · S. Agrawal · J. S. R. Alex
School of Electronics Engineering, VIT University, Chennai Campus,
Chennai, Tamil Nadu, India
e-mail: prithvish.n13@gmail.com

S. Agrawal
e-mail: shr7635@gmail.com

J. S. R. Alex
e-mail: jsranialex@vit.ac.in

D. Thalmann et al. (eds.), *Intelligent Embedded Systems*, Lecture Notes
in Electrical Engineering 492, https://doi.org/10.1007/978-981-10-8575-8_6

mortar retail stores. Our endeavor is to design an efficient product that can be piloted to the souk at this very instant. A queue-based billing system prevails in the existent scenario in most of the developing countries and few developed countries, demanding an intermediary labor intensive or conveyor belt-based billing of products. One of the measures adopted includes RFID tagging of products. However, high cost of both tag and reader makes item-level RFID tagging impractical except for high-value products [1, 2]. A simpler approach would be to embed RFID devices in consumers' shopper cards to identify them; such retail system essentially involves tradeoffs between enhanced functionality and privacy protection [3, 4]. Other than the RF-ID technology, the modern marketplace has also integrated other methods for shopping such as the NFC technology for smart payment [4]. In recent times, modern retail also uses smart phone and app technology [5] which might sound as a cost effective and a customer friendly but has its own disadvantages and dependencies which are addressed by the authors later in the paper. Instead of adapting to these, the authors' proposal is to come up with an advancement that utilizes the currently available barcode scanners and other related resources such as the shopping cart/trolleys.

The Internet of Things (IoT) by definition refers to a ever-growing network of physical objects including computing devices, mechanical and digital machines, animals or people or objects provided with unique identifiers having ability to transfer data over the same network without requiring human or computer interaction [6]. In today's world, IoT has been incorporated in every sector from industrial automation to home appliances to medicine, sports, and a lot more [7, 8]. The proposed smart cart contemplates to integrate the concept of IoT with a conventional barcode as the tagging identity. This also makes the design easy on the pocket. The system will efficiently automate the functionality of the store, reducing labor dependency to a great extent. This smart cart will keep updating the replenishment system for restocking, making the store's operations proficient. At the same time at the buyer's end, the current bill will be displayed on their smart devices keeping them well informed about their purchases and aiding them to take further purchase decisions.

Besides, the design also entails a security feedback unit which scrutinizes all the products in the cart. Based on that, discrepancies and falsehood can be alarmed, locking the cart at the very instant. At the cloud end, flexibility is provided as per the wish of the owner/management unit of the store. It facilitates in monitoring the consumer behavior, recognizing trends among various brands. The owner can supervise the functioning of the store at ease from any corner of the world. Further expansions based on the owner's desire can be implemented on the design as well.

2 System Design

The entire system is composed of two subsystems:

I. The smart cart database system
II. The smart cart feedback system

Figure 1 shows the design of the smart cart database system, which utilizes barcode scanner, the sensing element of the subsystem, that takes barcodes of different commodities as input to the system, checks in for the commodities into the database, further adding/removing it to the cart system. Following each scan, the data are directed to the cloud, revising the inventory. This smart cart will keep updating the restocking process so that no product is off the shelf. It will also lead customers to the location of their preferred products and assist them on the basis of their earlier purchases. The same smart cart will keep customers informed about their billing amount to aid them to make further purchase decisions. A customer shall contain his/her own login ID and password. The cart not only eases the work at the store but also adds technological aspects to a store [9].

Figure 2 shows the block diagram of the smart cart feedback subsystem which is a combination of both weight sensing unit and product imaging unit particularly added to make a flawless exemplary system. The customer may face contrarieties such as product not being scanned properly and added to the cart, product being scanned but not added to the cart, products with same weight but different price, and product description being added to the cart. Additionally, there may be instances where the customer attempts scanning of a single product but adds several products into the cart or attempts where high-priced commodity is placed into the cart, scanning the barcode of a low-priced commodity. Thus, feedback system ensures the error-free and theft-free supermarket in absence of manual security.

Fig. 1 Design for the smart cart database system

Fig. 2 Block diagram of the smart cart feedback system

3 Modules

3.1 Beagle Bone Black

Beagle Bone Black is an easy on pocket, community-supported platform for user development. It comes with a TI processor AM335x ARM Cortex-A8 supporting a processing speed of around 1 GHz [10], 512 MB DDR3 RAM, 4 GB 8-bit eMMC on-board flash storage and a USB-Client which abridges the task of interfacing the USB-based barcode scanners and transmission of data over the cloud. The Linux-based platform entitles the user to design on their favored programming language. Its pocket-sized structure provides an easy-to-fit-in feasibility onto the cart [11]. Figure 3 [12] shows the Beagle Bone Black.

Fig. 3 Beagle Bone Black

Fig. 4 Barcode scanner

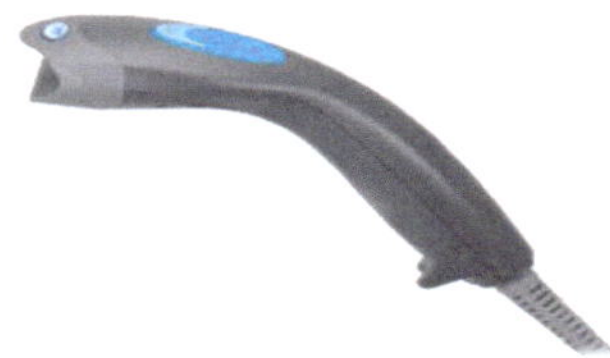

3.2 Barcode Scanner

TVS Electronics BSC101 Star is the barcode scanning device/module used in the prototype. It is a fast, high-performance USB-based CCD (charge coupled device) scanner which comes with a 32-bit microprocessor guaranteeing an unmatched scan rate of almost 500 scans in one second. Equipped with ultra-scan decoding technology, it can predominantly decode GS1 Data Bars along with simple 1D barcodes. The accuracy with which it can decodes makes it a pertinent choice for the applications involving both retail checkout and inventory management. It can also function as a hands-free scanner when configured in auto-scan mode and provided with a stand, making it an apt choice for the system [13]. Figure 4 [14] shows TVS BSC-101 Star (USB-based barcode scanner).

3.3 USB Wi-Fi Adapter

A 150 Mbps Wireless N Nano USB Adapter TL-WN725 N is a low cost, small-sized USB Wi-Fi Adapter which gives connectivity over the cloud. It supports IEEE 802.11 bgn standards with an operating frequency range of 2.400–2.4835 GHz. It has an RTL-8188EU real-teak chipset which supports infrastructure and ad hoc mode [15]. Figure 5 [16] shows Wireless N Nano USB Adapter TL-WN725.

3.4 Power Bank

A Xiao-mi 5 V 1.5 A 5200 mAh Power Bank provides power to the whole system, when not in emulation mode. The power bank is capable of powering the system for 4–5 h with full efficiency when fully charged. The rechargeable power bank enables the system to run in standalone mode. The push button and power level

Fig. 5 USB Wi-Fi adapter

Fig. 6 Xiao-mi power bank

indicating LEDs on the power bank shall notify the shopkeepers to recharge the battery when not in use. Figure 6 [17] show the Xiao-mi power bank.

3.5 *Parse Cloud*

A cloud-based free server which stores data in tabular form makes it a perfect choice for hosting a retail inventory. The added features include adding users (customers) to the database as they buy items from the shops. Analytics, the prime requisite, are facilitated by the Parse cloud itself. Further analysis can also be accomplished by downloading data and monitoring it on a regular basis. The cloud also provides two modes to access the server, master mode and user Mode, giving security and integrity to the inventory data. The server also provides reports on crashes and other performance parameters. For convenience, Parse also provides a REST API documentation which helps any Linux-based embedded hardware for easy integration [18]. The above-mentioned features make Parse the best choice for demonstrating the prototype Fig. 7 is the main window of the Parse cloud.

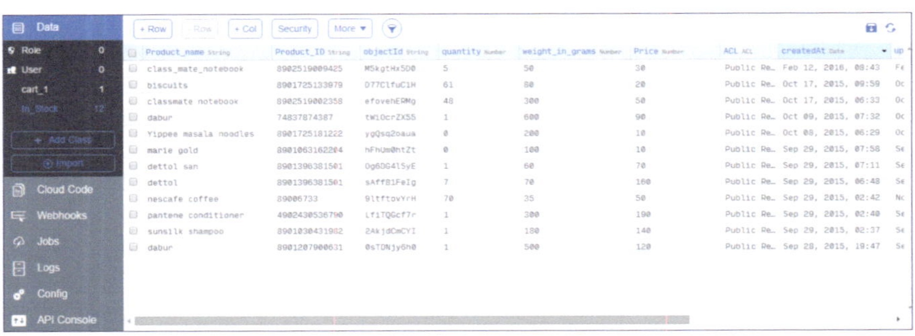

Fig. 7 Parse database

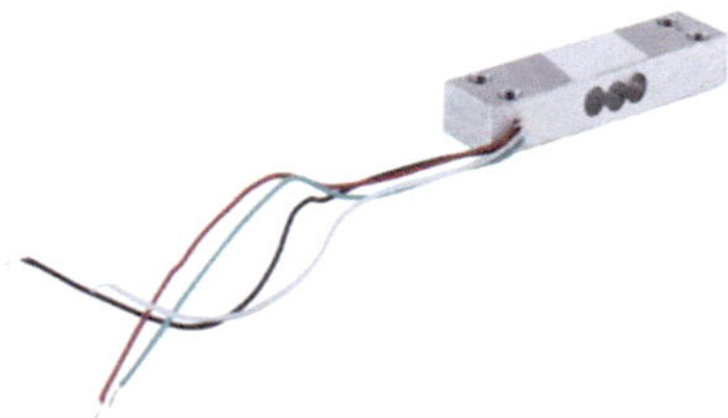

Fig. 8 Load cell CZL601

3.6 Load Cell

The load cells shall work as one of the evaluating module in the cases when the activities involving scanning and the product placing in the cart are not synchronous to each other. For instance, a customer scans a product but places it back on the shelf without removing it from the E-cart/E-bill. Further, it also handles dishonest customers and avoids larceny in the automated stores without the need of manual security. In few cases, the system shall fail when two products of same weight and different prices comes in the play. For such cases, the secondary feedback system comes into action. Figure 8 [19] shows the load cell CZL601 which was used to design a weighing mechanism (placed at the base of the trolley) with a load capacity of 5–60 kg and a rated output of 2.0 mV/V ±5% [20]. The required calculations were done for the same [21–23].

3.7 Product Imaging

Product imaging takes a real-time feed of the top layer of the smart cart all the time and cross verifies the products visible with the scanned products. In case if any additional item comes into picture which has not been scanned previously, the cart is locked and is not functional to the customer. Also, the owner is informed about the same. This is achieved by means of scale invariant feature transform (SIFT) algorithm [24] which acts as a product tracking mechanism for the cart. Based on the comparison between database images and the real-time images taken by an arbitrary camera, a match score is generated which when less than a particular value signifies that the object is not present in the cart and is scanned by the customer. The loopholes present in the system can be nullified by the above load cell weighing mechanism implemented in the cart. As such, both these feedback systems together eliminate the chance of any discrepancy. Figure 9 shows the working of the SIFT algorithm.

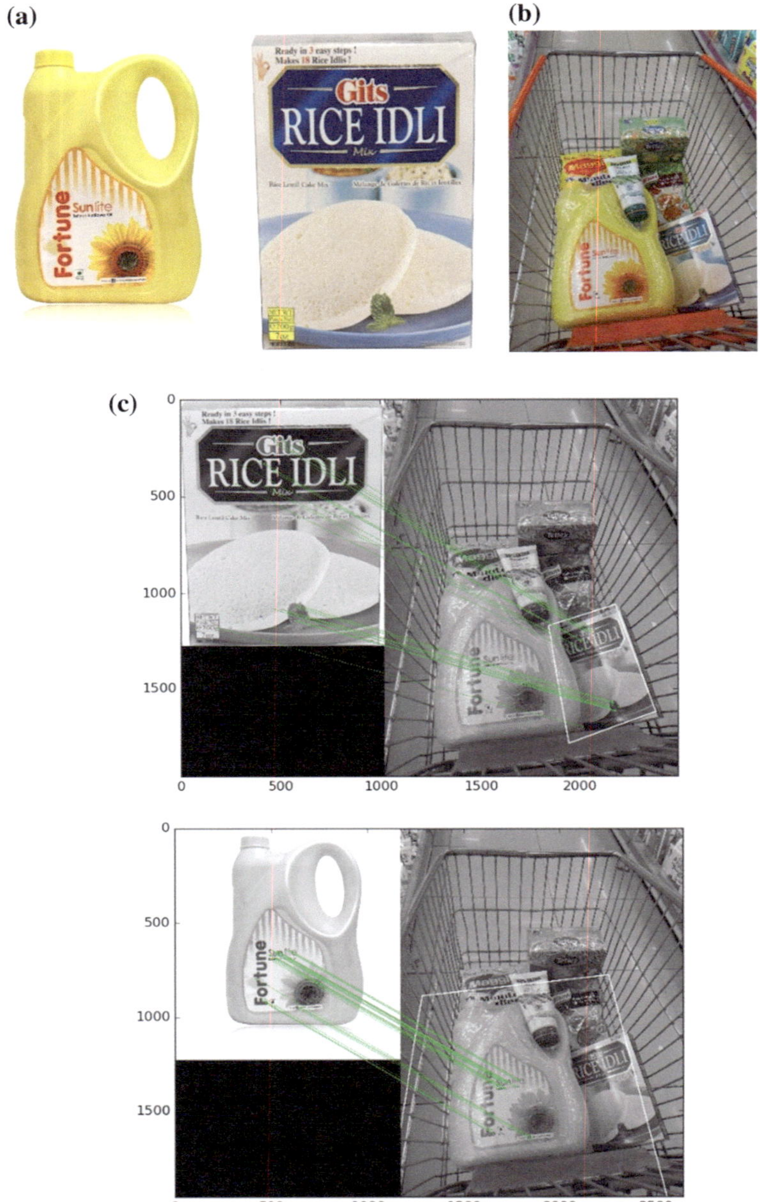

Fig. 9 **a** Dummy product images stored in the database, **b** dummy real-time image of the trolley, **c** SIFT algorithm implementation results

4 Simulation and Results

4.1 Hardware Implementation

The hardware platform for the proposed design has been built using Beagle Bone Black Rev C with Debian OS flashed on the eMMC [25]. The external modules comprised of USB port extender connected to the USB client of the Beagle Bone Black. The barcode scanner, Wi-Fi adapter, and USB CAM as mentioned above are connected to different ports of the port extender. The load cell is connected to one of the channels of the 12-bit ADC. A SPDT switch has been added to the prototype for simplifying the process of removing and adding items to the cart. LEDS are also added for proper guidance to the customer when products are added to the cart and removed from the cart. In case of any discrepancies, the cart generates an error signal indicated by a set of LEDs on the cart. The hardware does not comprise of dedicated display but provides a placeholder for the mobile which connects to the Internet/router provided, but the smart shop acts as a cost-efficient method of displaying the current bills. The power bank shall be connected when it runs as a standalone device. Figure 10a shows the placement of every component on the real cart. Figure 10b–d show the prototypes of the cart in emulation and stand-alone mode.

4.2 Software Environment

The software environment in the design draws the use of SSH (secure shell login) via any Linux-based OS onto the platform or via the Shell terminal Putty. Figure 11a shows the live window of the software. Beagle Bone Black also requires a couple of drivers and libraries in order to interface with the USB-based modules. It uses the RTL-8188eu driver for USB Wi-Fi adapter [26]. It uses the PyUSB library [27, 28] for interfacing with the raw barcode USB data. OpenCV is used for implementing the image processing segment of the project.

A program was also made for the owner to add or remove products, search for products, update fields like name price of the product and restock from his end when needed. Figure 11b shows the run screen of the program at the software end.

4.3 Results

Parse provides a facility on the cloud to analyze data up to a certain extent. If one needs to analyze data further, advanced data analytics software can be used. Parse

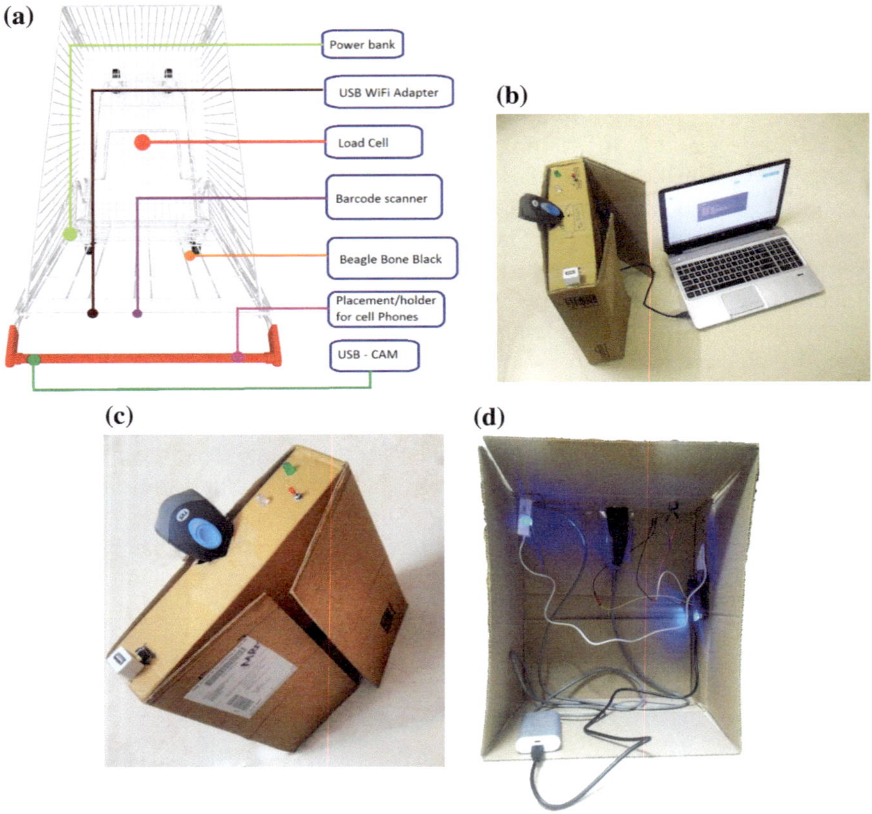

Fig. 10 **a** Placement of modules on the cart, **b** smart cart in emulation mode, **c** smart cart as a stand-alone device, **d** placement of modules in as a prototype cart

facilitates the owner showing him the daily in stock, alarming him when a product is out off shelf and needs immediate restocking. It also provides a graphical analysis of activities—adding/removing the products in a cart, restocking the shelves, etc., on daily basis (responsiveness). Figure 12a, b present the graphical analysis on the in stock and responsiveness on daily basis. Figure 12c shows the dummy in stock created to test the system.

(a)

(b)

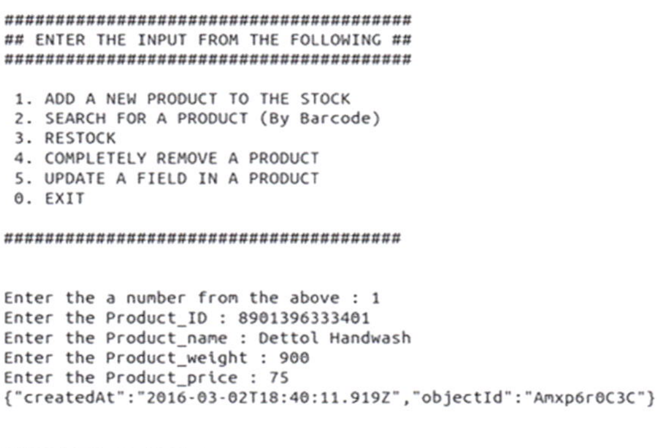

Fig. 11 **a** Simulation of the database update on Putty when cart in emulation mode, **b** simulation of the In-Stock update software at the owner end

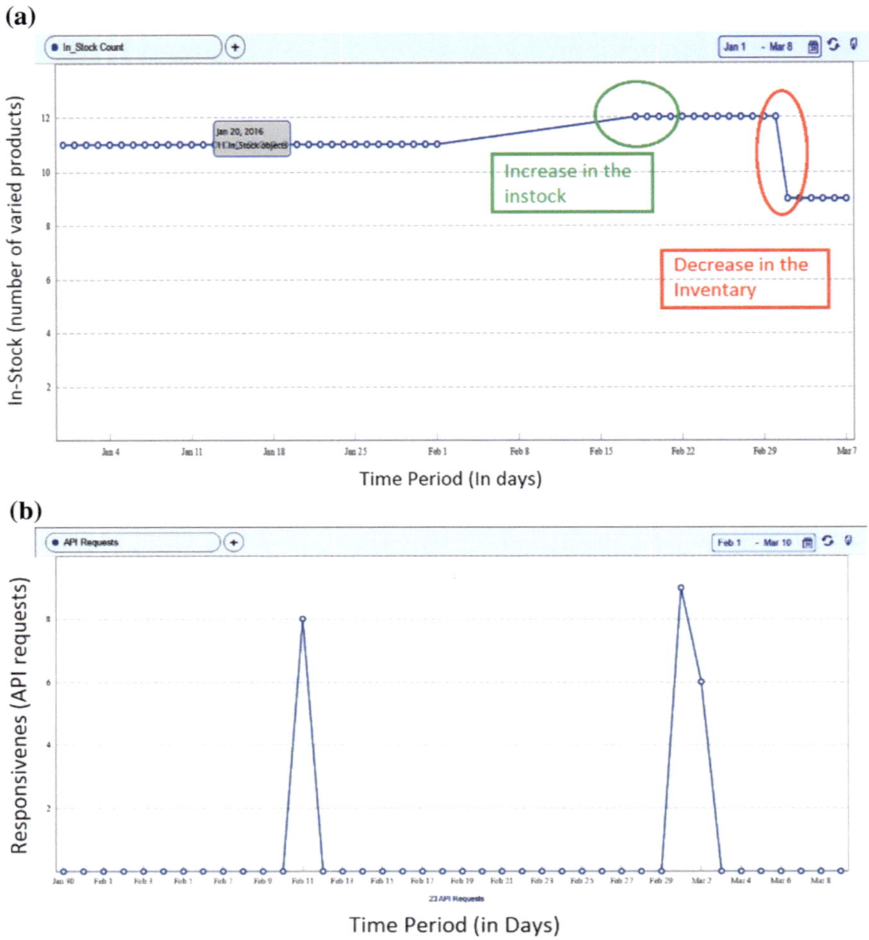

(c)

Product_name String	Product_ID String	objectId String	quantity Number	weight_in_grams Number	Price Number	ACL ACL	createdAt Date	up
class_mate_notebook	8902519009425	MSkgtHx5D0	5	50	30	Public Re…	Feb 12, 2016, 08:43	Fe
biscuits	8901725133979	D77ClfuC1H	61	80	20	Public Re…	Oct 17, 2015, 09:59	Oc
classmate notebook	8902519002358	efovehERMg	48	300	50	Public Re…	Oct 17, 2015, 06:33	Oc
dabur	74837874387	tWiOcrZX55	1	600	90	Public Re…	Oct 09, 2015, 07:32	Oc
Yippee masala noodles	8901725181222	ygQsq2oaua	0	200	10	Public Re…	Oct 08, 2015, 06:29	Oc
marie gold	8901063162204	hFhUm0htZt	0	100	10	Public Re…	Sep 29, 2015, 07:58	Se
dettol san	8901396381501	Og6DG41SyE	1	60	70	Public Re…	Sep 29, 2015, 07:11	Se
dettol	8901396381501	sAff81FeIg	7	70	160	Public Re…	Sep 29, 2015, 06:48	Se
nescafe coffee	89006733	9ltftovYrH	70	35	50	Public Re…	Sep 29, 2015, 02:42	Nc
pantene conditioner	4902430536790	LfiTQGcf7r	1	300	190	Public Re…	Sep 29, 2015, 02:40	Se
sunsilk shampoo	8901030431982	2AkjdCmCYI	1	180	140	Public Re…	Sep 29, 2015, 02:37	Se
dabur	8901207900631	0sTDNjy6h0	1	500	120	Public Re…	Sep 28, 2015, 19:47	Se

Fig. 12 **a** Analysis of the In-Stock on monthly basis, **b** analysis of the responsiveness of the store on daily basis, **c** dummy In-Stock

5 Conclusion

Parse cloud's tabular format of storing data provides an effectual, versatile and easy mechanism for implementing the proposed retail-based smart cart shopping system. This retail entity combines the advantages of cloud-based service and makes the auditing and management of not only one but multiple retail stores at ease. The proposed feedback unit adds reliability to marketplace and also reduces the labor security. The SIFT can be applied only in case of a constant light luminosity in the surrounding. For faster computation and real-time feedback, the images stored in the database must be comparatively small in size such that the output match score is also not affected. Limitations include aberrant behavior of SIFT in dim light environment and minimal investment in the production of customized central hardware resulting in an option a bit extravagant compared to the options such as smart phone app to scan barcodes and do all the processing. While on the other side, the above design proves advantages at places where security in retail comes into picture along with using the contemporary barcode scanners still present in most stores avoiding E-waste and dependency over smart phones. Further, the smart cart's payment system can be implemented by using either of the technologies RFID or NFC connected to credit/debit card swiping machines present at remote locations of the store, which read the corresponding tags present on the carts and gain information of the total items present in the cart. The total bill can further be generated and send them to their e-mails. Future scope includes adding more computation at the cloud end, alarming the owner whenever the product is out of stock and adding data analytics for analyzing customer behavior for profit making. Implementing such a system in retail shall bring down labor dependency, easy management at the owner's end, and in turn will improve the pre- and post-shopping experience.

Acknowledgements The following is a note stating that in this paper is a result of cumulative hardwork of all the authors, i.e., Prithvish, Shraddha, and Prof. J. S. Rani Alex as mentioned above. All the authors contributed equally in the research work resulting in the following paper. We would also like to thank Prof. J. S. Rani Alex for reviewing the paper and guiding our way through the research providing immense support.

References

1. Chandrasekar P, Sangeetha T (2014) Smart shopping cart with automatic billing system through RFID and ZigBee. IEEE. ISBN: 978-1-4799-3834-6/14
2. Ali Z, Sonkusare R (2014) RFID based smart shopping: an overview. In: 2014 international conference on advances in communication and computing technologies. IEEE. ISBN: 978-1-4799-7319-4/14
3. Roussos G. Enabling RFID in Retail. Brubeck College—University of London

4. Wenxing O, Lei W, Zhipeng J, Changhong Y (2015) Implementation of smart shopping system based on NFC Technology. In: Seventh international conference on measuring technology and mechatronics automation. IEEE. ISBN: 978-1-4673-7143-8/15
5. Gao JZ, Prakash L, Jagatesan R (2007) Understanding 2D-barcode technology and applications in m-commerce—design and implementation of a 2D barcode processing solution. 31st annual international of computer software and applications conference, 2007. COMPSAC 2007. IEEE. ISBN: 978-0-7695-2870-8, ISSN: 0730-3157
6. IoT Agenda (Online) (2015). http://internetofthingsagenda.techtarget.com/definition/Internet-of-Things-IoT. Accessed on 14 June 2015
7. Sinha N, Alex JSR (2015) IoT based iPower saver meter. Indian J Sci Technol 8(19) (IPL0142)
8. Sinha N, Pujitha KE, Alex JSR (2015) Xively based sensing and monitoring system for IoT. Accepted in 2015 international conference on computer communication and informatics (ICCCI-2015), Coimbatore, India, 08–10 Jan 2015
9. Blau J. Supermarket's futuristic outlet
10. TI Processor onboard on Beagle Bone Black (Online). http://www.ti.com/product/AM3359. Accessed on 14 June 2015
11. Beagle Bone Black (Online). https://beagleboard.org/black. Accessed on 14 June 2015
12. Beagle Bone Bloack JPEG image (Online). https://encrypted-tbn2.gstatic.com/images?q=tbn:ANd9GcSD5-qb_dxg1vJuk4apuEKMOUirgjIhY_W7h4gPp3uDEYstg_StUw. Accessed on 14 June 2015
13. TVS BS-C101 Barcode scanner (Online). http://www.tvs-e.in/product-detail-db.aspx?id=45&typ=btn_bt&pro=tb_4. Accessed on 14 June 2015
14. TVS BS-C101 Barcode scanner JPEG image (Online). http://bwindia.net/bwfiles/images/Barcode%20Scanner%20TVS%20CCD%20Based%20USB%20101%20Star%20Handheld.jpg. Accessed on 14 June 2015
15. USB Wi-Fi Adapter—TL-WN725N (Online). http://www.tp-link.com/il/products/details/cat-11_TL-WN725N.html. Accessed on 14 June 2015
16. USB WiFi Adaptor TL-WN725N image (Online). https://encrypted-tbn3.gstatic.com/images?q=tbn:ANd9GcSSNjRMJXw4uhjEp-pVBC5b_fcsCll9iTeAAB0abhA4Pm2-bbqhwczHkF0. Accessed on 14 June 2015
17. MI Power bank image (Online). https://encrypted-tbn2.gstatic.com/images?q=tbn:ANd9GcRxIVJsN2jejtFYYdO1k7cJX7JEr26kg05SfW7EordEh6UG0qDDvdeb7w. Accessed on 14 June 2015
18. Parse Cloud (Online Document). http://www.parse.com/. Accessed on 14 June 2015
19. Load cell image (Online). https://encrypted-tbn0.gstatic.com/images?q=tbn:ANd9GcSwy3Bhb94hKnAeIKJvKMrz0Bo2QqvoYEvE7NjqJi_QdoqpePOUc6j174k. Accessed on 14 June 2015
20. Load cell specification sheet (Online Document). www.kadcontrols.com/KadWebsite/LoadCell_Tech/CZL601.pdf. Accessed on 14 June 2015
21. Hernandez W (2006) Improving the response of a load cell by using optimal filtering. ISSN 1424-8220© 2006 by MDPI
22. Dahikar PB, Gaikwad KD (2013) Design and development of novel weighing scale system. Int J Eng Res Technol (IJERT) 2(5). ISSN: 2278-0181
23. Gangwal U, Roy S, Bapat J. Smart shopping cart for automated billing purpose using wireless sensor networks. ISBN: 978-1-61208-296-7
24. Lowe DG (2004) Distinctive image features from scale-invariant key-points. Int J Comput Vis
25. Beagle Bone Black Debian OS eMMC images (Online Document). https://beagleboard.org/latest-images. Accessed on 14 June 2015
26. Realteak Drivers (Online Document). http://www.realtek.com.tw/products/productsView.aspx?Langid=1&PNid=21&PFid=48&Level=5&Conn=4. Accessed on 14 June 2015
27. PyUSB (Online Document). https://github.com/walac/pyusb. Accessed on 14 June 2015
28. PyUSB (Online Document). http://walac.github.io/pyusb/. Accessed on 14 June 2015

Voting System for India

Shrikant Subhash Warghade and B. Karthikeyan

Abstract While many steps have been taken by the government, the voting percentage is not as per the expectations. In this paper, a new voting system is proposed for India. The proposed system is the combination of the currently working system as well the use of AADHAAR database. The voter's data, i.e., the voter identity number mentioned on the election card provided by Election Commission of India and AADHAAR details are saved. This stored data are used at the time of voting, and the status of the voter will be updated.

Keywords AADHAAR · EVM · Census · Migration

1 Introduction

In democratic nations, voting is done in order to exercise the right of the people, to whom the nation wants to see as their leader, to whom they want to take the decision for the benefit of the country. In India, the voting is done by two methods, the voting by ballot paper and the voting conducted by using electronic voting machine (EVM). The Election Commission of India tried their best to increase the voting percentage in India, but the results did not show the fruitfulness to their efforts. There can be many reasons behind this. But one of the reasons behind this can be migration of the people from one place to another place.

S. S. Warghade · B. Karthikeyan (✉)
School of Electronics Engineering, Vellore Institute of Technology, Vellore, Tamil Nadu, India
e-mail: bkarthikeyan@vit.ac.in

S. S. Warghade
e-mail: shrikantsubhash.warghade2015@vit.ac.in

D. Thalmann et al. (eds.), *Intelligent Embedded Systems*, Lecture Notes in Electrical Engineering 492, https://doi.org/10.1007/978-981-10-8575-8_7

2 Literature Survey

As per the available data, the average voting percentage in the 16th Lok Sabha was 66.38% [1]. This was the highest voting percentage in the history of India. In the 15th Lok Sabha, the average voting percentage was 59.7%. As per the census report of India [2], the total number of people 'migrant by place of birth' was 307,149,736 which are 29.9% of the total population. The reasons for migration can be different, but the important thing to be considered is that most of these migrants are not able to visit their area where they need to cast their vote. The percentage of the migrants is not small so as to be ignored. A holiday is declared on the voting day, but a holiday is for only one day. A single day is not enough for the people to travel from their present place to the place where they need to go for voting. Statistics regarding the migration take place in India is tabulated in Table 1 using census report.

Using statistics in Table 1, a new voting system is proposed in this paper. The new system will be composed of the voting by EVM as well as the online voting. People who are well aware and able to reach their area can use any of the available methods for voting. Those who are far away and not able to visit their area for voting can use the online voting option.

3 System Overview

There are already some proposed systems which completely rely on online voting technique [3–5]. These systems are proposed as an alternate method for the conventional voting system, but still, those systems are not implemented. In this paper, the proposed system is online but it will include the currently existing method, i.e., the voting by EVM. This will help to make people gradually aware about online voting systems as well as providing a way to resolve security issues associated with

Table 1 Census report on the migration by place of birth

	Category	Migrations by place of birth	Percentage
A	Total population	1,028,610,328	
B	Total migrants	307,149,376	29.9
B.1	Migrants within the state of enumeration	258,641,103	84.2
B.11	Migrants from within the district	181,799,637	70.3
B.12	Migrants from other districts of the state	76,841,466	29.7
B.2	Migrants from other states in India	42,341,703	13.8
B.3	Migrants from other countries	6,166,930	2.0

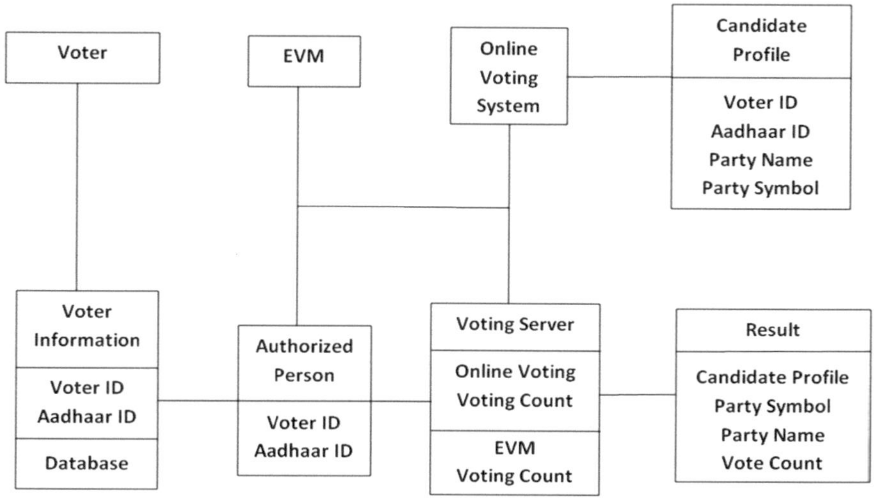

Fig. 1 Block diagram of the system

a purely online voting system. The block diagram explains the working of the system (Fig. 1).

The voter will have the choice to select one of the options: (1) voting by EVM or (2) online voting. The voter's information will be available in the database. The authorized person (field officer) can verify the information. The update from online voting system will be saved to voting server. For people who will choose the EVM voting option, the EVM status will be updated by authorized person. The candidate profile will be there in the database. The result from both the methods can be collected at the end point.

3.1 *The Voter Information*

As per the available information, 92% of the adult population in India above 18 years has acquired the AADHAAR [6]. The information of AADHAAR holders will be added to the database by the Election Commission of India. For those who have not received the AADHAAR unique identity number, their Voter ID number will be added to the database. Either of the details, i.e., Voter ID details or AADHAAR card details will be sufficient on their own (Fig. 2).

Voter Name	Image	AADHAAR ID	Voter ID	Voting Option	Status of Voting
Aakash koli		6523 4879 6987	PKS6478163	Online ● EVM ○	Done ● Not Done ○
Satish Solanke		2546 6398 7561	HFA8561247	Online ○ EVM ●	Done ● Not Done ○
Deepali Verma		6236 6978 6951	AYD8746952	Online ○ EVM ●	Done ● Not Done ○
Amit Sonurkar		2536 4536 6397	PSC7896541	Online ○ EVM ●	Done ● Not Done ○

Fig. 2 Voter database

3.2 *Candidate Information*

As per the area and ward, the information of the candidates those who are taking part in election will be added to the database. It will be same as that of the data which are saved in the EVM. Hence, the ward wise candidate information will be available online as well as in the EVM (Fig. 3).

Candidate Name	AADHAAR ID	Image	Party Name	Symbol	Select to Vote
A. Deshmukh	4552 8647 9875		NCP	NCP	○
A. Jagtap	3245 6389 7854		BJP	BJP	●
A. Shelke	6578 4896 3541		BSP	BSP	○
S. Patne	9658 4875 6325		Congress	Congress	○
A. Murade	2587 9648 3645		Shiv Sena	Symbol of Shiv Sena	○

Fig. 3 Candidates database

3.3 Way of Conduct

For people who are not able to travel to the voting booth from where they reside, the online option will be the most suitable one. The people who can visit their voting booth will have both the options of voting. For the people who are going to select the voting option by EVM, the field officer (who will be on duty at that booth) will update the status of that person as EVM and his voting will be marked by that method. This person will follow the usual method of voting by EVM. On the other hand, for the person who is not able to travel to his voting booth as per area and ward, they just need to visit the nearest voting booth. The information of all areas and wards will be available. They will select the online voting option. The person can cast their vote to the desire candidate. The flow chart will give the details of that as in Fig. 4.

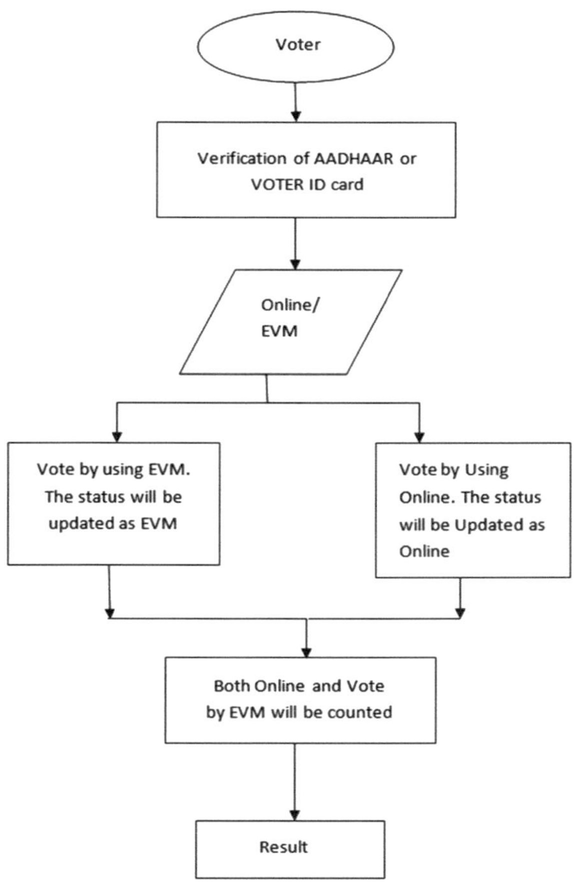

Fig. 4 Flow chart of voting process

4 Security Issues

Internet-based online voting system has some security issues [7]. The attack can be done at the following points:

- the server
- the client
- the communication path

In the penetration attack, the use of delivery mechanism transports the malicious payload to the client in the form of Trojan horse or control program. After the execution of this, it can spy on the ballots, stop voters from casting their votes, and sometimes even worst it can modify the votes. The threat is that this attack cannot be detected by security mechanisms such as encryption and authentication and secure hypertext transport protocol (https) [7].

The communication path refers to the path between the client and the server. This path must be secured when the vote from client end to server end is transferred. The current cryptographic technologies such as public key infrastructure are sufficient to preserve the confidentiality over the authenticated communication link between the client and the server [7].

5 Advantages of Proposed System Over Present System

Considering that India is the biggest democracy in the world, the following advantages are expected [7]:

Accessibility: With the present development in the field of communication, Internet is available throughout the globe. This gives the access to all the people to cast their vote from any corner of the world.
Accuracy and expedience: Many problems like rejection, mismarked or invalid votes can be avoided because of online voting system. The manual effort of vote counting can be avoided.
Cost-effectiveness: The cost of elections will reduce to great extent if the online voting system is implemented.
Transparency: The complete system can be transparent if it is managed by the third party.

Present system does not provide the facility for the people who are not in their area, so people from any place will be able to cast their vote. Those people who do not acquire Voter ID card will be able to vote if they have AADHAAR card [10, 11]. On duty people can also vote because of this proposed system. As it includes the present system also, people who are not well aware of the online system can use the conventional method.

6 Result and Future Scope

The online voting system has been already accepted by the countries like UK, Estonia and Switzerland. Also, the municipal elections in Canada and the party primary elections in USA are conducted as online [8]. The proposed system will be helpful for increasing the voting percentage in India. The present figure of voting percentage is expected to increase as it covers the maximum possibilities for voting.

In future, the voting by EVM can be replaced by this method. The AADHAAR database like fingerprints, facial details, retinal details can be used to make the system highly secure and reliable by including fingerprint scanning, face recognition [3, 9]. With the growing speed of technology, this can be achieved in India, when Internet will be available in remote as well as isolated places. A personal contact number can be added to database. Hence, one-time password (OTP) verification can also be used in order to make the system highly secure. A complete online system will save the time and documentation work. It will also help in faster declaration of results. Most importantly, it will help to increase the voting percentage.

References

1. https://en.wikipedia.org/wiki/Indian_general_election,_2014
2. http://censusindia.gov.in/Census_And_You/migrations.aspx
3. Patel R, Ghorpade V, Jain V, Kambli M (2015) Fingerprint based e-Voting system using Aadhaar database. Int J Res Emerg Sci Technol 2(3)
4. Gaike RL, Lokhande VP, Jadhav ST, Paulbudhe PN (2016) Aadhaar based electronic voting system. Int J Adv Sci Res Eng Trends 1(2). ISSN: 2456-0774(Online)
5. Raj RS, Raghavendra A, Madhushree KR, Bhargavi D (2014) An online voting system using biometric fingerprint and Aadhaar card. Int J Comput Technol 1(4). ISSN: 2348-6090
6. http://timesofindia.indiatimes.com/india/92-of-Indias-adult-population-has-Aadhaar-card/articleshow/50662611.cms
7. Aggrarwal K. Issues in implementing of online voting system. Int J Eng Sci Comput 6(5)
8. Kaliyamurthie KP, Udayakumar R, Parameswari D, Mugunthan SN. Highly secured online voting system over network. Indian J Sci Technol
9. Agarwal H, Pandey GN (2013) Online voting system for India based on AADHAAR ID. In: Eleventh international conference on ICT and knowledge engineering, 2013
10. Chakraborty S, Mukherjee S, Sadhukhan B, Yasmin KT (2016) Biometric voting system using Aadhaar card in India. Int J Innov Res Comput Commun Eng 4(4)
11. Khairnar S, Kharat R (2016) Survey on secure online voting system. Int J Comput Appl (0975–8887) 134(13)

Human–Robot Interaction Using Three-Dimensional Gestures

K. Ponmani and S. Sridharan

Abstract In the future of manufacturing industries, it is seen that the robots will replace most of the human workers to achieve faster production. All the industrial areas use automated robotic arms which do certain tasks assigned to them with amazing speeds and pinpoint accuracy. But areas such as medicine, space research, and military robotics require robot arms to be manually controlled to operate with objects that cannot be dealt with human hands. In order to achieve speed and accuracy like automated robots, a new approach, called the leap motion technology, a latest invention in human–computer interaction area is proposed. Hence, three-dimensional gestures are used and gesture recognition method using 3D pointing is schemed for interaction of human and robot using leap motion controller in real time. It facilitates the human pointing 3D gestures for the robot to understand in real time. The arm of robot captures the gestures shown by human partner to interact with the user's gestures and perform the task immediately. The position of the human's hand relative to the head is adjusted so that the speed of the robot is controlled. The status of the robot movement can be viewed in LCD. This proposal designs to advance a valuable 3D pointing gesture recognition system in real time for human–robot interaction.

Keywords Robotic arm · Leap motion controller · Human–computer interaction Three-dimensional pointing gesture recognition · Application programming interface (API) · Internet of Things (IoT)

K. Ponmani (✉) · S. Sridharan (✉)
Department of ECE, Easwari Engineering College, Ramapuram, Chennai, India
e-mail: ponmanikrishnamurthy@gmail.com

S. Sridharan
e-mail: sridharsubbiah@gmail.com

D. Thalmann et al. (eds.), *Intelligent Embedded Systems*, Lecture Notes in Electrical Engineering 492, https://doi.org/10.1007/978-981-10-8575-8_8

1 Introduction

In the beginning of a new era, the bridge was between man and machine. Humans and machines run no longer parallel to each other, but instead, they go hand in hand. To boost lifestyles, this new technology is helping in this scope. The technology used in robotics at the beginning was of a joystick and then came the touch screen, and now it is the advent of gestures [1].

Almost all areas of technology use robots these days. Starting from the popular field of robotics from manufacturing industries to the advanced researches on planet, space missions, robots are used everywhere taking place of humans to tackle the complex tasks quicker and easier. Most of the articulated robots used in manufacturing of cars and electronics are automated, which means they are preprogrammed to do specific set of tasks repeatedly, where human interaction is not required in such operation. However, in some areas, robots are to be manually controlled to do certain tasks that are not programmed earlier. Space and military missions require manual robotic arms to do the real-time operations. Here the need of a controller arises to manipulate the arm.

This paper proposes a new technology to trace and control the robotic arm for different activities. By moving the user's hand in any direction, the movement can be controlled. This paper provides an overview of leap motion technology used in controlling a robotic arm. In gesture recognition technology, a camera observes the various gestures and gives it as input to the computer for processing. A new scope of human–computer interaction which utilizes the computer vision and image processing techniques for detecting the gesture, understanding its objective and making it meaningful for the computer. The challenge lies in the accuracy of the gesture recognition software and device. How well the sensors detect the gesture and how quickly they respond to the user with the meaning of the gesture, somewhat similar to the sign language. Hand gesture recognition is emerging with high potential in research industry. In surgical methods, industrial applications, robotics, and gaming we have this technology as a boon for us.

A robotic arm is made up of numerous joints that replicate human arm joints, i.e., shoulder, elbow, wrist, and fingers. All these joints are driven by DC motors inside, which are operated by the controller. The controller manipulates each joint of the robotic arm. With the invention of the leap motion, we are able to track a whole human hand including joints (wrist, fingers) using infrared rays, accurate to millimeter. This paper walks through the research carried out on a robotic arm integrated with leap motion sensor to mimic the action of a human hand which makes real-time controlling more user friendly than ever before.

2 Related Work

Tremendous amount of work is done as of now, in order to recognize the motion of the hand. In observing the detection of motion of hand, collection of articles has been used. This technique which is used in various scopes contains automatic hand language, automatic sketching, computer graphics, and also industrial robots are used to replace the work of humans. This paper provides the most successful technique to use robotics. In these fields, two types of techniques are used:

(1) contact type and (2) noncontact type.

(1) Contact type: Contact type of devices contains data gloves, electromagnetic tracking system, exoskeleton, etc.
(2) Noncontact type: This type contains vision-based system, speech recognition, camera-based, etc. [2].

The technique used here comes under noncontact type since it uses a leap motion sensor to track the hands and fingers of humans. Unalike Microsoft Kinect, leap motion does not produce access to raw data in the form of a cloud of points. Captured data is handled by proprietary drivers supplied by vendor and accessible through API [2]. For human–computer interface, leap motion was proposed, not as general purpose 3D scanner; hence, it is enhanced for recognizing human hands and pointy objects.

3 Existing Method

In existing method, two-dimensional gestures were used to predict the movement of human–robot interaction. Existing system tracks only the movement of single hand of the user. The hand gestures of the human are predefined so that the system responds only to the predefined gestures. In previous system, the speed cannot be controlled according to the user's hand movement. Body gesture recognition and virtual reality concept are used to interact with 3D environments. Engagement recognition method is used in human–robot interacting game. Microsoft X-box Kinect was used to detect 3D perception of human's motion.

4 Proposed Method

Gesture recognition using 3D pointing method was introduced for interaction of human and robot using leap motion controller. It facilitates the human pointing 3D gestures for the robot to understand in real time. The vector calculations are computed using leap motion by extracting the joint coordinates of hand through which the pointing direction of the human partner is acquired. Thereby, the robot is capable to interpret the 3D pointing gesture shown by the human partner, depending on the coordinate information of human's arm and hand. Speed control of robot can be achieved by adjusting the position of the user's hand relative to the head. The status of the robot movement can be viewed in LCD.

5 Leap Motion Technology

The leap motion controller presents a new gesture and a system to track position with pinpoint accuracy, in the scope of hand gesture controlled human–computer interface. The controller operation is based on infrared optics and camera. In order to face upward, a small USB peripheral device [3] is designed to place on physical desktop. The accuracy in the detection of fingertip position is around 0.01 mm, with a frame rate of up to 300 fps as per the statement of manufacturer. The user has the opportunity to move hand in 3D, just like in real world, as the field view of the controller is up to 150°. 3D representation of leap motion sensor shows how a leap sensor tracks human's hand when placed above it is shown in Fig. 1.

Within it, two monochromatic IR cameras detect the heat signature from the hand and three IR LEDs occupy their position which keeps track of the movement of the hand [4]. Together they provide the data of positions of each joint in hand including the bone of fingers. Instantaneously, it can trace all 10 of the human fingers [1]. To a distance of about 1 m, the device observes a roughly hemispherical area. The leap motion sensor with its coordinate axis is shown in Fig. 2.

The sensor provides the position of user's hand and fingers represented in 3D coordinates i.e. (X, Y, Z) [4]. The data that is obtained from the leap motion sensor contains various details of hands such as position, orientation, frame ids, and finger bone information. The captured data [5] by the camera is fed to the leap motion

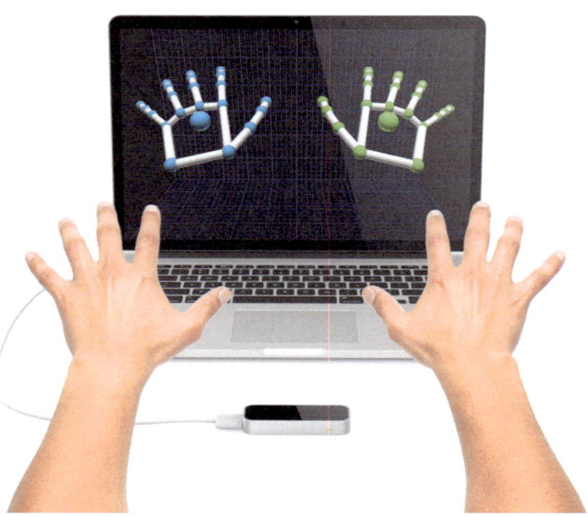

Fig. 1 3D representation of leap motion sensor shows how a leap sensor tracks your hand when placed above it

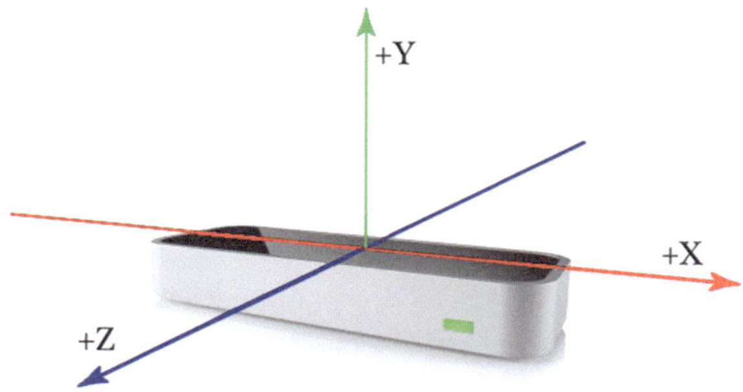

Fig. 2 Leap motion sensor with coordinate axis

controller software in the computer. After obtaining the position of the hand from the sensor it cannot directly send them to the robotic arm, because the arm does not know how to reach the position until its joints are turned to specific angles. This outstanding functionality is that it makes a good choice to integrate with a robotic arm.

6 Human–Robot Interaction

The human–robot interaction mechanism describes the interaction that occurs within the human and robot. The gestures performed by the user are recognized by the leap motion controller, and the task is performed by the robot using three-dimensional pointing recognition method. The function of each module presented within the human–robot interaction model is represented in Fig. 3.

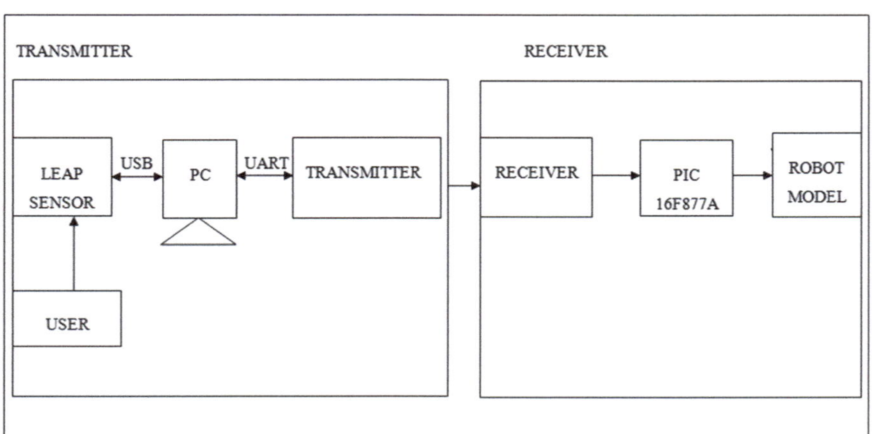

Fig. 3 Function of human–robot interaction

The block diagram represented here regarding the human–robot interaction describes the mechanism underwent during the process. This mechanism consists of two sections, namely transmitter section and receiver section.

When the user shows the hand gestures above the leap sensor, it can sense and detect the user's hand motion up to 2 ft above the sensor and track those images. The three-dimensional image sensing is developed using those tracked images from the leap motion controller, and the gesture done by the human can be viewed in the PC to which the leap motion controller is been connected. The gestures recognized by the image sensing system are transmitted to the receiver side of the robot system.

The gestures received are interfaced along with the PIC 16F877A microcontroller within the robot model. Thereby, according to the movement of the human, the action to be performed by the robot is displayed in LCD. Accordingly, the recognized image and from the instructions displayed in LCD the action is performed by the ARM of the robot.

The hand gestures from the leap motion controller are sensed and simulated in the PC through the Unity software for which the application is developed by SeeSharp coding. Similarly, on the receiver side, for the robot to interact with the human gesture the code is been developed using the Embedded C through the MPLAB IDE.

7 Software Modules

The hand gestures are recognized and captured using Unity software; using SeeSharp language the application is developed. Embedded C is used in coding the PIC microcontroller to interface with the robot model, and it is compiled using MPLAB IDE compiler.

7.1 Unity

The leap motion controller system comprises of a hardware device and a software component which runs as a service on the host computer. The images originated from the hardware are scrutinized by the software component and send tracking information to applications. In order to acquire data, the leap motion Unity plug-in is connected to this service. The scripts incorporated with the plug-in translate the leap motion coordinates to the Unity coordinate system.

7.2 MPLAB IDE

MPLAB IDE is a software program to expand applications for microchip microcontrollers which runs on a PC. Microchip PIC microcontrollers are used as a design desktop for the application progress of embedded designs with MPLAB Integrated Development Environment (IDE), which is a comprehensive editor and project manager. MPLAB IDE, that runs on a PC which incorporates all the components, is required to design and extend the applications of embedded systems. Simulators are assembled into MPLAB IDE so that a program can be verified without any further hardware.

8 Simulation

The source files are to be dumped, in order to create code that is executable by the target PIC microcontroller MCU. Further, the language tools (assemblers, compilers, linkers, etc.) can be selected so that the code can then be built into executable code. The project manager in MPLAB IDE controls this process. The hand gesture shown by the human is recognized using leap motion controller, and the simulated images in PC are shown in Fig. 4.

Fig. 4 Simulation of hand gesture

Fig. 5 Arm of the robot designed

9 Implementaion

The arm of robot model is designed to interact with the human and to perform the task or action of human. The robot designed by acrylic material using DC motors is shown in Fig. 5.

The leap motion controller is ported to the PC and synchronized to the Unity software. PIC microcontroller is interfaced with the designed arm of Robot. The prototype designed is implemented to perform the action of the human. The prototype implemented is shown in Figs. 6 and 7.

10 Advantages

- The position of the human's hand relative to the head is adjusted to control the speed of the robot.
- There are no predefined gestures.
- It can recognize both the hands simultaneously.
- It is applicable in the absence of human.

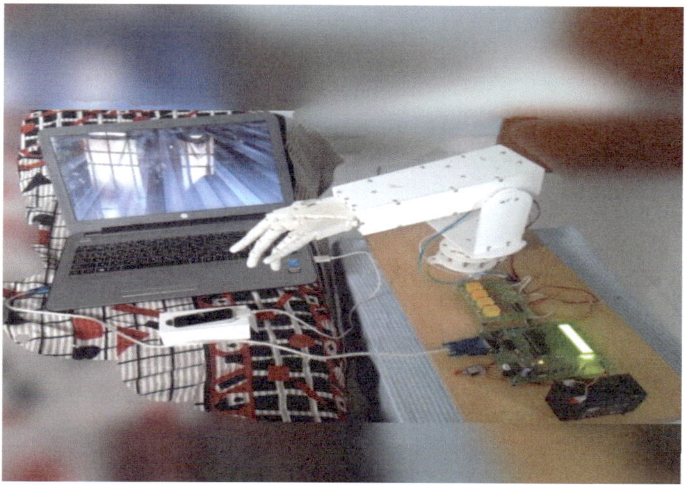

Fig. 6 Implementation of human–robot interaction

Fig. 7 Flow graph of human–robot interaction

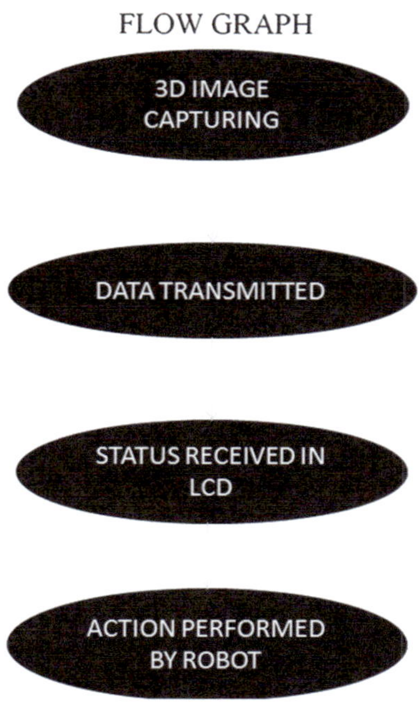

11 Conclusion

The three-dimensional pointing gesture recognition method is proposed, where the hand gestures shown by human are sensed by the leap motion controller and simulated through Unity software in PC. Here, the user can choose the gestures and the human gestures are shown preferably to the user's speed. Thereby these gestures are interpreted by the robot that is interfaced along with the leap controller, and the action to be performed by the robot is displayed in LCD. From the status shown, the robot performs the action. The system designed here is for the robot to perform any task according to the gestures interacted with the human.

12 Future Scope

The designed prototype, in future, can be implemented to perform the particular task in various fields, especially in the medical field. This can be utilized in medical applications as in case of absence of doctors in hospital at the required time; then this system can be implemented using Internet of Things (IoT) by fixing cameras to the robot. So that doctor from their own place can interact with the robot to complete the task.

References

1. Ugale A, Chandwadkar DM (2016) Overview on latest gesture controlled systems for robotic arm. International Journal of Computer Applications (0975–8887) 135(1)
2. Savatekar RD, Dum AA (2016) Design of control system for articulated robot using leap motion sensor. Int Res J Eng Technol (IRJET) 3
3. Sathiyanarayanan M, Mulling T, Nazir B (2016) Leap motion device for gesture controlling an unmanned ground vehicle (robot). IJEDR 4(4); ISSN: 2321-9939
4. Venna TVSN, Patel S (2015) Real-time robot control using leap motion a concept of human-robot interaction. In: 2015 ASEE Northeast section conference
5. Khelil B, Amiri H (2016) Hand gesture recognition using leap motion controller for recognition of Arabic sign language, University of Tunis El Manar, Tunis, Tunisia

Integration of the Smart Phone and IOT for Smart Public Toilet Hygiene Monitoring System

Prashant Namekar and B. Karthikeyan

Abstract As the number of inhabitants in the nation is expanding day by day, it gets to be distinctly hard to deal with the cleanliness of the urban areas. One vital reason for this issue is some time for individuals it is hard to find the public toilet and even if one finds it then there may be an issue of cleanliness with the one found. With help of the recent technology known as IOT, we are proposing a solution for this issue. Sensors will be conveyed in different public toilets which will accumulate the cleanliness parameters and this acquired information will be uploaded to the database. This information can likewise be utilized by maintenance office for further activities. Utilizing this information, one can without much of a stretch get the exact location and the cleanliness level of the toilet utilizing our android application on his smart cell/tablets.

Keywords IOT · Android · Raspberry Pi

1 Introduction

In urban areas, it becomes quite difficult to find public toilets and another problem with it is the cleanliness of it. The toilets are mainly affected by the toxic gasses like ammonia (NH_3), hydrogen sulfide (H_2S).

These gasses are heavier than air, virulent, corrosive, flammable, and explosive. It contains mixes dispensed by the body which is undesirable and can bother to skin

P. Namekar
Embedded Systems, Vellore Institute of Technology, Vellore, Tamil Nadu, India
e-mail: pjn.prashant@gmail.com

B. Karthikeyan (✉)
School of Electronics Engineering, Vellore Institute of Technology, Vellore, Tamil Nadu, India
e-mail: bkarthikeyan@gmail.com

D. Thalmann et al. (eds.), *Intelligent Embedded Systems*, Lecture Notes in Electrical Engineering 492, https://doi.org/10.1007/978-981-10-8575-8_9

and eyes and may lead to unhealthy conditions. In order to tackle this, we are developing a system which will be effective to all these issues.

For prototyping purpose, we are considering only single toilet with different sensors. But in reality, many such integrated systems can be created.

To detect the various toxic gasses and toilet humidity, we used the odor sensor and humidity sensors respectively. To know the water availability in the selected toiled, water-level detecting mechanism is used. An ultrasonic sensor is placed in the tank caring the wastage generated from the toilet to inform its level to maintenance authority.

The Raspberry Pi B+ is used as a smart central controller to collect the sensors data and provide an interface to the server. Hereby for the purpose of the demonstration, we have developed our own host Web server which will be able to serve many clients. All the data obtained from the RPi will be stored in the database via the server. At the client side which is the Web page and android application, we are providing different logins for the user and the maintenance authority.

IOT does not request any different Web switch establishment or any unique settings on the modem. It works over TCP/IP system alongside DHCP, HTTP, and MQTT like conventions at application layer which we use for the most part while browsing in our PC browser [1].

For the development of the android application, we are making use of the Android Studio. As we will be in need of the user's current location, for this build library are provided by the Google Play Services. Hence, it becomes quite easy to develop a high-end application.

2 Workflow

The complete system is divided into three different parts. The first part deals with the collection of the data from various sensors connected to the Raspberry Pi. The sensors are connected to the Raspberry Pi using the I2C bus.

In second section, each RPi placed in toilets will generate a data packet which will consist of a unique identification number which is assigned to each individual RPi Node, GPS coordinates (latitude, longitude), depending on the location name will be stored in each RPi node, and the data obtained from the various sensors connected to the RPi. This complete packet will be sent to the host server via which it will be stored in the database.

The third and final part is the application level which deals with accessing the data on the server using the user's android application and maintenance authority Web portal.

Figure 1 is showing the complete block diagram of the system. The Raspberry Pi will act as a smart controller which will collect all sensor data connected. Each toilet will be equipped with the GPS module and sensors.

The location (latitude, longitude) of each toilet equipped with this system is registered at the server with its location and a unique identification number assigned to it.

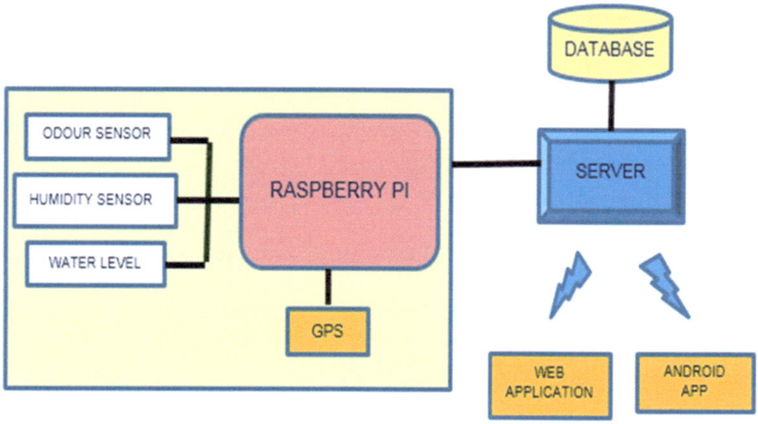

Fig. 1 Block diagram of the system

The user will provide its GPS location to the server using its smart phone, the server, on the other hand, will send back the information of the nearest matching location toilet with the various level of the hygiene corresponding with that toilet.

3 Design Detail

3.1 Sensor Interfacing

The system consists of the sensors like odor sensor, humidity sensor, presence of the water at the selected toilet can be found using the ultrasonic sensor, and even an ultrasonic sensor is placed in the tank caring the wastage to inform its level to maintenance authority.

The toilets mainly affected with the toxic gasses like ammonia (NH_3), hydrogen sulfide (H_2S). The odor sensor selected here is capable of sensing these gasses, and it is able to identify the level of presence of these toxic gasses.

To detect the presence of the water level, we are using the ultrasonic sensor which can inform us regarding the presence of the water level at the selected toilet.

Sometimes there are some cases where the tank caring the human wastage gets overflow which causes inconvenience to the person using the toilet. Hereby we are placing the ultrasonic sensor in the tank which will be able to inform the maintenance authority regarding the level before it gets filled [2].

3.2 Ethernet

An Ethernet cable is required to connect the Raspberry Pi to the cloud.

3.3 Web Application

This section is divided into two sections:

First section deals with the client side which is development of the Web page. The client usually is running on the Web browser. We have used HTML JAVASCRIPT/JQUERY for this purpose. The client-side generates an HTTP request to the server side, which is in the form of request message. The request consists of the method token after that an URI, as we are developing a local host server so we are placing request of the form http://www.localhost:8080/Smarttoilet/. In this case, usually the request consists of the GPS coordinates. We have provided a login page for the maintenance authority as well as for the users. Based on the received data, we have generated different graphical formats.

Second section is the server side, in which we have developed a Java Web hosting server, i.e., Apache Tomcat which is used depending on the client-side query and it will generate an appropriate document and send back to the client. The server authenticates the login information provided by the client side. Based on the requested coordinates by the client to the server side, we have developed an algorithm which will find the nearest matching coordinates and will send back the appropriate information back to the client side, which will be displayed on the Web page [3].

3.4 Mobile Application (Android App)

Web server can also be accessed by using our developed android application which will send its current location using its GPS. And retrieve of the list nearby toilets with their hygiene level [4, 5].

4 Discussion

Figure 2 shows the working web page in which it shows the data obtained for a particular location in different graphical formats. It shows the hygiene level on the Gauge bar. Additionally, it also displays the weather forecaster for the particular selected city.

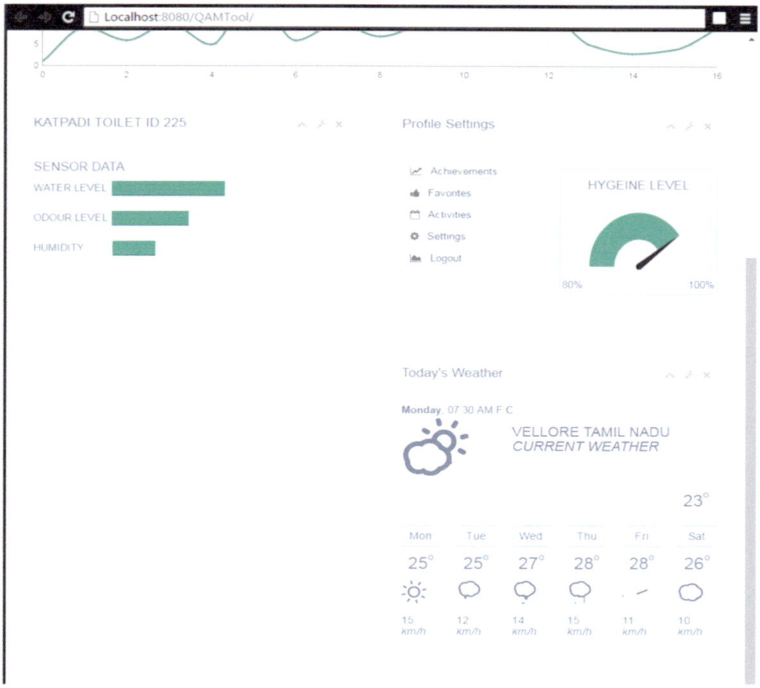

Fig. 2 Web page

The Raspberry Pi [6–8] will collect all data from the sensors after every 30 min. The GPS module GY-NEO6MV2 will also send its data to the RPi. Raspberry Pi operates on 5 V which is provided using a 5 V power supply for the demo purpose. The SD card connected to the RPi is loaded with the Raspbian OS. All the gathered information will be then forwarded to the server where it will first check for the toilet to which the device belong, this information is mentioned in the code dumped in the Raspberry Pi. Once the toilet is found, it checks if the device is registered in that organization by matching the device ID of the device with the device ID of the devices registered in that database. If it matches, it further checks for the authorization token provided at the time of the device registration. If the device is able to successfully verify itself then it is allowed to send and receive the data further to and from. The server will store all the obtained data in the database which can later be accessed by the user via server again.

5 Conclusion

Hence, we conclude that using integration of the smart phone and IOT for smart public toilet hygiene monitoring system; it has become easy for finding hygienic nearby public toilets.

This is a user-friendly system which will also help maintenance authority to analyze and monitor the hygiene of the toilet.

References

1. Internet of Things, European Research Cluster on the Internet of Things (Online), http://www.internet-of-things-research.eu/about_iot.htm
2. Kim B-K, Hong S-H, Jeong Y-S, Eom S-S (2008) The study of applying sensor networks to a smart home. In: Proceedings of fourth international conference on networked computing and advanced information management, pp 676–681
3. Tsai S-M, Wu S-S, Sun S-S, Yang P-C (2000) Integrated home service network on intelligent Intranet. IEEE Trans Consum Electron 46:499–504
4. Miorandi D, Sicarib S, De Pellegrinia F, Chlamtac I (2012) Internet of things: vision, applications and research challenges. Ad Hoc Netw 10:1497–1516
5. Yang LL, Yang SH, Yao F (2006) Safety and security of remote monitoring and control of intelligent home environments. In: IEEE international conference on systems, man, and cybernetics, vol 2, Taipei, Taiwan, 2006, pp 1149–1153
6. Choochotkaew S, Yamaguchi H, Higashino T, Shibuya M (2015) Data prioritization at multi-user IoT gateway with multiple sensor data attributes. DPSWS Bull 5:53–61
7. Bonomi F, Milito R, Zhu J, Addepalli S (2012) Fog computing and its role in the internet of things. In: Proceedings of the first edition of the MCC workshop on mobile cloud computing (MCC'12), pp 13–16
8. Vilros (2014) Raspberry PI user's gide. Vilros Inc., USA

Hyperelliptic Curve Cryptography-Based Lightweight Privacy-Aware Secure Authentication Scheme for Vehicular Ad Hoc Network

Kirti A. Yadav and P. Vijayakumar

Abstract Authentication is one the important parameters in designing security for vehicular ad hoc network (VANET). Authentication protocols based on public key infrastructure (PKI), elliptical curve digital signature algorithm (ECDSA) suffer from overhead problem at the road side unit (RSU) and extracting of certificates from the trusted authority. Traditional identity-based signature scheme is used to provide the privacy and authenticate the vehicle-to-vehicle communication and vehicle-to-road side unit of the VANET. ID-based signature scheme employs elliptic curve cryptosystem form authentication process and also provides batch message verification mechanism. ECC-based traditional scheme requires 160-bit key size to encrypt the message transferred between RSU and vehicle. It requires large memory space as well as consumes more energy and leads to high communication and computational overhead. In order to overcome the above-mentioned limitation, a new lightweight privacy-aware secure authentication scheme is proposed for VANET. The proposed authentication scheme employs the advantage of hyperelliptic curve cryptosystem to sign the message transferred between RSU and vehicle. Since HECC requires lesser 80-bit key size than the traditional signature scheme, it provides greater level of security with less communication and computational overhead. It also traces the real identity of the vehicle using pseudo-identities which eliminates the process of extracting the certificates from the trusted authority. Finally, compare the performance analysis of both traditional and proposed ID-based signature scheme using the different simulation parameters.

Keywords Public key infrastructure · Elliptic curve cryptography
Hyperelliptic curve cryptography · ID-based signature scheme · Communication and computational complexity · Pseudo-identity

---

K. A. Yadav (✉) · P. Vijayakumar
School of Electronics Engineering, VIT University, Chennai Campus,
Chennai 600127, Tamil Nadu, India
e-mail: yadavkirti.ankush2015@vit.ac.in

P. Vijayakumar
e-mail: vijaya.kumar@vit.ac.in

D. Thalmann et al. (eds.), *Intelligent Embedded Systems*, Lecture Notes
in Electrical Engineering 492, https://doi.org/10.1007/978-981-10-8575-8_10

1 Introduction

Nowadays, people are more frequently moving to city for better livelihood with all the basic facilities like medical, security and may be due to job satisfaction, many of them shift to city. Not only this, but due to rapid movement of crowd to city areas, villages are under development so that they could be in the count of smart cities. With this increased demand of the city areas, government has raised many schemes and funds for the projects under smart cities. When considering this development, traffic management comes to major issue that is to be resolved due to the increase in the quantity of vehicles. Traffic management many a times encounters failure due to basic reasons like large amount of vehicles, inadequate space for accommodation of these vehicles. Many a time during natural calamities like floods, landslides, it becomes difficult for management of traffic and moves it to a safe area. For an efficient measure for management of traffic, a new system has been introduced which is called as intelligent transport system (ITS). It is one of the most developing and fast-emerging technologies which is trying to make traffic system smart enough and manage traffic so that hazardous situations may be avoided. Under ITS, there is a system specifically made for vehicles which is called as vehicular ad hoc network (VANET). It is a type embedded system equipped with basic units that are capable of communicating with the nearby vehicles and road side unit (RSU). It mainly consists of on-board unit (OBU) mounted with the vehicle and RSU which is stationary at the road side. With the help of this and some other peripherals, vehicles can easily communicate with each other and also transfer any message related to safety or traffic. VANET emphasizes the aim for sharing emergency message and current weather forecast to avoid traffic and provide passenger safety. While sharing these messages, there are chances that the original message may get altered or even may not be delivered to the destination vehicle or RSU. This may result in entry of a malicious vehicle to create false activities in the system and cause illegitimate functions. Hence, it is necessary to secure the messages while establishing communication. The driver's privacy needs also to be considered in this case. All the equipments, i.e., RSU, control unit, key generation mechanism unit except the user vehicle are registered with government authorities [1]. So it is difficult to cause the illegitimate activity, but the vehicle is a private user which may be easily compromised by an attacker. This paper tries to propose hyperelliptic curve cryptography-based lightweight privacy-aware secure authentication scheme for VANET which will be able to provide lightweight mechanism in terms of encryption of data and also satisfy reduced energy consumption and time consumption as compared to traditional ID-based signature scheme with batch verification.

2 Traditional ID-Based Signature Scheme with Batch Verification

ID-based signature scheme with batch verification consists of five phases namely setup, extract, sign, verification, and batch verification to check the valid signature. In the setup phase, private key generator (PKG) computes the system parameters, i.e., public key for the master private key, two one-way hash functions, and elliptic curve parameters, i.e., the generator point of curve and the prime order of it. In extract phase, user, i.e., the vehicle registers with the PKG using its unique identity. PKG computes corresponding private key and sends to the user through a secure channel. Whenever the user wants to transmit a message, it needs to encrypt it with its identity. For this, in sign phase, the message is encrypted with the unique ID of the user and corresponding private key. The signature of received message is checked for its valid signature in verification phase. Here, if hash of the received message along with hash of its user's identity and computed values satisfies the equation, then the signature is valid. For n-different message verification simultaneously, batch verification is computed wherein the summation of all the received hash composed signatures and hash composed identities must satisfy the computed value to obtain a valid signatures.

Authentication provided in this mechanism is two-way communications, i.e., from vehicle to RSU and RSU to vehicle. Here, pseudo-identities are preloaded for specific duration, and private keys are loaded into the vehicle from the trusted authority during the ID generation and extraction phase.

Vehicle-to-RSU authentication consists of four phases: system setup, pseudo-ID generation and private key extraction, vehicle message signing, and verification of the traffic information messages. In system setup phase, essential system parameters include hash functions and trusted authority public key. In private key phase, trusted authority generates pseudo-identities for respective vehicles and then, the PKG computes the respective private key for the vehicle. During these two phases, all vehicles in the network get registered with the trusted authority and get their corresponding private key. When a message needs to be end, the vehicle sends its message with signature and time stamp. Time stamp helps to provide information whether the pseudo-ID generated is still valid for mentioned particular duration. In batch verification process, n numbers of traffic information messages are confirmed by the RSU by computing the corresponding hash values and validate the signatures.

For RSU-to-vehicle authentication, the phases involved are same; however, system setup phase and pseudo-ID generation are not required. RSU-to-vehicle communication does not require privacy preservation as in case of required for vehicle-to-RSU communication.

This traditional scheme of ID-based signature proposed in [1], which is carried out using ECC, require 160-bit key size to encrypt the message that is to be transferred between RSU and vehicle. Due to this, the communication overhead and computation overhead are increased as the number of encoding bits is increased. In terms of energy consumption also this algorithm requires more power for its computation of 160-bit encryption key [1].

3 Related Work

Generation of pseudonyms is required while providing authenticity for every vehicle. Li et al. [2] proposed an algorithm that utilized RSA, which provides faster computational rate. This scheme employs ECC-based and ID-based online or offline signatures. During the generation of pseudonyms, it does not affect the efficiency of this authentication scheme. Together, this speeds up the authentication process for vehicular communication. In an identity (ID)-based signature proposed in [3] process, a pair of users has to verify each other's signatures without exchanging public key certificates. However, the signer's key is shared with the private key generator authority. The sharing of secret key by various entities can be made using threshold scheme. Different ID-based scheme with threshold can be implemented using pairings with elliptic curves. Using ID-based signature, multiple signatures cannot be verified at the same time. Yoon et al. [3] proposed a scheme which is able to verify multiple messages at the same time and is referred to as batch verification. Biswas et al. [4] proposed an ID-based proxy signature framework with the standard ECDSA for VANETs, with efficient and bandwidth friendly way of signature forwarding. Here, if the proxy signature forwarded from the RSU is verified and accepted by an OBU, OBU then can forward the contents. Biswas et al. [5] proposed an elliptic curve digital signature algorithm (ECDSA) in combination with the identity-based (ID-based) signature. Here, the current position information on a signer vehicle and each individual receiver are used for identification (ID) corresponding vehicle. This eliminates the need for a third-party public key certificate for message authentication in VANETs. Lin et al. provide secure and privacy-preserving protocol based on group signature and identity. It also guarantees the requirements of security and privacy but can also provide the desired traceability of each vehicle in the case where the ID of the message sender has to be revealed by the authority for any dispute event (ID)-based signature techniques. However, the overhead parameter is increased in this approach. So to further enhance the performance and reduce the communication overhead by using a more efficient broadcast authentication protocol is also suggested in this paper [6]. Malhi and Batra used digital signature scheme and aggregate verification scheme for vehicular communications, and the ID-based signature scheme is used for vehicle to RSU. They designed privacy-preserving pseudonymous based security framework, which employs the use of bloom filters in message verification. Password-protected OBU is one of the methods which is suggested here for the faithful consideration of less overhead in maintaining the keys [7].

To ensure security in VANETS, this paper tries to present a comprehensive authentication approach with lightweight encryption using hyperelliptic curve cryptography. Vehicles as a node and road side unit (RSU) (fixed access point) need to share the application data along with a private data from the authentication unit (AU). AU confirms with the authenticity of the vehicle and the security during the communication. Whenever a vehicle needs to send a message, it creates an event with some of the parameters drawn for communication.

$$\text{Event} = \{\text{VID}, \text{AUID}, \text{message}, \text{message ID}, \text{Sign of the vehicle}, \text{Time}\}\ \text{Sign of AU}$$

where VID = vehicle identity, AUID = authentication units identity. The period of validity for a particular event can be calculated through its time stamp determined by AU. Every time a new event is created and a message is to be transferred, vehicle needs to be get authenticated from the AU. This affects the communication delay and also increases communication overhead as every time vehicle has to wait till it gets authenticate for communication. Our proposed method combines the asymmetric encryption using HECC (80 bits) and digital signature with unique identity key each along with proxy server. This proposed infrastructure is an attempt to reduce the delays due to involvement of PKI and AU authentication process

4 Hyperelliptic Curve Cryptography-Based Lightweight Authentication

Authentication is a cryptographic process where receiver of a message gets to know about the source messenger identity and then encrypts message if source identity is legitimate. There are different symmetric and asymmetric ways of encrypting the message which is sent to RSU and other vehicles in network. In symmetric cryptography, also known as private key cryptography, same key is used for both plain text encryption and ciphertext decryption. In asymmetric cryptography, also referred as public key cryptography, pair of keys (private and public) is used. Public key is used to encrypt plain text, and private key is used to decrypt the received ciphertext. Asymmetric encryption is preferred over symmetric as here private key is not known to any other section than the receiver. Whenever a message is to be forwarded, a vehicle signs the message with its own private key obtained from the PKI and attaches its certificate if required. On the other end, receiver vehicle or the RSU compares the key used and verifies the message. To provide security with minimum overhead, common approaches used are elliptic curve cryptography (ECC), RSA, Diffie–Hellman, and hyperelliptic curve cryptosystem (HECC). But the size of the key in (HECC) is less than ECC, Diffie–Hellman, and RSA with same level of security. Therefore, HECC is most compact and safe cryptographic mechanism which offers better performance as compared to ECC and RSA, such as high efficiency, shortest key length, and bandwidth. 80-bit HECC has the same security strength as 160-bit ECC and 1024-bit RSA.

RSU at a time requires more than 2500 to 5000 message to be verified within one second [8]. It becomes difficult to manage at the RSU alone for verifying these messages and confirm its authenticity.

Proposed method makes use of proxy vehicle in order to authenticate the vehicle and reduce burden on nearby RSU. This method not only reduces the time consumption at the RSU, but also using HECC encryption, it tries to secure the message with less number of encrypted key. The important thing required here is that vehicle which is used as proxy must be authenticated first by the trusted authority. For this, the ID of the vehicles that are to be used for proxy authentication must sign in with the trusted authority and then, it can authenticate other mobile vehicles in their area of coverage. Figure 1 shows the brief idea of Proxy ID-based authentication. Here, we can form a group of vehicles presented in a geographical area of about 100 m.

If in case a highway scenario is considered, the average speed of the vehicle with a very small variation is calculated. In this group of vehicles, a proxy vehicle is formed for reducing the burden of authentication. A group of vehicles in the range of 100 m is capable enough to establish faithful communication without a packet loss. A proxy server is formed from the group itself who will be responsible for authentication of the group members. For this, the vehicle that first enters the group or the vehicle who first initiates the communication can be chosen as a proxy

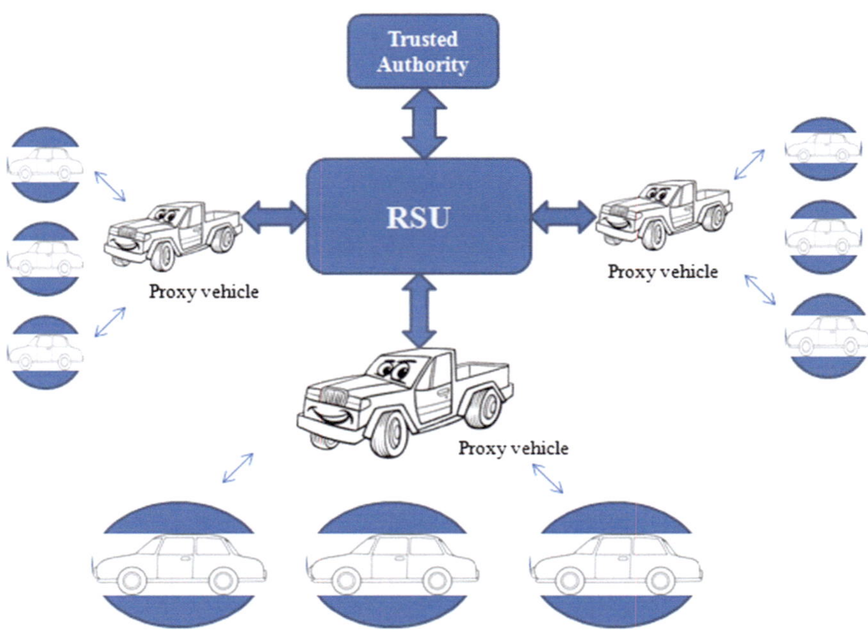

Fig. 1 Proxy ID-based authentication using lightweight HECC encryption

vehicle. Every time when an event is created the message is encrypted with K_{gr} and concatenated with K_{gr} then encrypted with P_{ux}.

$$\text{Event} = \{(\text{message}, K_{gr}|K_{gr}, P_{ux})\}P_{rp}$$

where K_{gr} = symmetric encryption key of the group, P_{ux} = public key of the proxy vehicle, and P_{rp} = private key of the proxy vehicle.

The digital signature algorithm (DSA) is the basis of the digital NIST signature standard. This algorithm can be adapted for elliptic and hyperelliptic curves. The hyperelliptic curve algorithm has not yet been appeared a lot in current literature survey for encryption of the message. So the proposed mechanism tries to inculcate HECC-based encryption along with proxy ID-based signature.

4.1 Algorithm

Deploy the vehicles in the node (Example): 40 vehicles, 5 proxy servers (vehicles), 2 RSU and 1 trusted authority (TA)

1. Choose a vehicle that needs to send message in the network
   a. Vehicle sends its identity (UID) to the TA
   b. TA then validates this identity and sends private key to the respective user (example for vehicle A-private key is P_A)
   c. Vehicle A uses this private key P_A for further message communication in group
2. Vehicle A can act as a proxy server authenticated by the AU to verify the message from other user vehicles. This reduces the burden of RSU to verify the messages.
3. Suppose m is the message to be send by the vehicle A
4. Choose a hyperelliptic curve for encryption of message
5. Take a message for encryption
6. Message along with proxy vehicle identity with group identity is encrypted using HECC
7. The characters of message (m) are converted to its ASCII value, w
8. For signing a message with users identity, digital signature is used
   a. Signing a message w by a user A so that any user can verify the signature: $d_A(w)$
   b. Signing a message w by a user A so that only user B can verify the signature: $e_B(d_A(w))$
   c. Sending a message w and a signed message digest of w obtained by using a hash function standard h: $(w, d_A(h(w)))$

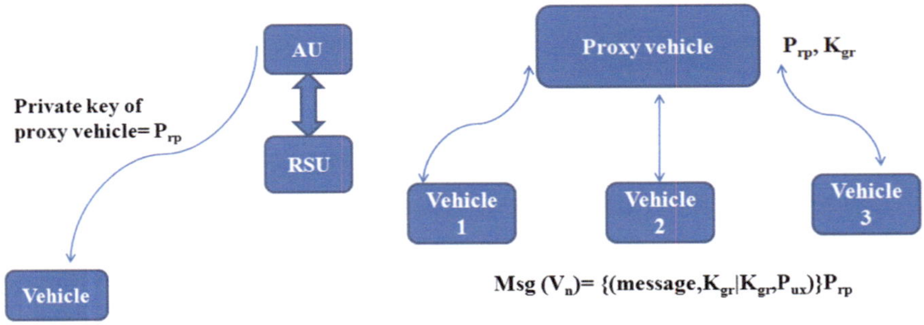

a. Creation of proxy vehicle b. Communication with proxy vehicle

Fig. 2 Establishment and communication with the proxy server (range of 100 m)

The proposed algorithm can reduce the computational and communication overhead with the help of HECC-based encryption (Fig. 2).

5 Conclusion and Future Scope

Key size required for HECC is very much less as compared to that traditional ECC. HECC requires just 80-bits key size. Thus, a less amount of data is required for transfer of message created during an event. With the help of this, communication overhead and the computation time can thus be reduced maintaining the same level of security. For future scope, implementation of the proposed scheme is to be done using NS2 or NS3 simulation.

References

1. Lo N-W, Tsai J-L (2016) An efficient conditional privacy-preserving authentication scheme for vehicular sensor networks without pairings. IEEE Trans Intell Transp Syst 17.5:1319–1328
2. Li et al (2015) ACPN: a novel authentication framework with conditional privacy-preservation and non-repudiation for VANETs. IEEE Trans Parallel Distrib Syst 26(4)
3. Yoon HJ et al (2015) Batch verifications with ID-based signatures
4. Biswas S (2011) ID-based safety message authentication for security and trust in vehicular networks. In: 2011 31st international conference on distributed computing systems workshops
5. Biswas S, Jelena Misic (2013) A cross-layer approach to privacy-preserving authentication in wave-enabled vanets. IEEE Trans Veh Technol 62.5:2182–2192
6. Lin X et al (2007) GSIS: a secure and privacy-preserving protocol for vehicular communications. IEEE Trans Veh Technol 56.6:3442–3456
7. Malhi A, Batra S (2015) Privacy-preserving authentication framework using bloom filter for secure vehicular communications. Int J Inf Secur 15.4:433–453
8. Kudlikar GH, Nagaraj U (2015) A survey on various security schemes in vehicular ad hoc network. Int J Innovative Res Comput Commun Eng 3(11)

A Dynamic Approach of Energy Saving Control Strategy in Smart Homes

S. Sofana Reka and S. K. Pranesh

Abstract This paper focusses on the control analysis with energy saving strategy based on loss aversion analysis in energy management. In smart homes, need of smart meter roles down as an important element. Home area management is considered vital when considering the proper strategy inside the smart meters. Appliances strategy with smart meters at homes is considered in this work using Markovian stochastic Petri nets. The control approaches in the smart meters are analyzed using meta-heuristic algorithm for energy scheduling for appliances. This approach helps in analyzing the control analysis in the smart meters at the operational level building infrastructure. Further, the fulfillment of the consumers at smart homes using this dynamic approach is done. The loss aversion analysis is also considered at energy-level approach for the run time evolution in smart meters.

Keywords HEMS · Smart buildings · Smart meter · Stochastic petri nets
Smart homes

1 Introduction

Home energy management system (HEMS) establishes the basic focus of today modern smart buildings. As the future moves on to the era of smart buildings, sustainability of the energy management schemes invoking the need of proper energy saving strategy is vital. Sustainable energy grids play a key role among the modern infrastructures in various sectors from residential, commercial, and

S. Sofana Reka (✉) · S. K. Pranesh
School of Electronics Engineering, VIT University, Chennai Campus,
Chennai, India
e-mail: sofanareka.s@vit.ac.in

S. K. Pranesh
e-mail: skpran@gmail.com

D. Thalmann et al. (eds.), *Intelligent Embedded Systems*, Lecture Notes
in Electrical Engineering 492, https://doi.org/10.1007/978-981-10-8575-8_11

industrial. This paper exhibits the dynamic control logic strategy in smart meters. Smart metering approaches in modern age smart homes are mandatory [1–7]. The smart meter showcases different components at various levels starting from smart buildings to energy forecast. The control analysis inside the smart meter with energy saving analysis forecasts at a tremendous outset in smart homes.

The understanding of home area management system involves different categories. Moving to the era of the modern electric power systems, the need of new technology dynamically with different control strategies is really important [3]. The smart meter fixation at the modern residential homes eases the need of different analysis

- Reduces the prospectus of energy fixation.
- Customer and the utility foremost interaction is been clearly underwent.
- Centralized structure with communication infrastructure, proper controlling strategy.
- The control strategy involves the proper structure of energy saving scheme along with necessary parameter structures [8–10].
- Loss aversion analysis with proper energy saving schemes is to be involved.

The general outlook of HEMS is described in Fig. 1. The structure invokes the parametric diversion of the two schemes involved as control analysis and the energy saving schemes. The energy saving schemes involves the intervention of the users and the utility [11] and the further need toward interaction.

The control center determines the proper scheduling of the appliances, and the analysis can be done by different algorithmic approach. Further for a smart grid environment, smart meter involvement is needed at the various levels.

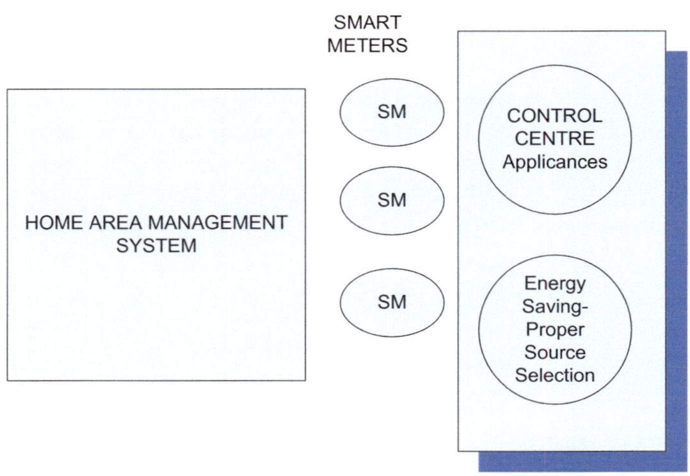

Fig. 1 General outlook of HEMS

This paper is subdivided into different approaches. The dynamic approach of the new control strategy with harmony search meta-heuristic algorithm forms a basic framework in loss aversion approach. The entire module is separated into three approaches by modelling the framework with the new analysis in smart meter. Evaluation and controlling the logic are done at the second level using stochastic Petri nets. Further, proper monitoring is done in the smart meter using run time execution considering at different smart homes. The process control with the new energy saving approach and loss aversion analysis is done in DECIS-based solver and Java programming. The run time execution approaches at the different levels are exhibited using MATLAB.

2 Background and Survey

The work implemented in the previous literature depicts different control analysis based on the energy scheduling and energy consumption in a smart grid environment [3–7, 12]. The work shown in [5] explains the control analysis in a home structure bearing the hybrid-based algorithm [13–15]. The control analysis and the customer interaction are explained. The model analysis or the schematic setup had to be established in this paper.

Further, the work approached in [4–12, 16] clearly explains the different analysis of energy consumption using renewable energy approaches and the need of the energy consumption with setting up different schedule-based algorithms. The paper [6] explains the demand response modelling using game theory approaches pertaining with energy scheduling and classification made. The smart grid infrastructure with energy saving brings in renewable energy as a need of control strategy together with model designed.

The paper establishes different modelling approaches which are explained with the proper approaches of stochastic Petri nets [8]. The stochastic Petri nets are used to model the stochastic modelling using Markov stochastic dynamic programming. This process evolves [9, 10, 13] the need of the dynamic control using the different stochastic Petri nets modelling and the use of meta-heuristic algorithm.

3 Proposed Model

The proposed model set up is made with the control strategy and general analysis is done by considering the loss aversion analysis. The control logic model is incorporated with stochastic Petri nets with particle swarm optimization—meta-heuristic algorithm for the energy saving scheme. The model is designed considering different factors and subject to constraints in stochastic model.

The main aspect of the residential building is subjected to the control center which controls all smart meters in neighborhood are network (NAN). The NAN

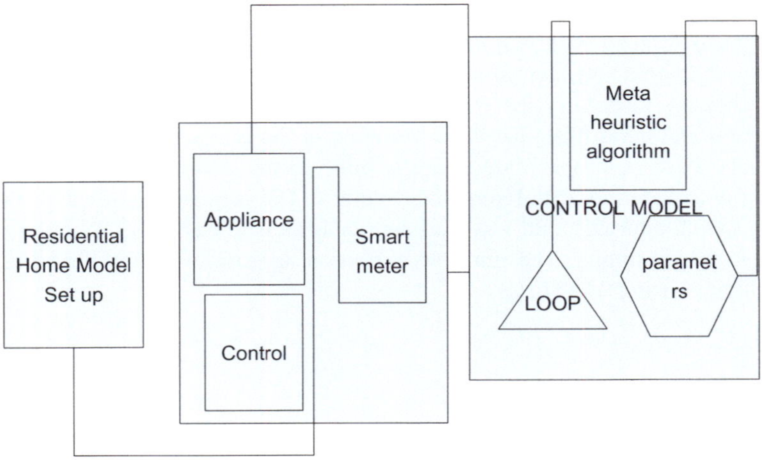

Fig. 2 Proposed scenario

network furnishes the logic and data stored using an algorithm communicated between the user and the utility. The consideration of the energy obtained from the battery source and the renewable sources. The model explained in Fig. 2 shows the connection between the smart meters with the control parameters which are obtained in the loop formats. The entire data obtained from the NANs are collected in Petri nets mathematically using stochastic formulation.

The collected data from the NAN network for an efficient cyber security analysis is created using DECIS solver and JAVA programming with different appliances. The structure is differentiated with the following analysis.

Figure 3 explains the control logic used in the proposed model, and the steps are explained in the following section.

- Modelling of the parameters of smart meter data is collected from each of NAN network, and the data are considered as U nets in the total parameters.
- The total parameters are evaluated depending on the hourly basis tariff for the different scheduling appliances considered as P nets.
- The evaluation structure is considered totally as the probability of the P in the total sets in U network. The scheduling of appliances is considered as two-level stochastic process $F = U \cap P$.
- Further, the control logic of the data is obtained in the stochastic Petri nets with two-level multi-objective function.
- The function P considers all the networks and final logic model is created using the PSO model for the data entry from the appliances.

The evaluation model for the entire logic is done by the stochastic Petri nets considering all the users, and the logic is created using PSO. Further in the paper,

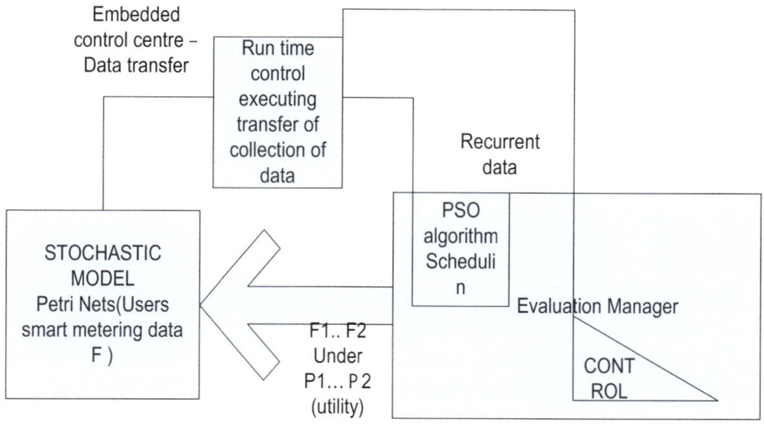

Fig. 3 Control logic strategy

the loss aversion is also introduced for an efficient energy saving scheme. The Petri nets are modelled with five processes in the proposed scheme. New approach of scheduling the appliances and the characteristic is mathematically modelled in the further sections.

4 Mathematical Formulation

The mathematical formulation is separated based on the collection of data with control logic separation at every level of smart metering data. Considering the users at the smart meter as F to the set of the different control centers as $P1$ to Pn. P is the center of the different logic centers at F level of the users to reach the Markov model at each level as

$$p = p * v + y1 * f1(p - x) + y2 * f2(g - p) \tag{1}$$

$$F = U \cap P. \tag{2}$$

$$\begin{bmatrix} F1 \\ F2 \end{bmatrix} = [g1 \ldots gn] \tag{3}$$

$$P\{f_{k^{1\ldots n}}\}/U_{i\ldots n} = \frac{g_{k(F_N)}}{\sum_{i \in F(U_i)} g_k(F_N)} \tag{4}$$

Table 1 Calculation of variables considered

Total appliances considered	Petri nets calculation	Algorithm fitness function	Energy saving (%)
5	5	3.55	3
10	3	4.55	4
12	2	6.3	5
15	5	2.4	5

The list of p and v parameters are linked with every smart meter for the controlling of appliances, and the scheduling is done based on the final fitness function obtained using stochastic model and particle swarm optimization. The x and g variables are the level of multi-objective function used at each level at the NAN. The final fitness equation considering the loss aversion analysis and modelling of Petri nets in a local editor in JAVA with PSO technique for local scheduling of appliances is given as the above equations with Petri nets are user interaction.

5 Results and Inference

The dynamic approach of the new control strategy with meta-heuristic algorithm forms a basic framework in loss aversion approach. The entire module is separated into three approaches by modelling the framework with the new analysis in smart meter. Evaluation and controlling the logic are done at the second level using stochastic Petri nets. Further, proper monitoring is done in the smart meter using run time execution considering at different smart homes. The process control with the new energy saving approach and loss aversion analysis is done in DECIS-based solver and Java programming. The run time execution approaches at the different levels are exhibited using MATLAB and discussed in Table 1.

6 Conclusion

This approach model is done as a control parameter with reduction of energy consumption tariff-based on two logistics of generalized stochastic Petri nets for the collection of data, and the mathematical formulation is done using DECIS solver. The loss aversion management is made with the control logic for scheduling the appliances using meta heuristic algorithm. The paper enhances the new logic template using an evaluation manager, and further, the process can be extended using control logic analysis for an entire smart building approach. The dynamics invoked can be extended for a platform in smart building.

References

1. Farhangi H (2010) The path of the smart grid. IEEE Power Energy Mag 8(1)
2. Nguyen TA, Aiello M (2013) Energy intelligent buildings based on user activity: a survey. Energy Build 56:244–257
3. Reinisch C, Kofler M, Iglesias F, Kastner W (2011) Thinkhome energy efficiency in future smart homes. EURASIP J Embed Syst 2011(1):104617
4. Shaikh PH, Nor NBM, Nallagownden P, Elamvazuthi I, Ibrahim T (2014) A review on optimized control systems for building energy and comfort management of smart sustainable buildings. Renew Sustain Energy Rev 34:409–429
5. Klein L, Kwak JY, Kavulya G, Jazizadeh F, Becerik-Gerber B, Varakantham P, Tambe M (2012) Coordinating occupant behavior for building energy and comfort management using multi-agent systems. Autom Constr 22:525–536
6. Reka SS, Ramesh V (2016) A demand response modeling for residential consumers in smart grid environment using game theory based energy scheduling algorithm. Ain Shams Eng J 7(2):835–845
7. Calderaro V, Hadjicostis CN, Piccolo A, Siano P (2011) Failure identification in smart grids based on petri net modeling. IEEE Trans Industr Electron 58(10):4613–4623
8. Brezovan M, Badica C (2013, May). A review on vision surveillance techniques in smart home environments. In: 2013 19th International conference on control systems and computer science (CSCS). IEEE, pp 471–478
9. Martins J, Platzer A, Leite J (2011, Oct) Statistical model checking for distributed probabilistic-control hybrid automata with smart grid applications. In: International conference on formal engineering methods. Springer, Berlin, Heidelberg, pp 131–146
10. Ghasemieh H, Haverkort BR, Jongerden MR, Remke A (2015, June) Energy resilience modelling for smart houses. In: 2015 45th Annual IEEE/IFIP international conference on dependable systems and networks (DSN), pp 275–286
11. Chen TM, Sanchez-Aarnoutse JC, Buford J (2011) Petri net modeling of cyber-physical attacks on smart grid. IEEE Trans Smart Grid 2(4):741–749; Elissa K Title of paper if known (unpublished)
12. GridLAB-D (2011) Gridlab-d simulation software. http://www.gridlabd.org/
13. Shi Y (2001) Particle swarm optimization: developments, applications and resources. In: Proceedings of the 2001 congress on evolutionary computation, vol 1. IEEE, pp 81–86
14. Faria P, Soares J, Vale Z, Morais H, Sousa T (2013) Modified particle swarm optimization applied to integrated demand response and DG resources scheduling. IEEE Trans Smart Grid 4(1):606–616
15. Marinaki M, Marinakis Y, Stavroulakis GE (2011) Fuzzy control optimized by a multi-objective particle swarm optimization algorithm for vibration suppression of smart structures. Struct Multidisciplinary Optim 43(1):29–42
16. Van Kasteren TLM, Englebienne G, Kröse BJ (2010) Activity recognition using semi-markov models on real world smart home datasets. J Ambient Intell Smart Environ 2(3):311–325

A Novel Approach for Night-Time Vehicle Detection in Real-Time Scenario

M. Aswin and G. Suganthi Brindha

Abstract Intelligent transportation systems and computer vision have intensively been studied in the past decades since the need for intelligent transportation system increases as the growth of vehicles is increasing rapidly than ever before. Traffic surveillance is the important feature in intelligent transportation system which involves in detection and counting of vehicles. In this paper, a night-time vehicle detection system is proposed for effective detection and counting of the vehicles during night-time with the help of the headlights and also classification of the vehicles is carried out in this system. This system involves in processing the night-time traffic video in MATLAB, where image segmentation, background subtraction, blob analysis are done to detect and pair the vehicle headlights for the counting of vehicles. Different templates of headlights are created, and comparison is carried out with the headlight of the vehicles present in the traffic video to classify the types of vehicles present. The above algorithm for night-time detection and counting of vehicles is also implemented in system on chipboards, so that it can be effectively used for intelligent transportation system purposes.

1 Introduction

Nowadays, the intelligent transport systems has been widely practiced in most of the countries, and intra-vehicular communication (IVC) is a fast-growing technology under intelligent transport system which involves in sharing information about traffic and road conditions among vehicles; in order to get the traffic information in a area which has to be shared, the vehicle detection has to be carried out for

M. Aswin (✉) · G. Suganthi Brindha
Department of Electronics and Communication, SRM University,
Chennai 603201, India
e-mail: aswinsaras@gmail.com

G. Suganthi Brindha
e-mail: brindhasrm2013@gmail.com

D. Thalmann et al. (eds.), *Intelligent Embedded Systems*, Lecture Notes
in Electrical Engineering 492, https://doi.org/10.1007/978-981-10-8575-8_12

counting the vehicles passing that area. Further, the vehicle detection is also used in calculating speed [1] and in autonomous driver assistance system. In proposed system, vehicle detection can be carried out in night-time scenario under various illumination conditions by identifying the headlights of the vehicles. In this system, a night-time traffic video is processed using MATLAB, where pre-processing is done for the given video after which threshold segmentation and background subtraction of the video are done so that the headlights can be detected and paired; blob analysis is done for the detected headlamp, and counting is done based on the number of paired headlights. In order to classify the types of vehicles in the given video, each frame of the video is processed to compare with the headlight templates. The above same algorithm for counting is also implemented in a Raspberry Pi 3 system on chipboard for effective implementation in intelligent transportation system.

2 System Overview

2.1 Need for Vehicle Detection

Vehicle detection becomes necessary for monitoring traffic. This traffic monitoring involves in counting and classifying the types of vehicles passing in that area, where this information can be used for many purposes like intra-vehicular communication (IVC) which helps in finding fastest route to the destination by avoiding traffic, and it is one of the fast-growing technology in intelligent transportation system.

2.2 Existing System Drawbacks

Most of the system involves in daytime vehicle detection [2–4]. Daytime vehicle detection is easy because of its proper visibility of the vehicle images, whereas incase of night-time it is difficult due to poor visibility. Some system involves only in counting of vehicles [5, 6]. In this proposed system, both vehicle detection and classification are carried out in night-time and also counting of vehicles is implemented both in MATLAB software and Raspberry Pi 3 system on chipboard.

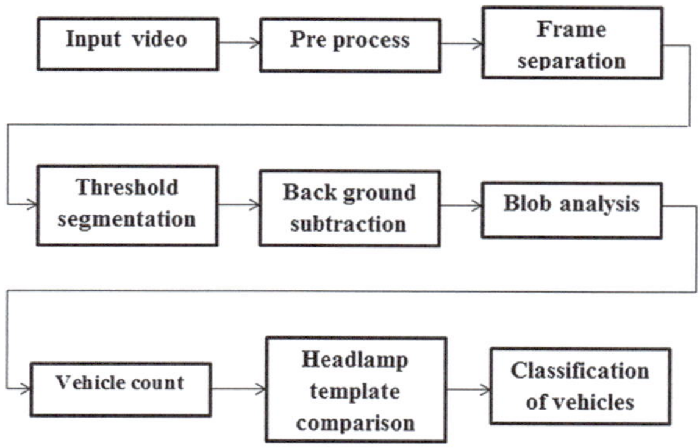

Fig. 1 Flow of the system

3 System Architecture

This system involves in processing the night-time traffic video using MATLAB for the detection, counting and classification of vehicles as shown in Fig. 1. This video processing system processes one or more frame at a time from a continuous stream of video frames.

A traffic video captured during night-time is fed as a input to the MATLAB, which is shown in Fig. 2. Pre-processing is done for the given input video; it involves in removing the unwanted noise in the image using filters and converting it from RGB to grayscale.

Fig. 2 Input video

Fig. 3 Threshold segmentation

Grayscale image is an intensity image where the hue and saturation information is removed from the colour image, and grayscale image is formed by the weighted means of the red, green and blue image channels; this is done in order to reduce the computation requirement of the system.

After pre-processing image segmentation is carried out, there are different segmentation methods available; here, we use the threshold segmentation method.

Threshold segmentation method involves in setting a threshold value and converting the grayscale image to binary images which makes the computation process easy. By doing this process, the vehicle headlights can be easily detected, which is shown in Fig. 3.

After segmentation, background subtraction is carried out, and background subtraction involves in extracting the foreground object from the background of the image which helps in removing the other light sources like street lights; this involves in better differentiation of vehicle headlamp from other light sources. The paired headlights are continuously tracked till the second reference line, to confirm that the paired headlights belong to the same vehicle by eliminating the light reflections on the road. The detection and tracking of headlight are done through blob analysis as shown in Fig. 4.

The red colour lines present in the Fig. 4 are the reference lines. These two lines form the region of interest within which the vehicle headlight is detected, paired and tracked. The black colour rectangular box around the headlight in Fig. 4 implies the blob analysis, which performs the detection and tracking of headlight. This is composed of three stages, the first stage involves in the detection of headlight, where the detected headlight is surrounded by black rectangular box, the second stage involves in tracking of headlight which is done by examining whether the same headlight is found on the successive frames if the same headlight is not

Fig. 4 Tracking and pairing of headlights

present on the successive frames, and then, it is considered as false detection, and the third stage involves in updating the count of the vehicle if the paired headlight remains same on all the successive frames till it successfully leaves the region of interest. When the vehicle leaves the region of interest, the tracking of headlight is stopped.

The classification of the vehicle is done by extracting the frames from the video and processing one frame at a time to compare with the headlight templates created. The number of frames obtained from the video depends on the frame rate of the CMOS sensor. Headlight templates for truck and car are compared with the vehicle headlights in the selected frame, and after finding the match, the type of the vehicle is displayed. The same algorithm for vehicle counting in night-time scenario was implemented in Raspberry Pi 3 system on chipboard, which processes the night-time traffic video obtained by the CMOS sensor and the same output is obtained.

4 Result

The count of the vehicles is shown in Fig. 5. Figure 5 shows the number of vehicles which have passed through the region of interest.

The counting of the vehicles is done based on the tracking of vehicle headlamps. The tracked headlights are indicated by a black rectangular box.

Figure 6 shows the classification the vehicles, which is done based on template matching. The headlights of the vehicle present in the processed frame are

Fig. 5 Vehicle count output

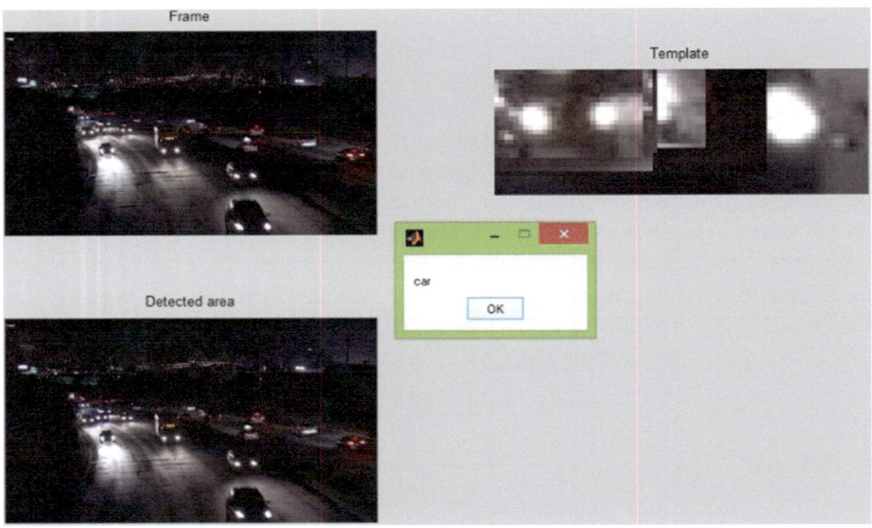

Fig. 6 Vehicle classification output

compared with default headlight templates of vehicles which are predefined in the system. Figure 6 shows the selected frame and processing of that frame; the template region shows the default headlight templates. These templates are compared with the headlights present in the frame; if the match is found, the corresponding vehicle type is displayed.

5 Conclusion

Thus, the effective detection and counting of vehicles during night-time under various night-time illumination conditions are implemented successfully. Counting is done based on the detection and pairing of headlights of the vehicles, and this shows effective detection during night-time when compared to many other systems which involves in vehicle detection during the daytime, and also, this system does the classification of vehicles during night-time by template matching method. This system was successfully implemented both in hardware and software, where most of the system is implemented only through software. The software implementation is carried through MATLAB, where the counting of vehicles is hardware implemented through Raspberry Pi 3 system on chipboard. The information collected through system is used for many applications like energy saving purposes which involves in smart on/off of street lights based on the number of vehicles passing through that area and also in intra-vehicular communication, which is fast-growing technology in intelligent transportation system. The challenges faced in this system are if the input video is of high quality, it takes more amount of computational time. This can be rectified in future by using system on chipboard containing powerful processor of high clock speeds.

References

1. Zhang W, Wu QMJ, Wang G, You X (2012) Tracking and pairing vehicle headlight in night scenes. IEEE Trans Intell Transp Syst 13(1):140–153
2. Sivaraman S, Trivedi MM (2012) Real-time vehicle detection by parts for urban driver assistance. In: Proceedings of IEEE intelligent transportation system conference, pp 1519–1524
3. Zhou J, Gao D, Zhang D (2007) Moving vehicle detection for automatic traffic monitoring. IEEE Trans Veh Technol 56(1):51–59
4. Sun RMZ, Bebis G (2006) Monocular precrash vehicle detection: features and classifiers. IEEE Trans Image Process 15(7):2019–2034
5. Guo J-M, Hsia C-H, Wong K, Wu J-Y, Wu Y-T, Wang N-J (2016) Night time vehicle lamp detection and tracking with adaptive mask training. IEEE Trans Veh Technol 65(6):4023–4032
6. Salvi G (2014) An automated night time vehicle counting and detection system for traffic surveillance. In: Proceedings of IEEE international conference on computer science and computational intelligence, vol 1, pp 131–136

Local Visualization for Troubleshooting the RF Mesh Network in a Wireless Metering System

Parvathi L. Prabhakar, Kiran Thomas, S. Sreekumar and S. Muthulakshmi

Abstract **Objectives**: To design a visualization of a network containing wireless meters installed with RF modules which uses RPL protocol for routing. **Methods/ Statistical analysis**: Meters are linked wirelessly for obtaining the packets containing its details and other parameters, by using a Constrained Application Protocol (CoAP). Thus, details collected from each meter are used for the construction of the visualization, by dynamically assigning the position of each node. **Findings**: RPL, routing protocol, is used for the construction of the network. The advantage of its self-healing nature makes the whole system stable by removing the irresponsive meters, when detected, from the network and reconstructing it from the starting. **Application/Improvements**: Maintenance of such a network which switches its neighbors at different instances may be complicated. In such cases, these visualization methods can be included. In future, the firmware update options for each meter can be implemented along with visualization.

Keywords Constrained application protocol (CoAP) · Dissector
RPL · Wireshark

P. L. Prabhakar · K. Thomas · S. Sreekumar · S. Muthulakshmi (✉)
Vellore Institute of Technology, Chennai, India
e-mail: muthulakshmi.s@vit.ac.in

P. L. Prabhakar
e-mail: parvathi.lp@gmail.com

K. Thomas
e-mail: kithos@gmail.com

S. Sreekumar
e-mail: sreeku.sp@gmail.com

P. L. Prabhakar · K. Thomas · S. Sreekumar · S. Muthulakshmi
Vellore Institute of Technology, Chennai, India

D. Thalmann et al. (eds.), *Intelligent Embedded Systems*, Lecture Notes
in Electrical Engineering 492, https://doi.org/10.1007/978-981-10-8575-8_13

1 Introduction

Smart grid technology is the most advanced and fast-growing innovation which can be used in the meter reading system. Conventional meter reading system leads to more wastage of energy and other resources. This can be avoided by using this latest technology so that we can connect the meters or any other home appliances so as to increase reliability and also to save the energy and cost.

Since for billing operators, while reading the meters, there may be chances of errors due to the placed meter position, and can be miserable by worst climate conditions. This smart grid idea includes crossing-over data and communication techniques with energy optimization to allow two-way control stream, to accomplish consistent operation, for end usage and delivery, and to empower a wide variety of renewable resources. Even in the initial phase of development of this smart grid technology, its present implementation has few advantages, for instance wireless metering system to gather information from electric meters, utilizing various other wireless technologies like ZigBee, RF modules and so forth. Wireless metering framework is fit for giving restricted interchanges from meters to an information perusing gadget (by means of a gateway). Thus, contrasted with the ordinary power grid, the utilization of wireless metering [1] framework spares utility suppliers the cost of occasional trips to those places for meter reading.

During the further development of smart grid technology, wireless metering system may catch up with new features added to it such as the advanced metering infrastructure (AMI) [2] which will be capable of sending and receiving in a bidirectional way such as informing each consumer about their recent power cost along with the monitoring of consumers on real time.

This can be solved by using a multi-hop network which associates a substantial number of the meters connected to a portal/gateway later being connected to a main controlling central unit which organizes a wide range of management.

In particular, such a mesh-based AMI system ought to give legitimate routing functionalities guaranteeing higher reliability and low inactivity conveyance for the traffic types from meter to the gateway meter reading, and utility administration information from the gateway to each meter. All the meters will be connected to the server using a data control unit (DCU), which acts as the gateway.

To fulfill the quality and also the reduced latency necessities of AMI, the routing convention for AMI must be able to adapt the continuous changes in the state of each node by giving quick and efficient route calculation techniques and in the meantime should just deliver moderate measure of protocol.

Here, wireless metering system [3] is used where they use IPv6 Routing Protocol for Low Power and Lossy Networks (RPL) [4]. RPL is nothing but a routing protocol [5] which is being developed by the Internet Engineering Task Force (IETF) [6] specifically for the low power and lossy networks (LLN), thus helping in various low-cost networking applications. Directed acyclic graphs (DAG) are used by RPL as a reflection of the network topology, and every node keeps up its position in a DAG structure by utilizing a rank property to decide its relations with different nodes in the DAG.

In this paper, we present a visualization for the DAG constructed by the RPL in the wireless metering system. The block diagram of the system is given in Fig. 3. The visualization includes various nodes, represented as each meter, and how the connections are between the nodes will be plotted as a hierarchical graph. The RPL routing protocol is good enough in self-healing the network.

This is one among the major advantages of RPL. Because of this quality, the meters' connections may vary from time to time. This is managed and can be represented in the form of a graph which may help for the easiness of troubleshooting and development.

2 RPL Framework

As mentioned, RPL is a protocol which helps in routing the different nodes in a network efficiently. This is done by constructing a directed acyclic graphs (DAG) which focus on connecting nodes neglecting any formation of a cyclic path. Another arrangement preferred over DAG is the destination-oriented DAG (DODAG). This is because a single border router is used as the gateway for the whole network. Figure 1 shows the DAG and DODAG structure.

As a starting step for constructing this DODAG [7], a DODAG Information Object (DIO) message, which is a control message, has to be initiated and broadcasted from the border router. Those meters which receive these DIO messages will approve itself to be a child node for the border router. Thus, those nodes will register their parent as the border router.

Similarly, the each and every meter which is not identified as the leaf node will send such control messages and register their corresponding parent. The DIO message has the details of the DODAG ID which defines which particular DODAG the meters are connected, rank of the node, and the objective function which gives the idea of how to optimize the connection.

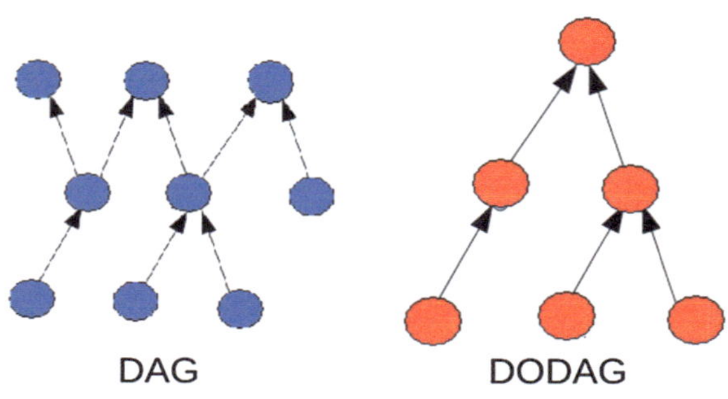

Fig. 1 Difference between DAG and DODAG

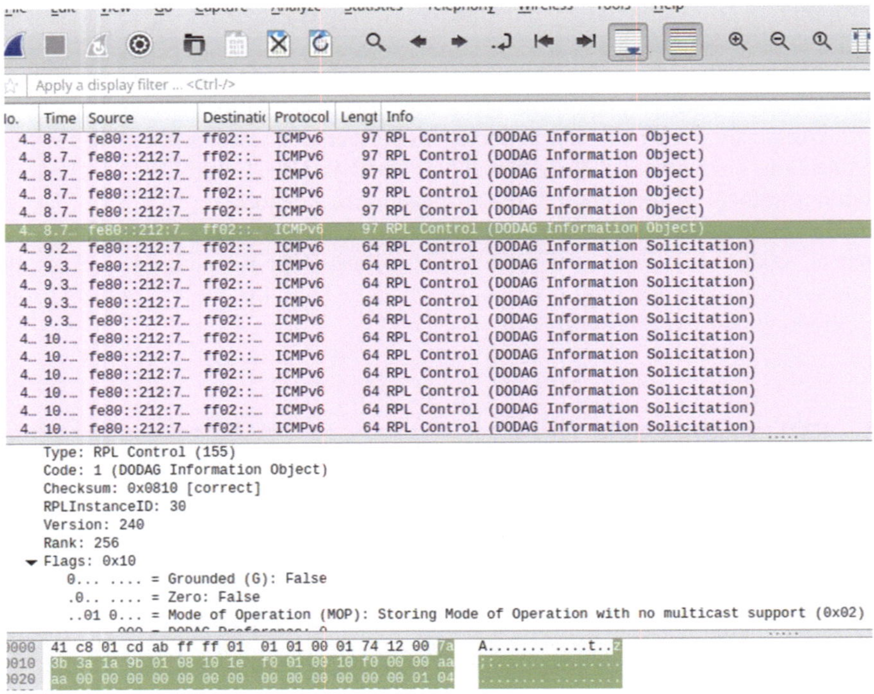

Fig. 2 Wireshark output of dissecting RPL protocol

In the same way, in order to obtain the reverse hierarchy, the Destination Advertisement Object (DAO) message will be sent to the parent. This is being sent by each meter so that it can make sure that they are connected with their previous parent itself.

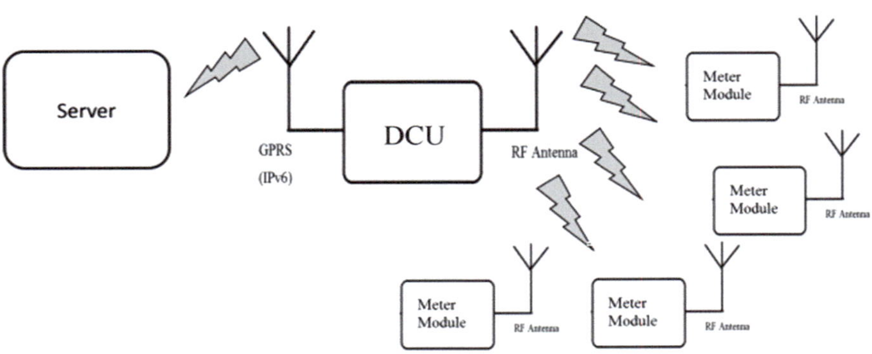

Fig. 3 Block diagram of wireless metering system

This parent confirmation is done when an acknowledgment (DAO-ACK) is received. Once if the acknowledgment is not received then the parent switching occurs [8]. Figure 2 shows the Wireshark window where the RPL protocol is dissected (Fig. 3).

3 Required System

The main elements in Fig. 4 are the Desktop and the DCU. DCU and meters have their own unique IPv6 ID. These meters are connected in a network where all of them will be advanced toward the Data Controlled Unit (DCU). Thus, it is the DCU which will collect all the information from the meters and provide it to the user.

The DCU has a Linux OS and a Contiki OS [9]. Contiki OS is installed in the RF card and is used for the low-cost RF network, and the packets from the meters are received into this, which is later transferred serially to the Linux OS. A CoAP [3] server is implemented in the DCU and each meter which has resources required by the user. A CoAP client program is written so as to fetch the required information from the meters through the border router. This information is later used for the construction of the mesh visualization. Figure 5 shows a Copper (Cu) client extension screen that displays the information by fetching the MAC ID of a meter.

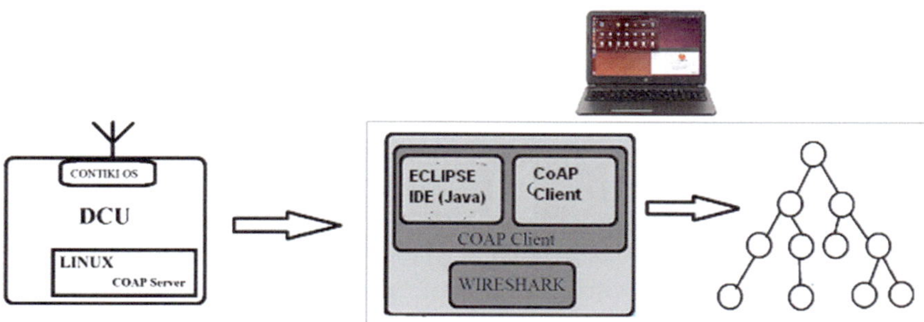

Fig. 4 Block diagram for displaying the network hierarchy

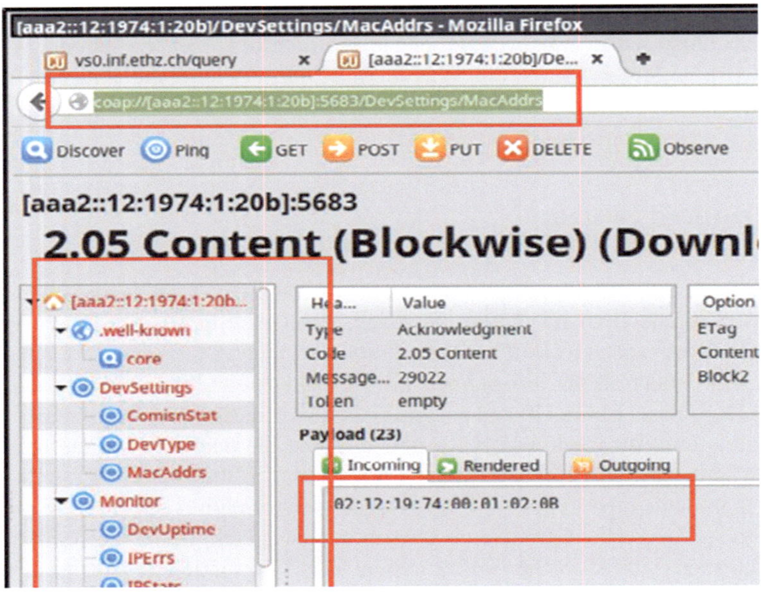

Fig. 5 Copper (Cu) client extension screen that displays the information by fetching the MAC ID of a meter

4 Constrained Application Protocol (CoAP)

CoAP is the most recently and commonly used Web protocol in the applications where there are restricted resources available like RAM, ROM. The advantage of CoAP over the other protocols like TCP is that before establishing the connection for communication, it requires a three-way handshaking, while when using CoAP the data can be sent along with the initial packet irrespective of the handshaking.

A CoAP server and a CoAP client are required for such a communication to happen. The resources can be implemented in the server (here meters or DCU) such that the required information can be easily fetched by the client. And if we check for the size of the transmitted information it will be in few bytes. There are many methods to access the information in the meters, i.e., by using GET/PUT/POST/OBSERVE. GET and OBSERVE are used for obtaining the data from the server, while PUT and POST are used for altering the data or the firmware update of the meters.

A widely used CoAP client which is more user-friendly is a tool known as Copper. It is a downloadable extension for Mozilla Firefox, which has a DISCOVER option used for viewing the resources available at the server. The server resources can be accessed using a URL which has a format of "coap://" followed by the IP address in square bracket, port number, and the resource path [10].

For example, coap://[aaaa::1111]:5683/resource

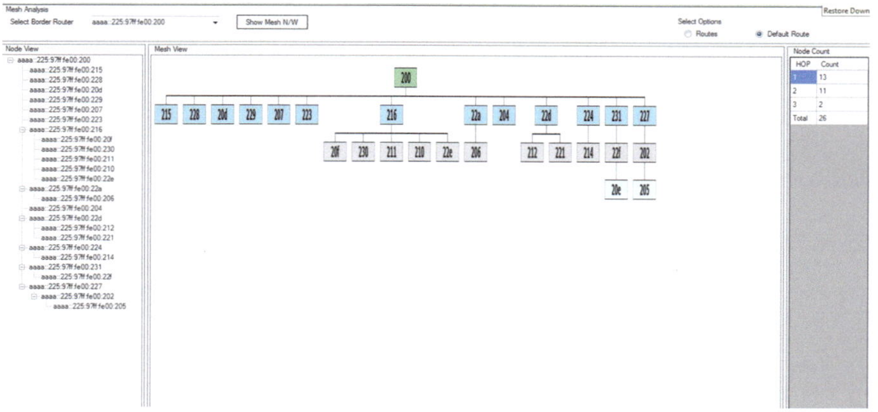

Fig. 6 Network hierarchical view of a meter

5 Visualization of RF Network

The meters and the DCU are implemented with the CoAP server which is capable of providing the data required by the client for constructing the hierarchy.

Due to the self-healing utility of the RPL routing protocol, the connection between the meters may vary from time to time and thus the visual representation also should alter itself. This is done by monitoring the parent switching event and thereby polling the node again whenever a change in it is identified.

Figure 6 shows the hierarchical view of the wireless metering mesh. Each level is categorized by using different color codes.

6 Conclusion

This work proposes a troubleshooting tool for a wireless metering system that is implemented using a DODAG graph with the help of RPL routing protocol.

For the easiness of troubleshooting the mesh network, the visual representations are more helpful than any other form. Thus, the whole network elements can be plotted graphically. The elements can be of the border router (DCU) or a meter. The root node will be the border router and the other elements that follow the border router will be the meters. Each of them will have a unique IPv6 address.

Acknowledgments I would like to extend my sincere gratitude to many individuals and organizations. Without their kind support this would not have been possible.

I am highly indebted to Mr. Fathah Rehman and Prof. S. Muthulakshmi for their ultimate support, guidance, and patience as well as in helping me achieve the knowledge while the course of this project. I would like to express my gratitude toward all the employees of Kalki

Communication Systems for their kind cooperation and encouragement which helped me in the completion of this work. And I would like to sincerely honor the industry persons for giving me such attention and time.

References

1. Akyildiz IF, Wang X, Wang W (2005) Wireless mesh networks: a survey. Comput Netw 47:445–487
2. Wang D, Tao Z, Zhang J, Abouzeid AA (2010) RPL based routing for advanced metering infrastructure in smart grid. In: IEEE smart grid communication workshop, July 2010
3. Article title. http://www.indjst.org/index.php/indjst/article/view/89919. Accessed 11 Aug 2016
4. RPL. https://tools.ietf.org/html/rfc6552
5. Ancillotti E, Bruno R, Conti M (2013) The role of the RPL routing protocol for smart grid communications. IEEE Commun Mag 51(1)
6. Ancillotti E, Bruno R, Conti M (2014) Reliable data delivery With the IETF routing protocol for low-power and lossy networks. IEEE Commun Mag 51(1)
7. Accettura N, Grieco LA, Boggia G, Camarda P (2011) Performance analysis of the RPL routing protocol. In: 2011 IEEE international conference on mechatronics (ICM)
8. Ho QD, Gao Y, Rajalingham G, Le-Ngoc T (2014) Performance and applicability of candidate routing protocols for smart grid's wireless mesh neighbor area networks. In: 2014 IEEE international conference on I communications (ICC)
9. Tang CM, Zhang Y, Wu YP (2012) The P2P-RPL routing protocol research and implementation in Contiki operating system. In: 2012 second international conference on instrumentation, measurement, computer, communication and control (IMCCC)
10. CoAP. http://coap.technology/impls.html

Train Collision Avoidance System for Automatic Train Protection Using Internet of Things

Mohit Savner and G. Gugapriya

Abstract Objective: The train collision avoidance system (TCAS) is a synergic approach for intervention of train in danger zone using Internet of things. **Methods**: The proposed system surveillance the signal condition and observe the disobedience of the track signal. A sensor is placed beneath the track, which senses the pressure, temperature, altitude of the track. The presence of train is observed by heavy pressure, change in temperature, and altitude on the track. Eventually, the system notifies the intervention of train being in danger zone using Internet of things by ThingSpeak IoT platform. An alarm is given to the train side that blow through Beaglebone black using Wi-fi module and if locomotive pilot does not respond, then the train stops by controlling braking system. **Findings**: Track signal plays a paramount role in cognizance of train's direction. The disobedience of signal results in collision, derailment, and conflict. It is mainly occurred due to human error or rigid climate conditions like fog, rain and affects thousands of lives. TCAS-IoT prevails the problem of overshooting the track signal and acknowledging the locomotive pilot about intervention of train in danger zone. **Conclusion/ Improvement**: The proposed TCAS-IoT can avoid train collision and capable of taking action before collision or derailment of trains. It is a unique way to synthesize the intervention of train in danger zone and can save lot of precious lives.

Keywords Train collision avoidance system (TCAS) · Internet of things (IoT)
Beaglebone black · Global positioning system (GPS) · ThingSpeak
ESP8266 NodeMCU · BMP180 (pressure sensor)

M. Savner · G. Gugapriya (✉)
School of Electronics Engineering, VIT University, Chennai 600127, Tamil Nadu, India
e-mail: gugapriya.G@vit.ac.in

M. Savner
e-mail: mohit.savner2014@vit.ac.in

D. Thalmann et al. (eds.), *Intelligent Embedded Systems*, Lecture Notes
in Electrical Engineering 492, https://doi.org/10.1007/978-981-10-8575-8_14

1 Introduction

Train collision avoidance system (TCAS) has been aimed as an automatic train protection system using "Internet of things." It is an indigenous way of saving precious lives and protecting form disastrous calamities. Train collision mainly occurred due to misinterpretation of signal by train's pilot, and as a result, train enters into danger or interlocked zone. The signal misinterpretation is due to bad weather condition like fog, rain. This paper proposes a method to inform the pilot about intervention of train occurred by alarming and if pilot does not take any action then surveillance controller can apply an emergency brake through Internet of things.

The classification of railway accidents is human errors, signal error, and mechanical faults, more than 80% of the accident in Indian railways is due to human error that led to loss of thousands of lives. The proposed system is capable of preventing or protecting the train before the collision since it informs the presence/intervention of train in danger zone to the train's pilot. Effective pilot can take necessary action and can protect from a vast damage of lives of people. An alarm is given to the train's engine side, if after intervention train's pilot does not respond then train stops by controlling brakes through Beaglebone black.

RFID and Optocoupler which were used in earlier days were complex, sophisticated and costly. Nevertheless, here is the simplest, smartest, and cost-friendly way of analyzing the presence of train in danger zone. Along with these, GPS installed at the train's engine to detect the location of train.

1.1 Technical Background

According to Wang [2], the trains using parallel monitoring systems equipped with next generation of train control systems (NGTCS) shows the fail-safe status of complex control systems. This proposed idea implemented mainly for Chinese high-speed trains from 250 to 350 km/h. The implementation of an independent parallel monitoring system performed using data acquisition technologies and typical software application to avoid conflict of errors.

GPS-based real-time train tracking system by Immanuel Rajkumar [3] introduced the different way of tracking the train using GPS and Ethernet. The proposed system implements the real-time location of the train by GPS on Google map, the speed of the train is identified and the alcoholic detection of pilot status shown.

Anjali Bissa [4] proposed the TCAS using vibration sensor and Zigbee interfaced with LPC2148 microcontroller. The proposed system was based on vibration of track which is not much accurate due to unwanted interventions.

1.2 Proposed System

Figure 1 shows the block diagram of TCAS-IoT, the synergy of the proposed system is a unique, accomplished, and cost-effective method for train collision avoidance system using Internet of things (IoT). The proposed system works on the principle of real-time monitoring of trains and keeping track of train in safe zone, if train being in danger zone due to disobedience of signal then system identifies the incident and immediately an alarm given to train side. If train's pilot does not respond or take any action then an emergency brake is applied using internet of things. Beaglebone black and ESP8266 NodeMCU microcontroller accomplish the communication between train and track signal.

Figure 2 shows the system design flow of track side and locomotive side, track side an 8-bit, Wi-fi enable ESP8266 NodeMCU is installed. BMP180 pressure sensor is placed beneath the track and the clock signal. NodeMCU takes the record of signal in each interval (i.e., red, yellow, and green) conventionally whenever signal is in red state train has to stop behind the signal if not that lead to disobedience of signal and eventually that causes train collision or conflict. Track side a Beaglebone black is used which simulate GPS location, train braking, and buzzer. Wi-fi is connected to the Beaglebone black, which establishes wireless communication between train and track signal.

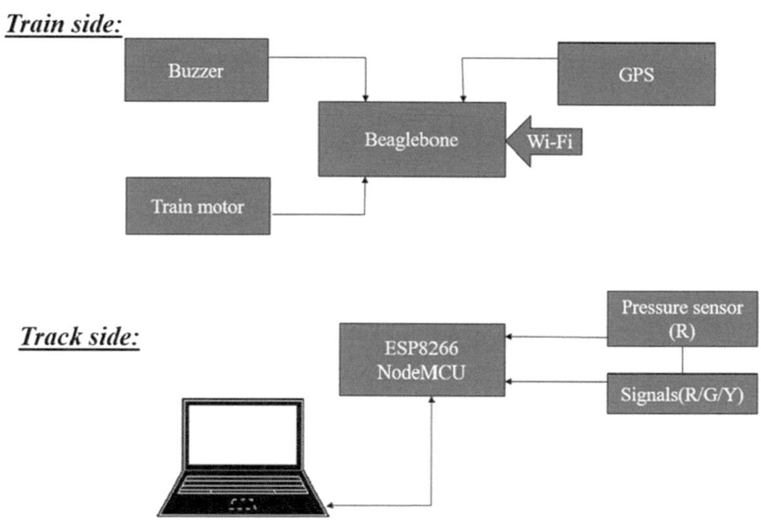

Fig. 1 Block diagram of TCAS-IoT

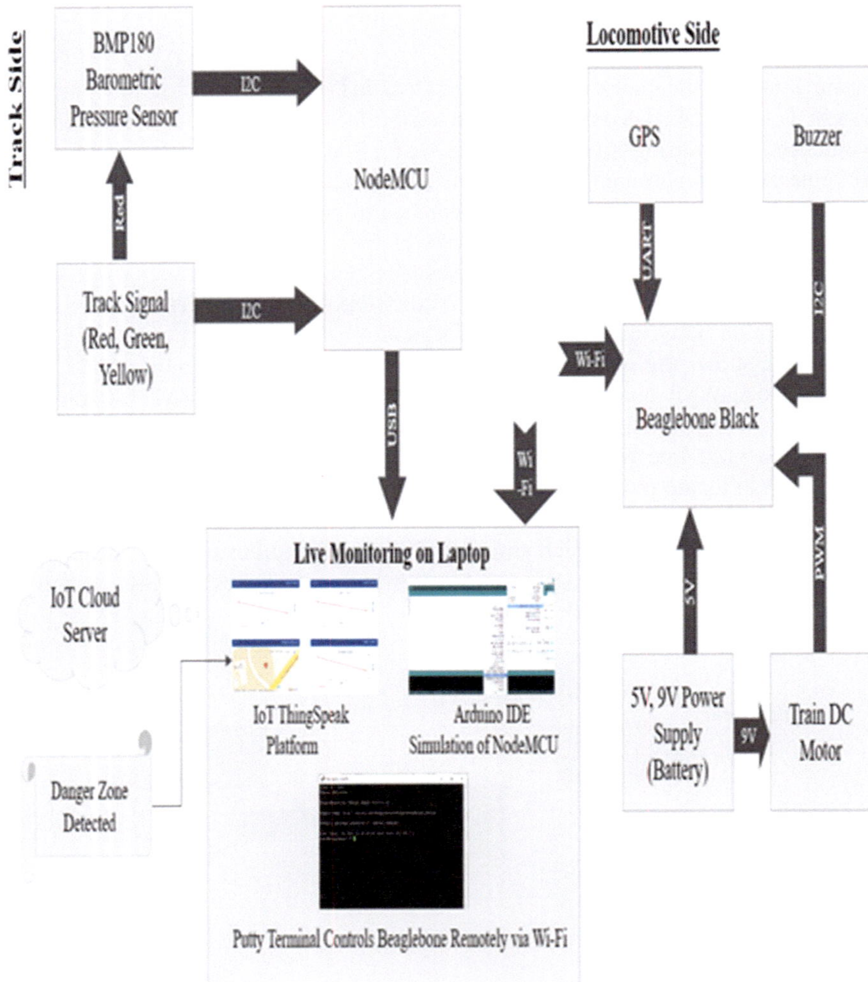

Fig. 2 Proposed system design flow

2 TCAS-IoT Architecture

TCAS-IoT provides a pre-acknowledgment to the locomotive about intervention of train in danger zone via alarming buzzer. In addition, for worst-case condition system stops the train using Wi-fi [5] communication between controller and train. TCAS-IoT assembled in two parts:

1. Track side
2. Train side

2.1 Track IoT Installation

Figure 3 shows the track signal installation which consists of ESP8266 NodeMCU, BMP180 pressure sensor, LED for track signal representation. Whenever the pressure is observed on red signal, condition the values of high pressure observed is fetched into cloud server using NodeMCU. ThingSpeak IoT platform shows the track pressure values continuously in graphical and form, whenever a high-density pressure is observed, i.e., intervention of train being in danger zone. NodeMCU's Wi-fi connects with locomotive Wi-fi via Beaglebone black and an alarm blown; train stopped by controlling locomotive DC motors for brake.

2.1.1 BMP180 Pressure Sensor

BMP180 [6] pressure sensor senses the pressure, temperature, altitude, respective parameters play a vital role in finding the track pressure and train recognition in danger zone. It is based on piezo-resistive technology for EMC robustness, high accuracy, and linearity. Pressure range is from 300 to 1100 hPa with I2C interface; it also enhances the performance of GPS navigation since it shows altitude.

Relation between Altitude and Pressure

$$\text{Altitude} = 44{,}330 * \left(1 - (p/p_o)^{1/5.255}\right) \tag{1}$$

$$P_o = P/(1 - \text{altitude}/44{,}330)^{5.255} \tag{2}$$

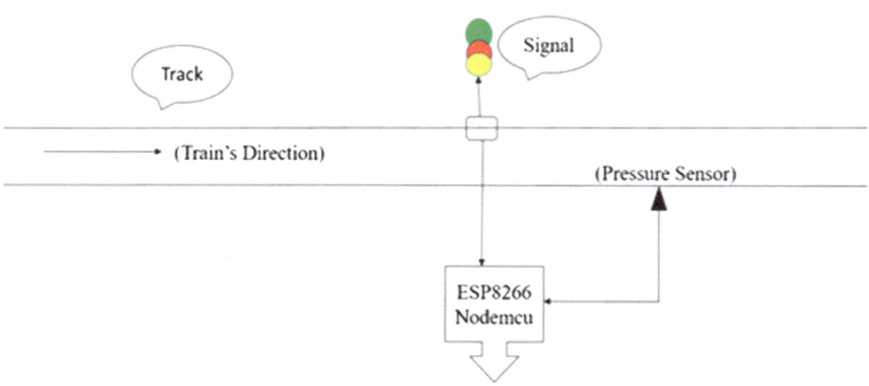

Fig. 3 Track signal installation

2.2 Locomotive Module Installation

The practical implementation of locomotive model is shown in Fig. 4 which consists of two DC motors, Beaglebone black, buzzer, GPS module, Wi-fi, batteries, L293D motor driver. Beaglebone black and DC motor interfaced using L293D motor driver IC [7] controls the breaking of train at emergency. In addition, a GPS module is installed to fetch the present location of train. Buzzer is used to invoke pilot about the intervention of train being in danger zone.

3 Implementation

3.1 Global Positioning System

Global positioning system is a constellation system of 24 satellites to trace the location of an object. It communicates with the satellites and generates parameters (latitude, longitude, altitude) of the respective location. It produces the NMEA

Fig. 4 Implemented locomotive and system modules

Fig. 5 Putty snapshot of GPS

sentences that generate a desired latitude, longitude, altitude, and speed. GPS satellite incessantly transmits current location and time at the receiver side, receiver invigilate multiple satellites, and compare equation to figure out the real-time location. GPS works on the principle of trilateration; it is the process of adjudging the locus or point by measuring the distance between the geometry of circle, spheres, or triangle.

GPS uses this technique to find the exact location with at least four satellites at a time. The satellites measure the distance from the respective location and simplify trilateration geometry to adjudge the exact location of the object.

Figure 5 shows the putty terminal snapshot illustrates the real-time train's GPS [8]—latitude, longitude, altitude, speed, universal time, and number of tracking satellites. GPS NMEA sentences provide these parameters, which is essential for mapping real-time location of train. The whole scenario was accomplished using Beaglebone black microcontroller. Figure 6 shows that each location has its own unique latitude, longitude with reference to this parameter the real-time location is

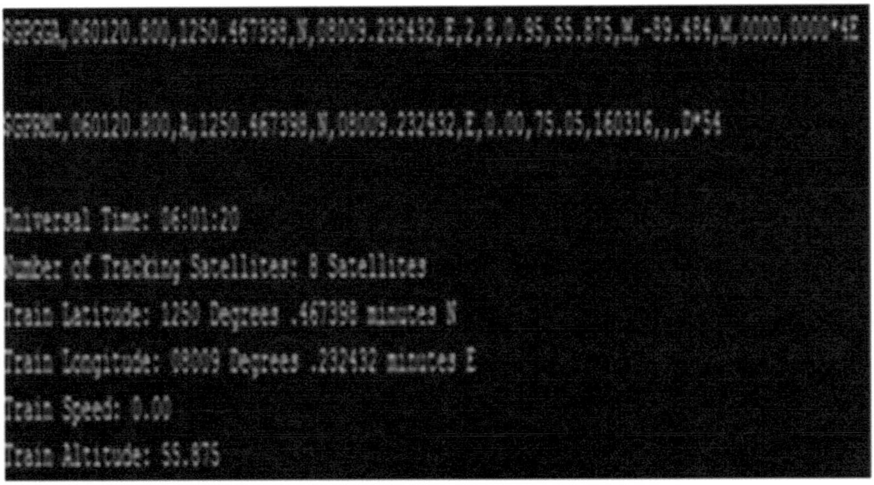

Fig. 6 Snapshot of GPS output

mapped out. National Marine Electronic Association (NMEA) gave the standards for definition of GPS NMEA sentence that is traced into latitude, longitude, altitude.

3.2 ThingSpeak IoT Platform

ThingSpeak [9] is an IoT platform where electronic devices (sensors, gadgets, etc.) data is fed into cloud server and the respective object can be controlled remotely. It is an emerging field of engineering and possesses enormous opportunity.

Figure 7 shows the track pressure graphically representation was observed with ThingSpeak IoT platform, where all three parameters, i.e., pressure, temperature, altitude of BMP180 sensor was analyzed. This gives privilege to track real-time location of subject on map. Therefore, the track signal has a particular track ID awarded by ThingSpeak, and also this has unique API that is useful for the synthesis of scenario. The presence of train observed heavy pressure on track, i.e., status train present at that location.

4 Results

The proposed system is illustrated in Fig. 8. The experiment was implemented with a locomotive model consist Beaglebone black, GPS, two DC motor, Wi-fi, Buzzer, L293D motor driver. Pressure sensor was placed ahead of track signal and if red signal is given then the pressure sensor passes the data to NodeMCU, which was

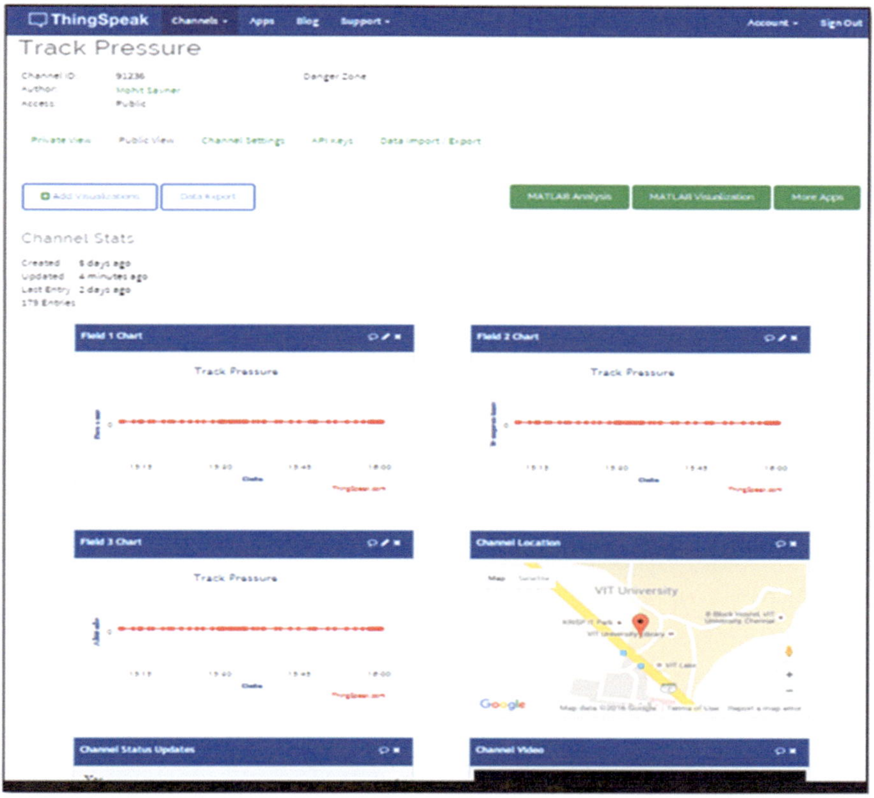

Fig. 7 ThingSpeak IoT platform graphical presentation

connected to Laptop for experimental observation. Figure 10 shows the real-time pressure values graphically on ThingSpeak IoT platform and pressure real-time values are shown in Fig. 11. The disobedience of signal observed then the locomotive was connected to laptop through putty terminal and alarming buzzer was initiated to the locomotive side. Further, locomotive was stopped using putty terminal by controlling two DC motors using Beaglebone and L293D motor driver (Fig. 9).

The disobedience of signal was observed then the locomotive Wi-fi connected to laptop through putty terminal shown in Fig. 9 and alarming buzzer was initiated to the locomotive side. Further, locomotive was stopped using putty terminal by controlling two DC motors using Beaglebone and L293D motor driver.

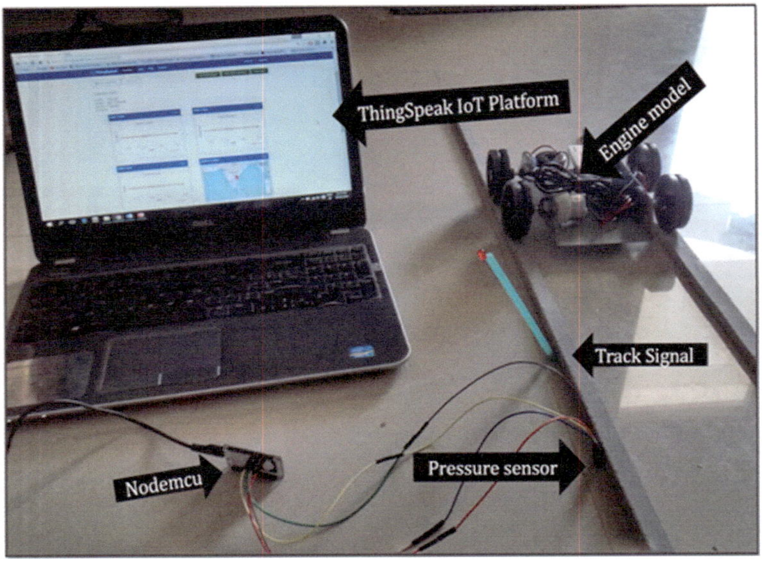

Fig. 8 TCAS-IoT implementation model

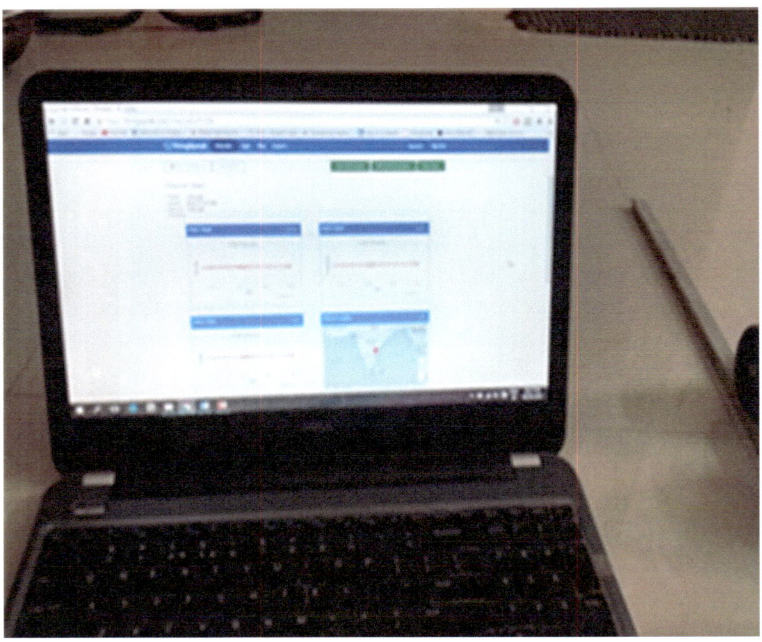

Fig. 9 ThingSpeak laptop remote observation

Fig. 10 ThingSpeak IoT snapshot updated data of pressure, temperature, altitude

5 Conclusion and Future Scope

TCAS-IoT is a synergy of Internet of things, which is a future's predominant technology and possesses enormous opportunity in every field of engineering. The proposed system can play a key role in avoiding train collision, conflict of trains, and can protect precious lives. It is a diligent and cost-effective way to avoid collision of trains in proactive manner, irrespective of weather conditions like fog, rain if train passes track signal in red state then an automatic alarming given to the locomotive, which plays a vital role in avoiding collision of trains, i.e., locomotive pilot gets acknowledgment being in danger zone. One-step further if pilot does not respond then an automatic brake is applied using IoT protocol (HTTP) via Wi-fi communication.

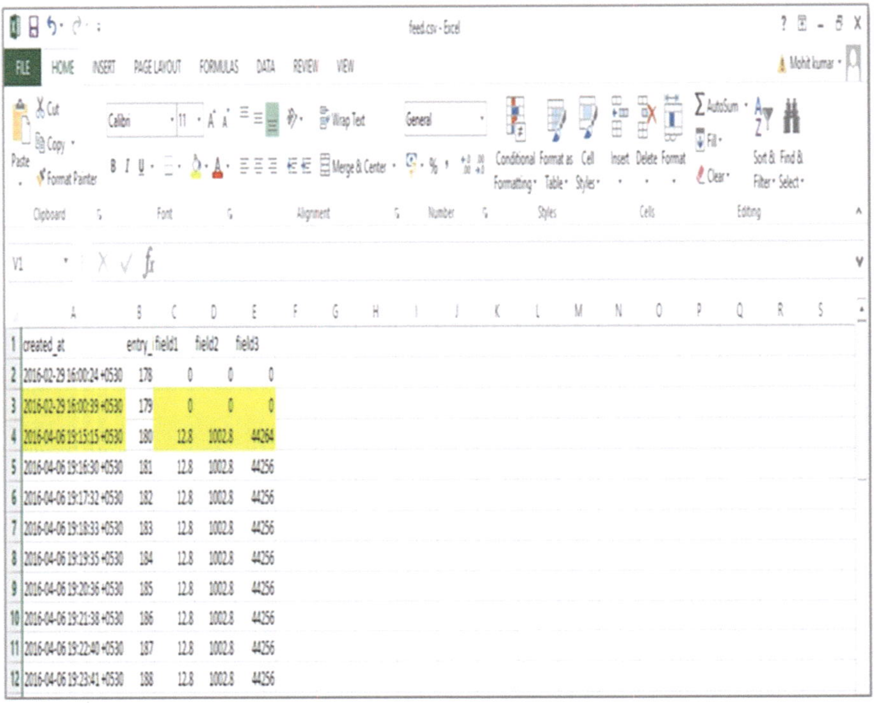

created_at	entry_	field1	field2	field3
2016-02-29 16:00:24 +0530	178	0	0	0
2016-02-29 16:00:39 +0530	179	0	0	0
2016-04-06 19:15:15 +0530	180	12.8	1002.8	44264
2016-04-06 19:16:30 +0530	181	12.8	1002.8	44256
2016-04-06 19:17:32 +0530	182	12.8	1002.8	44256
2016-04-06 19:18:33 +0530	183	12.8	1002.8	44256
2016-04-06 19:19:35 +0530	184	12.8	1002.8	44256
2016-04-06 19:20:36 +0530	185	12.8	1002.8	44256
2016-04-06 19:21:38 +0530	186	12.8	1002.8	44256
2016-04-06 19:22:40 +0530	187	12.8	1002.8	44256
2016-04-06 19:23:41 +0530	188	12.8	1002.8	44256

Fig. 11 BMP180 pressure sensor's data updated on ThingSpeak

Table 1 Locomotive weight difference types

Locomotive types	Weight (tonnes)
Diesel locomotive	112.8
Chair car (empty)	47
Parcel van (empty)	54
General bogie	49

The future scope of this project is practical implementation of an Earth pressure cell sensor instead of BMP180 pressure sensor since the locomotives weight ranges in tonnes. Thus, in practical implementation need to keep pressure of train/ locomotive with Earth pressure cell [10] sensor. The proposed work is an indigenous, efficient, and cost effective with reference to present working model for collision avoidance system.

Table 1 shows the difference between the heavy weights of locomotives [11] only can be sensed by earth pressure sensor that can bear up to 10–30 MPa of pressure.

References

1. Pavithradevi V, Roopini D, Aleem RVA, Solai Shree V (2014) Identification of the presence of train and its formation. In: IJIRSET, second national conference on trends in automotive parts systems and applications (TAPSA), vol 3, special issue 2, Apr 2014, pp 136–142
2. Wang J, Wang J, Roberts C, Chen L (2015) Parallel monitoring for the next generation of train control system. IEEE Trans Intell Transp Syst 16(1):330–338
3. Immanuel Rajkumar R, Sankaranarayanan PE, Sundari G (2013) GPS and ethernet based real time train tracking system. In: IEEE computer society, international conference on advanced electronic systems (ICAES), 2013, pp 282–286
4. Anjali Bissa G, Jayasudha S, Narmatha R, Rajmohan B (2013) Train collision avoidance system using vibration sensors and Zigbee technology. IJREAT Int J Res Eng Adv Technol 1(1)
5. Kapsch Group, European Train Control Systems (ETCS), White Paper by European Railway Traffic Management Systems—ERTMS. http://www.kapsch.net
6. Data Sheet—BMP180 Digital Pressure Sensor. http://www.bosch-sensortec.com. Accessed on 04/02/2016
7. Araari T, Jaziri I, Charaabi L, Jelassi K (2014) DC motor control based on embedded Linux. In: International conferences on electrical sciences an technologies in Maghreb (CISTEM), Nov 2014, pp 1–6
8. Sharma A, Balamurugan MS (2015) RF control based mobile robotic system for search mission with GPS tracker. Indian J Sci Technol (IJST) 8(19)
9. Maureira GAM, Oldenhof D, Teernstra L (2016) ThingSpeak—an API and web service for the internet of things. http://www.mediatechnology.leiden.edu/images/uploads/docs/wt2014_thingspeak.pdf. Accessed on 10/03/2016
10. Earth Pressure Cell. http://www.geokon.com. Accessed on 11/03/2016
11. Indian Railways Institute of Signal Engineering & Telecommunications (IRISET). Secunderabad, http://www.iriset.indanrailways.gov.in. Accessed on 05/02/2016

Automatic Driver and Vehicle Safety Monitoring System

S. Vijay Kumar and Abraham Sudharson Ponraj

Abstract **Background/Objectives**: Accident is increasing rapidly and is mostly due to drunk driving and drowsiness of drivers. A novel approach is presented in this paper to monitor the physical condition of driver and alert system to reduce human errors by means of an embedded system. **Methods**: To find a particular person's blood alcohol concentration in their bloodstream, breathalyzer which is also known as breath alcohol detector is used. IR sensor and biomedical sensors are used to detect the drowsiness by monitoring eyes pattern, EMG and EOG signals. These were the basic inspiration for creating the embedded system introduced here. The proposed system includes microcontrollers, alcohol sensor, biomedical sensor, IR sensor, camera modules, GPS, Wi-fi adapter, ZigBee module, solenoid valve and ignition switch circuitry. **Findings**: The proposed vehicle safety monitoring system will detect blood alcohol concentration (BAC) percentage from the particular person's breath samples. The EOG, EMG and IR sensors are used to monitor driver condition. The warning alarm and ignition switch circuitry are enabled and disabled based on the input values of the monitoring system. Thus, the condition of the driver is continuously monitored and calibrated on a scale. When it exceeds a particular limit, the alarm unit is enabled. GPS and camera module are used to trace the vehicle and driver. The monitored values are serially transferred to the master controller and further upload the data to cloud space by means of Internet which can be accessed or streamed. This data is used to track the vehicle and detect who is driving the car at any instant of time. **Conclusion/Improvements**: Non-intrusive nature of this system helps to detect driver's physical condition without making inconvenience to him. Real-time and online data transfer system of this system helps the respective authority to take immediate actions. This device can be very helpful for law enforcement department to solve vehicle theft, accident and hit-and-run cases.

S. Vijay Kumar (✉) · A. S. Ponraj
School of Electronics Engineering (S.E.N.S.E), VIT University,
Chennai 600127, Tamil Nadu, India
e-mail: vijaykumarchoudarys@gmail.com

A. S. Ponraj
e-mail: abrahamsudharson.ponraj@vit.ac.in

D. Thalmann et al. (eds.), *Intelligent Embedded Systems*, Lecture Notes
in Electrical Engineering 492, https://doi.org/10.1007/978-981-10-8575-8_15

Keywords Blood alcohol concentration (BAC) · Drowsiness · Biopotential
Eye movement detection · Electromyography (EMG) · Electrooculography (EOG)

1 Introduction

In today's world drunk driving [1, 2] and drowsiness [3, 4] had been recognized as a major point for large portions of accidents due to diminishing driver's perception for danger and also corrupt vehicle-handling abilities because of fatigue. Sleepiness includes physical and additionally physiological progressions. Driver's fatigue [5] not only affects those alertness and reaction time of the driver but also increments the possibilities for mishaps. The point when driver dives drowsy, his/her capacity levels, driving behaviour, proficiencies and choices would adversely influenced, and in these situations, accident rate builds likewise the subject neglects should take right movements former with an impact. Despite the fact that driver safety need been made throughout way what's more vehicle design, the number of genuine crashes may be even now increasing, demonstrating a requirement for programmed identification frameworks. Driver's sleepiness could make distinguished through different measures such as eye movement identification [6], yawning monitoring, driver–vehicle communication and so forth throughout this way, observing and stock arrangement of all instrumentation may be enhance.

In this paper, a breath alcohol identifier [7] which controls ignition framework [8, 9] and also fuel stream utilizing microcontroller will be created. Rather than simply demonstrating and displaying the monitored BAC percentage, these analysers are capable to control the vehicle starting loop mechanism with the help of programmed controller [10], solenoid valve and in addition an alert. Those basic parts for this ignition loop interlock framework need aid the MQ3 alcohol sensor, microcontroller unit, ignition framework circuit also fuel stream control solenoid valve. The framework will have the capacity to identify alcohol fixation over a person's breath and recognize the measure BAC (Blood Alcohol Concentration) rate.

To identifying driver's drowsiness, an IR sensor and biomedical sensors [11, 12] are used to identify the sleepiness by checking eyes pattern, EMG [13] and EOG [14, 15] signals. These sensors were interfaced for microcontroller and programmed to screen the driver physical states, for example, eyes open and shut state and biopotential [16, 17] for eye movement.

The warning alarm and ignition switch circuitry are enabled and disabled based on the input values of the monitoring system. Thus, the driver physical state is monitored uninterruptedly and calibrated on a scale. When the threshold limit is exceeded, the system gives an alarm. The implemented system also includes camera module, GPS, Wi-fi adapter, ZigBee module. The monitored values are serially transferred to the master controller and further upload the data to cloud space by means of Internet which can be accessed or streamed. This data is used to track the vehicle and detect who is driving the car at any instant of time. This device can be very helpful for law enforcement department to solve vehicle theft [18], accident and hit-and-run cases.

Compared to the already available alcohol and drowsiness detection systems in the market, the vehicle safety monitoring system offers high precession at less expense and offers a simple system which is versatile.

2 Proposed Algorithm

2.1 Ignition Interlock Device

The proposed algorithm for ignition interlock device is as shown in Fig. 1; after switching ON the power supply, the sensors begin working. The MQ3 alcohol sensor detects the alcohol content from driver breath and sends the comparing

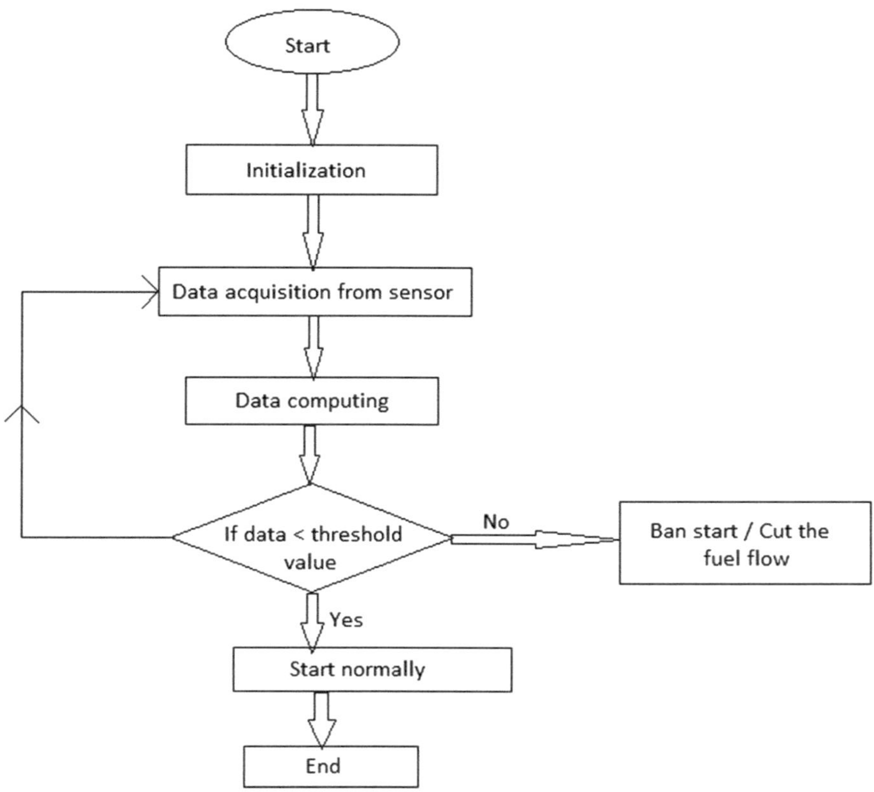

Fig. 1 Proposed algorithm for ignition interlock device

qualities to the microcontroller unit. The motivation behind the microcontroller unit is to check the information values which were sent by the sensors and are contrasted and the predefined limit which was customized in the microcontroller unit.

At the point when the sensor information quality is more than the edge esteem, then the microcontroller ought not to permit the ignition framework to start the engine and in the meantime, it stops the fuel stream and an alarm will be initiated to warn the individual.

2.2 Drowsiness Detection System

The proposed algorithm for drowsiness detection system is as shown in Fig. 2; after switching ON the power supply, the sensors begin working. The IR sensor detects the open and shut condition of driver's eyes and sends the comparing qualities to the microcontroller unit; comparably, the biomedical sensor detects the biopotential relating to EMG and EOG signs of driver eyes. The principal point of the microcontroller unit is to check the information values which were sent by the sensors and are contrasted and the predefined limit which was customized in the

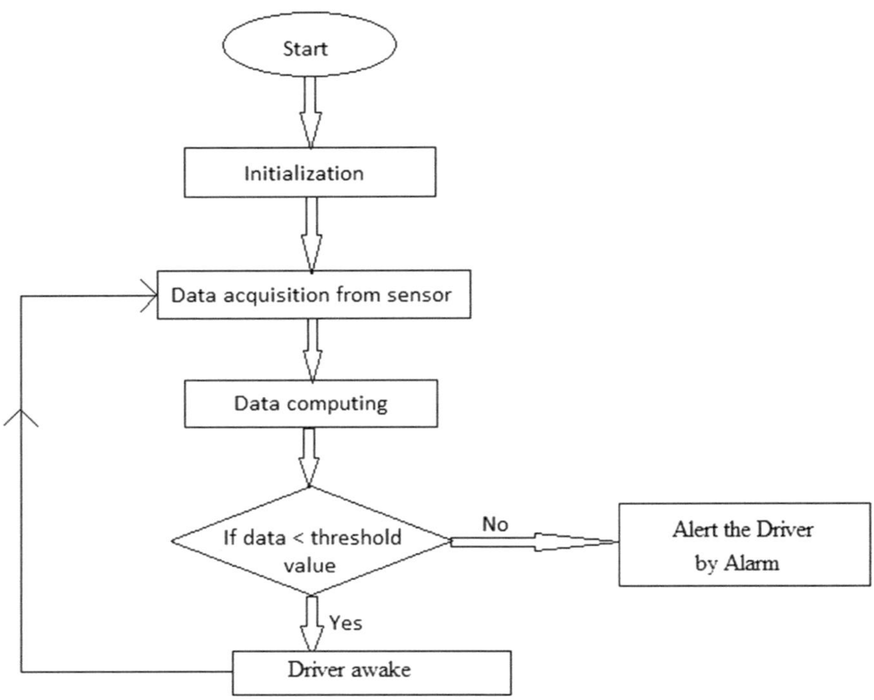

Fig. 2 Proposed algorithm for drowsiness detection system

microcontroller unit. At the point when the sensor information qualities are not in the scope of edge qualities, then the microcontroller enacts an alarm to caution the driver.

3 Block Diagram

3.1 Dashboard Device Module

The dashboard module of this project consists of ignition interlock device and driver, vehicle tracking system with GPS module, Wi-fi adapter, ZigBee adapter and camera module. All these components are interfaced and installed to Raspberry Pi microcontroller board to read the vehicle and driver data and upload to cloud space by means of Internet gateway. Fig. 3 shows the block diagram of dashboard module.

The dashboard module comprises of Raspberry Pi board and which acts as a master controller, and a ZigBee module is used to receive the data from the wearable device. The data from the master controller is uploaded to cloud space by means of Internet gateway. A Wi-fi adapter is used to establish Internet connection to this module.

3.2 Ignition Interlock Device

A breath alcohol ignition interlock gadget is an embedded framework gadget which is installed in a vehicle's dashboard. Prior to the vehicle can be begun, the driver

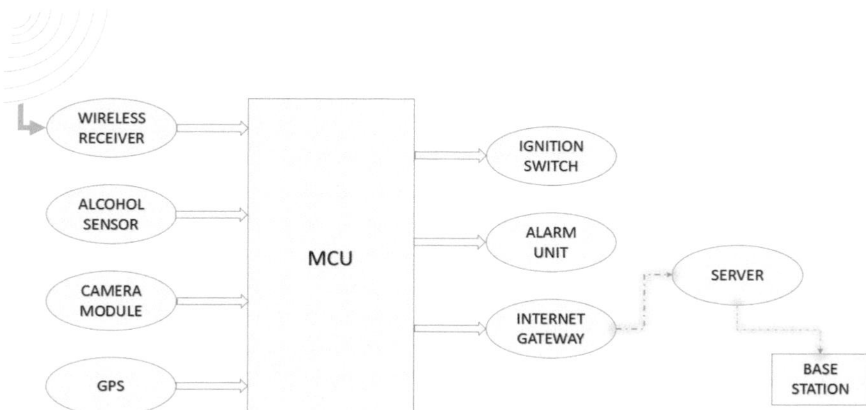

Fig. 3 Block diagram of dashboard device

must inhale into the gadget. In the event that the sensor value is over a limit blood alcohol fixation, the vehicle will not start.

All together, keep up its proficiency and precision. Occasional adjustment is required. This can be performed utilizing either a pressurized liquor blend at a known liquor focus arrangement.

Modern ignition interlock gadgets utilize an ethanol particular power module for a sensor. An energy unit sensor is an electrochemical gadget in which alcohol experiences a compound oxidation response at a reactant cathode surface (platinum) to create an electrical current. This current is then measured and changed over to an alcohol proportionate reading.

3.3 *Blood Alcohol Concentration (BAC)*

It is important to devise a strategy to make test gas fixations to mimic different BAC values. The blood alcohol fixation is characterized to be the rate of alcohol in grams per 100 mL of blood. In this manner, 0.08% BAC is 80 mg of alcohol in 100 mL of blood. MQ3 sensor recognizes the nearness of alcohol in air not blood. By this way, a moderately steady proportion of 2100:1 was actualized to make diverse false arrangements. This proportion is considered from a logically settled upon thought that the BAC is characterized as the measure of alcohol in grams per 210 L of air.

Ethanol has a specific gravity of 0.79. This implies 1 mL of ethanol weighs around 0.79 g. Figuring to discover the measure of ethanol required in every arrangement was defeated the accompanying focus levels of 0.02, 0.04, 0.06, 0.08, 0.10, 0.12 and 0.20.

With these amounts now inferred, it was then simple to make these concentrations by putting the measure of ethanol required into a 1 L cup. These jars were named and fixed to permit them to achieve harmony. This regularly took around an hour since the measure of ethanol was so low. When they were made and prepared, calibration testing was finished. Table 1 shows the alcohol concentrations in breath and their corresponding BAC value.

Table 1 Alcohol concentrations in breath and their corresponding BAC value

Alcohol concentrations in breath (g/L)	BAC (%)
0	0
9.5×10^{-6}	0.02
1.9×10^{-5}	0.04
2.8×10^{-5}	0.06
3.8×10^{-5}	0.08 (maximum legal limit)
4.7×10^{-5}	0.10
5.7×10^{-5}	0.12
9.5×10^{-5}	0.20

3.4 Drowsiness Detection Working Principle

A proximity sensor is a sensor used to recognize the nearness of adjacent articles with no physical contact. A proximity sensor regularly emits an electromagnetic field or a light emission radiation (infrared) and searches for changes in the field or return signal. The article being detected is referred as proximity sensor's target. The greatest separation that this sensor can recognize is characterized as "nominal range".

Proximity sensors can have a high unwavering quality and long useful life because of the absence of mechanical parts and absence of physical contact in the middle of sensor and the detected item.

Proximity sensors are generally utilized on cell phones to identify incidental touchscreen taps when held to the ear amid a call. They are likewise utilized as a part of machine vibration checking to quantify the variety in separation between a pole and its bolster bearing.

This is the basic thought made me to utilize a proximity sensor to recognize drowsiness by method for aligning human eye squint. Human eyeball and eyelid reflects IR light diversely so blinking will trigger a fast perusing change.

An Ag/Agcl sensor cable, AD8232 break out board and sensor pads are utilized to gauge the electrical action of the eye, muscles. This electrical movement can be outlined as an EOG or an EMG and yield as a simple perusing. EMGs can be extremely noisy, and the AD8232 goes about as an operation amp to acquire an unmistakable sign from zone of contact of individual's body.

The AD8232 is an incorporated sign moulding hinder for EMG, EOG and other biopotential estimation applications. It extracts, amplifies and filters very small biopotential signals in noisy conditions, for example, those made by movement or remote electrode placement. Figure 4 shows the block diagram of wearable device.

There is a fast transition in the reading while winking an eye as the human eyeball and eyelid reflects IR light differently. Whenever a person closes or opens his eyes, there is a change in voltage. According to these output voltages, we can detect whether eye is open or close. Let the open state of eye is 0 and closed state of eye is 1.

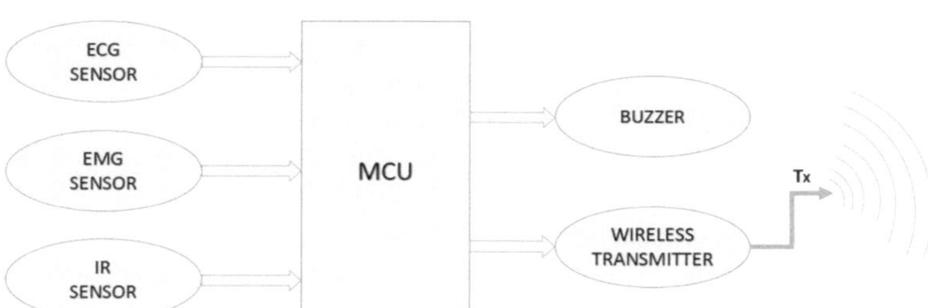

Fig. 4 Block diagram of wearable device (goggles)

Table 2 Sleep detection goggles output logic

IR output (A)	Biomedical output (B)	Goggles output (A and B)
0	0	0
0	1	0
1	0	0
1	1	1

Similarly by using biomedical sensors, we can detect the EOG and EMG signals from eye. When a person blinks or moves his eyes, there is a change in biopotential among the electrode cables. By analysing those signals from different samples and conditions, we can determine a threshold value for minimum and maximum change in voltages. Let the minimum threshold is 1 and maximum threshold is 1.

By considering both the outputs from IR sensor and biomedical sensor together, we can detect drowsy and sleep state of person. The following table gives how the logic works to determine drowsiness and sleep of a person.

Here in Table 2, the output value 0 determines the awake and the output value 1 determines sleep states of person. The goggle determines the drowsy or sleep state of person; only both the outputs from IR sensor and biomedical sensor are high.

4 System Testing and Calibration

4.1 Ignition Interlock Device

The entire framework will breakdown if the power supply falls flat. Henceforth, appropriate testing of the voltage controller is crucial. The finished circuit of the voltage controller is tried and its relating yield voltage over the Drove is measured utilizing multimeter.

The most essential assignment in this area is the calibration of the alcohol sensor. The input voltage that is sent to the microcontroller is not so much direct over all concentrations of ethanol. Consequently, watchful concentrations must be set linear or close to linear area. Concentration can even damage the sensor if too high a fixation was set on the sensor. Each readied ethanol fixation was put against the sensor, and the voltage level was seen on the multimeter and recorded. A normal worth for each particular concentration was taken.

In Fig. 5, the finished circuit without the alcohol sensor was put through useful testing. The sensor voltage was supplanted by a variable resistor. When this alignment was done, the sensor was associated with whatever is left of the circuit. The entire gadget was tried utilizing the arrangements that were set up for the calibration of the sensor.

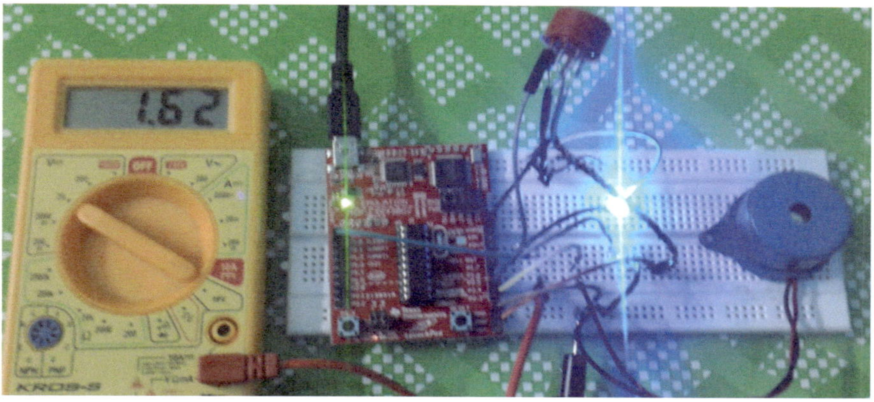

Fig. 5 Calibration of alcohol sensor

Table 3 BAC sensor value and its respective change in output voltage values at an operating voltage of 5 V

Voltage (V)	BAC (%)
0–2.5	0.00
3.0–3.1	0.02
3.1–3.2	0.04
3.25	0.06
3.30	0.08
3.42	0.10
3.75	0.20

After fixing the sensor to the driver and associating it to the motherboard, yield voltage from the sensor at typical room temperature is taken. This yield voltage is set as the reference voltage for the calibration process. At that point, human breath test is blown at the sensor and the consequent yield voltage is taken. At last, the sensor is then tried utilizing alcohol substance and its yield voltage is taken. The voltages set in the writing computer programs are then straightened out as indicated by the voltage measured amid calibration, and the equipment is on the other hand tried to check the BAC rate produced.

Table 3, Figs. 6 and 7 show the change in sensor value for various alcohol concentrations and its respective change in voltage.

4.2 Drowsiness Detection Device

As the human eyeball and eyelid reflects IR light differently, a wink of an eye will give rise to a fast change in the reading. Figure 8 shows the hardware implementation of eye blink detection goggles.

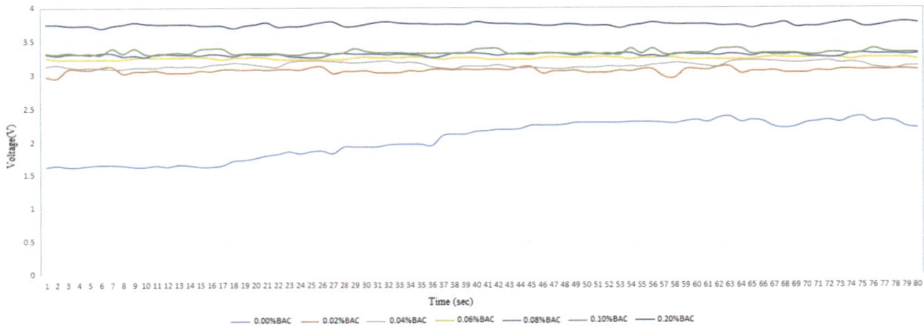

Fig. 6 Output voltages for different BAC values at operating voltage of 5 V

Fig. 7 Output voltages at corresponding BAC values for 5 V input

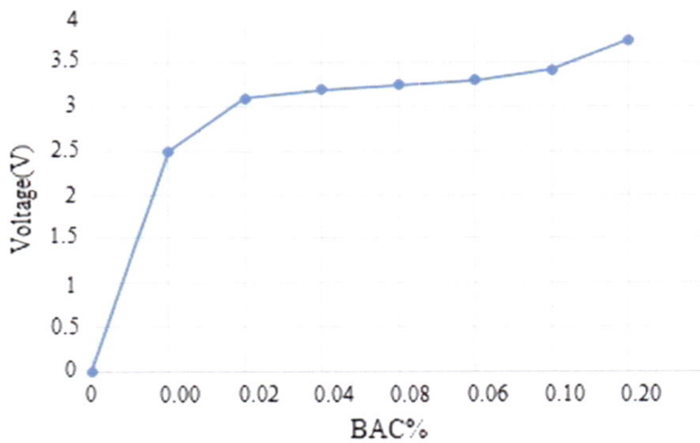

Figure 9 shows the output voltage ranges for different state of eyes. The voltage ranges present inside the red box determine voltage range of closed state of eye.

Myofibrils generate an electrical signal on the surface of the body, and it is recorded as the electromyography (EMG), and electrooculography (EOG) is the recording on the body surface constituted by flicker of an eye. A differential recording between two points on the body is made to record electrical activity. Traditionally, each differential recording is referred to as a lead. In this project, a three lead electrode sensor is used to detect ECG and EOG signals from human eyes. Three leads indicated as A, B and C. They are defined as:

$$A = V_{RE} - V_{LE},\ B = V_{RL} - V_{RE},\ C = V_{RL} - V_{LE.}$$

where RE = right eye, LE = left eye and RL = reference lead. Note that these were not independent, and the following relationship holds: B = A + C.

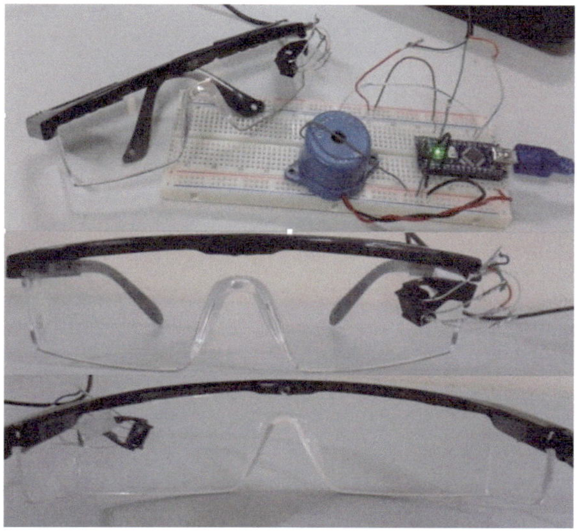

Fig. 8 Blink detection goggles

Fig. 9 Output voltages of IR sensor for different state of eye

Figures 10 and 11 show the experimental setup of EOG and EMG signal acquisition system to detect drowsiness by means of its output values.

As shown in Fig. 12, the voltage range present with in blue box represents eyes closed for a long time, green and red boxes represents normal blinking and moving of eyes.

Based on these voltages or signal variations, we can easily find out the drowsiness of the driver.

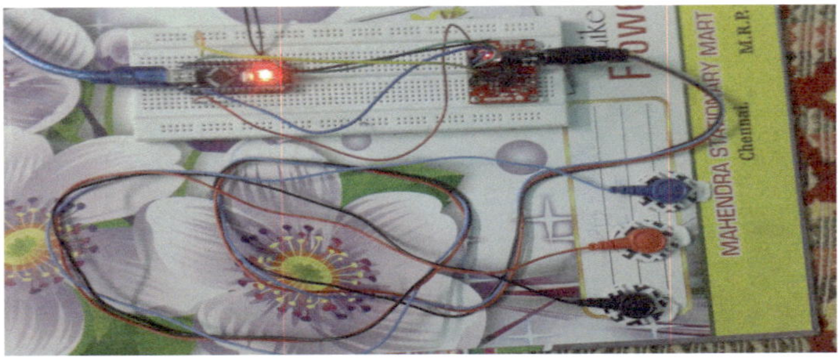

Fig. 10 EMG and EOG signal acquisition setup

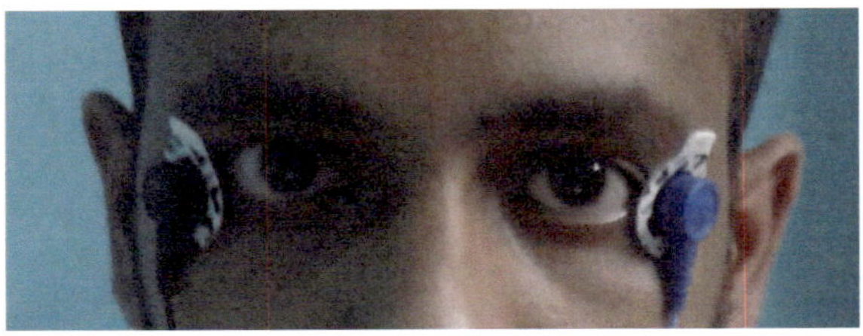

Fig. 11 Sensor cables attached to extract EMG and EOG signals

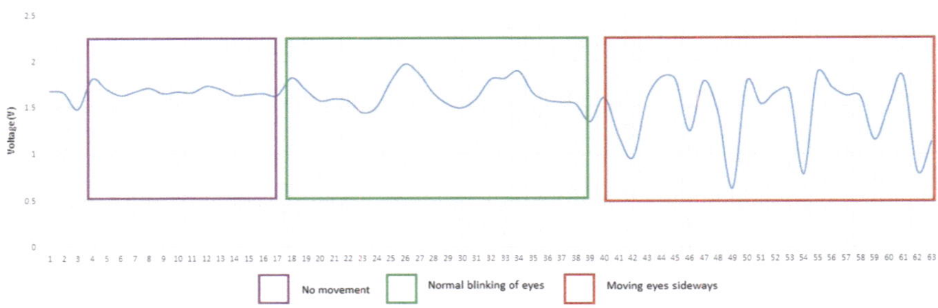

Fig. 12 Output voltages of EMG and EOG signals

5 Implementation and Results

5.1 Ignition Interlock Device

Figure 13 shows the implemented hardware of anti-drunk-drive system. The purpose of this embedded system is for alcohol detection and ignition interlock for vehicle. Figure 14 shows the output on PuTTY terminal for high levels of BAC, and Fig. 15 shows the output on PuTTY terminal for low levels of BAC.

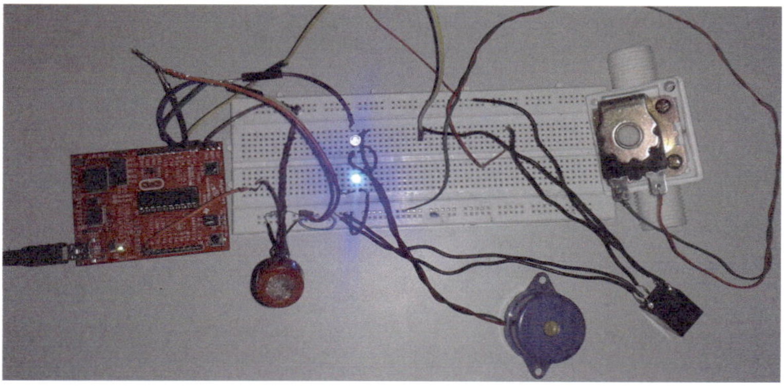

Fig. 13 Proposed ignition interlock device hardware

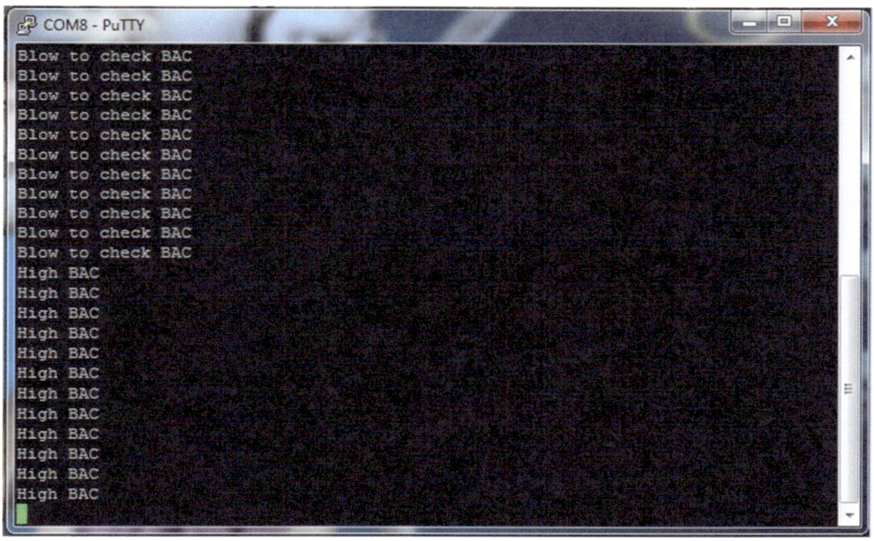

Fig. 14 Anti-drunk-drive system output to high BAC%

5.2 Drowsiness Detection System

The implemented hardware of drowsiness detection system and goggles are shown in Figs. 16 and 17.

There is a fast transition in the reading while winking an eye as the human eyeball and eyelid reflects IR light differently. Whenever a person closes or opens his eyes, there is a change in voltage. According to this output voltages, we can whether eye is open or close. Let the open state of eye is 0 and closed state of eye is 1.

Fig. 15 Anti-drunk-drive system output to low BAC%

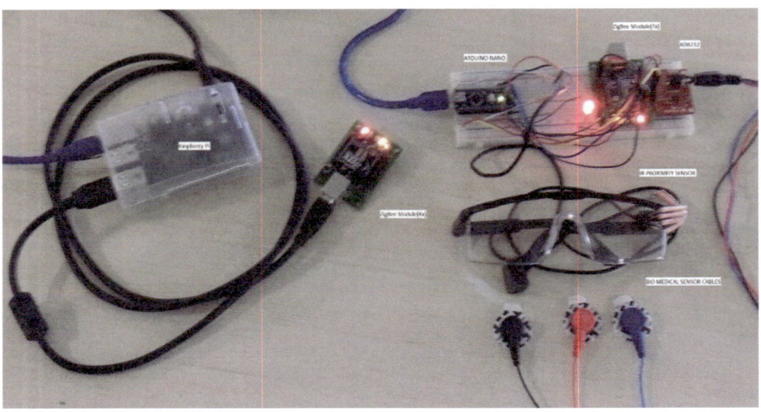

Fig. 16 Implemented drowsiness detection system

Fig. 17 Sleep detection goggles

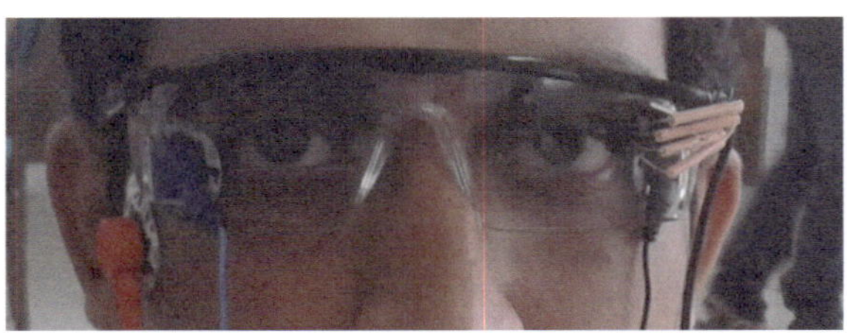

Similarly by using biomedical sensors, we can detect the EOG and EMG signals from eye. When a person blinks or moves his eyes, there is a change in biopotential among the electrode cables. By analysing those signals from different samples and conditions, we can determine a threshold value for minimum and maximum change in voltages. Let the minimum threshold is 1 and maximum threshold is 1.

By considering both the outputs from IR sensor and biomedical sensor together, we can detect drowsy and sleep state of person. The goggle determines the drowsy or sleep state of person; only both the outputs from IR sensor and biomedical sensor are 1. Output from wearable device on pi terminal is shown in Fig. 18.

The dashboard module comprises of Raspberry Pi board which acts as a master controller and a ZigBee module which is used to receive the data from the wearable device. The data from the master controller is uploaded to cloud space by means of Internet gateway. A Wi-fi adapter is used to establish Internet connection to this module.

For vehicle tracking and identifying the person who is driving the car, a GPS module and a camera module are also installed in the dashboard. Figure 19 shows the implemented hardware of automatic driver and vehicle safety system. Figure 20 shows the GPS data uploaded into the cloud space for real-time vehicle tracking.

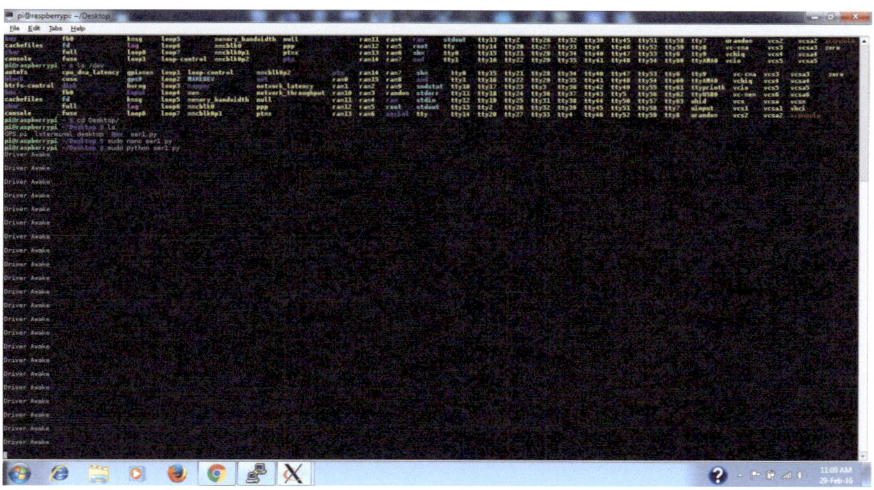

Fig. 18 Output from wearable device on pi terminal

Fig. 19 Implemented hardware of automatic driver and vehicle safety system

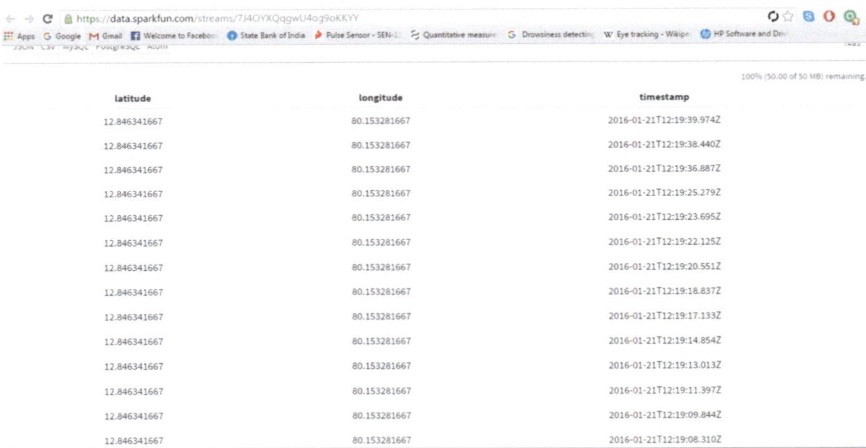

https://data.sparkfun.com/streams/7J4OYXQqgwU4og9oKKYY

Apps Google Gmail Welcome to Faceboo State Bank of India Pulse Sensor - SEN-1 Quantitative measure Drowsiness detectin Eye tracking - Wikipe HP Software and Dri

100% (50.00 of 50 MB) remaining.

latitude	longitude	timestamp
12.846341667	80.153281667	2016-01-21T12:19:39.974Z
12.846341667	80.153281667	2016-01-21T12:19:38.440Z
12.846341667	80.153281667	2016-01-21T12:19:36.887Z
12.846341667	80.153281667	2016-01-21T12:19:25.279Z
12.846341667	80.153281667	2016-01-21T12:19:23.695Z
12.846341667	80.153281667	2016-01-21T12:19:22.125Z
12.846341667	80.153281667	2016-01-21T12:19:20.551Z
12.846341667	80.153281667	2016-01-21T12:19:18.837Z
12.846341667	80.153281667	2016-01-21T12:19:17.133Z
12.846341667	80.153281667	2016-01-21T12:19:14.854Z
12.846341667	80.153281667	2016-01-21T12:19:13.013Z
12.846341667	80.153281667	2016-01-21T12:19:11.397Z
12.846341667	80.153281667	2016-01-21T12:19:09.844Z
12.846341667	80.153281667	2016-01-21T12:19:08.310Z

Fig. 20 GPS data uploaded into cloud space

6 Conclusion

In contrast to other sort of breath alcohol identifier, the proposed IC-based model is much cheaper and very vividly powerful also. Besides, it can be interfaced to embedded controllers with much ease as the output is in voltages. It is found that the estimation of the fixation gets to be saturation for BAC (%) more than 0.20

despite the fact that the alcohol concentration utilized is more than that. At this juncture when loss power comes into concern, the power dissipated by the semiconductor sensor model is very low despite the fact that it works in presence of heat. Likewise, the sensitivity of the sensor is rather high when put up against other alternate models. The semiconductor sensor is used for identifying different substances and is capable of misinterpreting them as alcohol. The sensitivity to alcohol is much higher which makes it a perfect solid liquor identifying gadget.

Drowsiness detection system implemented in this project is very cheap compared to existing systems with high-resolution cameras. Even though with high-resolution cameras, it is quite difficult to capture and process the driver eyes in night times, whereas the system implemented in this project uses low-cost and low power consuming IR sensor and biomedical electrode sensors which give output with high precision. While testing the system in real time, it gives 93% positive response to the given situation.

While coming to vehicle tracking and driver tracing, a GPS module and camera module are used. These data from GPS and camera are uploaded to cloud space by means of Wi-fi adapter. These data are used to tack the vehicle, to law enforcement department to solve the cases like vehicle theft, hit and run.

The actualized framework can be utilized as a part of a wide range of vehicles and can also monitor the alcohol substance of drivers and drowsiness of driver and guarantee the driver's life and property well-being. The guideline of gadget determination in this framework is stable performance and cheap, so the expense is lower. It is appropriate to the applications of high cost-performance ratio.

References

1. Murata K, Fujita E, Kojima S, Maeda S, Ogura Y, Kamei T, Tsuji T, Kaneko S, Yoshizumi M, Suzuki N (2011) Noninvasive biological sensor system for detection of drunk driving. IEEE Trans Inform Technol Biomed 15(1)
2. Sakairi M (2012) Water-cluster-detecting breath sensor and applications in cars for detecting drunk or drowsy driving. IEEE Sens J 12(5)
3. Chacon-Murguia MI, Prieto-Resendiz C (2015) Detecting driver drowsiness. A survey of system designs and technology. IEEE Consum Electron Mag
4. Roy R, Venkatasubramanian K (2015) EKG/ECG based driver alert system for long haul drive. Indian J Sci Technol 8(19). https://doi.org/10.17485/ijst/2015/v8i19/77014
5. Yang JH, Mao ZH, Tijerina L, Pilutti T, Coughlin JF, Feron E (2009) Detection of driver fatigue caused by sleep deprivation. IEEE Trans Syst Man Cybern A: Syst Humans 39(4)
6. Mbouna RO, Kong SG, Chun MG (2013) Visual analysis of eye state and head pose for driver alertness monitoring. IEEE Trans Intell Transp Syst 14(3)
7. Anghel MA (2012) Statistical inference of breath alcohol concentration measurement. In: 2012 international conference and exposition on electrical and power engineering (EPE 2012), 25–27 Oct, Iasi, Romania
8. Ignition Interlock System (2008). www.wikipedia.com

9. International Council on Alcohol, Drugs and Traffic Safety (ICADTS) Working Group on Alcohol Ignition Interlocks (2001) Alcohol ignition interlock devices. Volume I: position paper
10. Abdul Rahim H, Hassan SDS (2010) Breathalyzer enabled ignition switch system. In: 6th international colloquium on signal processing and its applications (CSPA)
11. Lin CT, Chang CJ, Lin BS, Hung SH, Chao CF, Wang IJ (2010) A real-time wireless brain–computer interface system for drowsiness detection. IEEE Trans Biomed Circ Syst 4(4)
12. Sun Y, Yu XB (2014) An innovative nonintrusive driver assistance system for vital signal monitoring. IEEE J Biomed Health Inform 18(6)
13. Venkatesh K, Geetha S (2015) Sleep stages classification using artificial neural network. Indian J Sci Technol 8(31). https://doi.org/10.17485/ijst/2015/v8i31/87271
14. Subbulakshmi K (2015) Computer human interface control by electro-oculography (EOG) signal. Indian J Sci Technol 8(31). https://doi.org/10.17485/ijst/2015/v8i31/84221
15. Hu S, Zheng G, Peters B (2012) Driver fatigue detection from electroencephalogram spectrum after electrooculography artefact removal. IET Intell Transp Syst. https://doi.org/10.1049/iet-its.2012.0045
16. Visu P, Varunkumar KA, Srinivasan R, Kumar RV (2016) Brainwave based accident avoidance system for drowsy drivers. Indian J Sci Technol 9(3). https://doi.org/10.17485/ijst/2016/v9i3/86381
17. Jung SJ, Shin HS, Chung WY (2012) Driver fatigue and drowsiness monitoring system with embedded electrocardiogram sensor on steering wheel. IET Intell Transp Syst. https://doi.org/10.1049/iet-its.2012.0032
18. Shinde PA, Mane YB (2015) Adavanced vehicle monitoring and tracking system based on Raspberry Pi. Intelligent systems and control (ISCO). In: 2015 IEEE 9th international conference. pp 1–6. https://doi.org/10.1109/isco.2015.7282250

Emergency and Traffic Congestion Avoidance Using Vehicle-to-Vehicle Communication

Anirban Das, Mahadev Desai, Nilkanth Mugatkar and Abraham Sudharson Ponraj

Abstract Intelligent transportation systems (ITSs), entitled with wireless communication, are supposed to improve the safety of passengers and effectiveness of our transportation network. Transceiver is used in vehicle to exchange critical information from host vehicle to preceding vehicle. Wireless short-range communication technology is very much fastest technology which is used for transferring information. Enabling vehicle to communicate with each other, information such as traffic environment, emergency situation can be shared with the help of wireless transceiver (Zigbee) in each vehicle and an embedded microcontroller. This leads to create a safer and comfortable drive, and also GPS module is used for detecting the position of the particular vehicle. Traffic congestion is one of the major issues in the urban areas. For traffic avoidance, when a vehicle passes the traffic congestion area, driver passes that message to the vehicle approaching to that area. In this case, driver can type the message using keypad and broadcast it through Zigbee, and the vehicle approaching to that area will get pre-alert.

1 Introduction

Time is very crucial when ambulances are utilized to save people lives but when an ambulance pass through a junction, its speed often must be reduced due to traffic. In this critical situations, the patient in the ambulance may need urgent treatment that

A. Das · M. Desai · N. Mugatkar · A. S. Ponraj (✉)
School of Electronics Engineering (SENSE), VIT University, Chennai 600127, India
e-mail: abrahamsudharson.ponraj@vit.ac.in

A. Das
e-mail: anirbandas02@gmail.com

M. Desai
e-mail: desaimv037@gmail.com

N. Mugatkar
e-mail: nilkanthmugatkar@gmail.com

D. Thalmann et al. (eds.), *Intelligent Embedded Systems*, Lecture Notes in Electrical Engineering 492, https://doi.org/10.1007/978-981-10-8575-8_16

can be administered only at a hospital. Due to the unavailability of advanced procedures in an ambulance, there is the possibility to suffer a loss of patient life. In this work, an emergency vehicle (ambulance) can give its presence to the vehicle driving ahead as a warning to make way for it. Also a vehicle that have observed obstacles or traffic conjunction on a particular route can communicate the information to other vehicle passing by and helping them to be better prepared or take an alternative route. This system uses the V2V communication method to avoid traffic and makes it easy for an emergency vehicle to reach its destination [1–4]. Traffic control is a challenging issue in India, as is the case in much of the rest of the world. Zigbee technology is a wireless standard designed to operate low-power wireless sensor networks, and it can aid emergency vehicles in dealing with traffic congestion. This module sends an emergency alert to the nearest vehicles for giving way to the ambulance. Nowadays, traffic congestion is one of the major issues in the urban areas. Basically, this project is proposed to do a traffic congestion alert system which is controlled by vehicle (driver). When a vehicle passes the traffic congestion area, the driver passes that message to the vehicle approaching to that area. In this case, driver can type the message using keypad and broadcast it through Zigbee, and the vehicle approaching to that area will get pre-alert [5–10].

2 Proposed System

The emergency signal is being transmitted to the nearest vehicle by using Zigbee technology and to avoid traffic congestion, the message which is typed by vehicle driver is passed to the nearest vehicle which is approaching that area. Latitude and Longitude readings are obtained using GPS module Skytra S1216R8 and successfully transmitted those data to the Cloud agent Think Spark (Figs. 1 and 2).

Hardware:

- Zigbee module.
- LCD.
- Skytra S1216R8 (GPS)
- Arduino Uno.

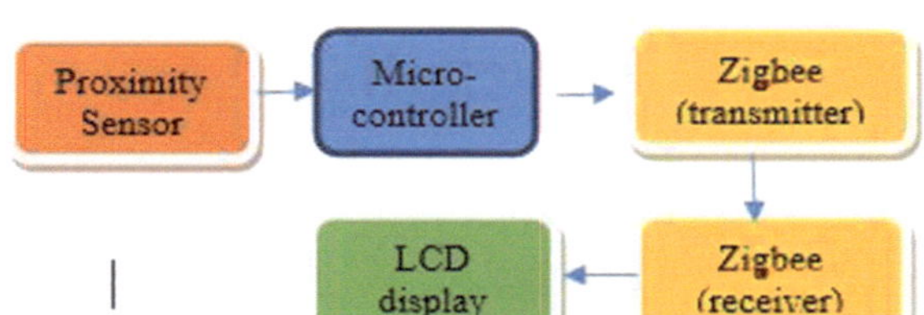

Fig. 1 Block diagram of vehicle-to-vehicle communication for traffic emergency purpose

Fig. 2 Block diagram of track position of the vehicle

<u>Software:</u>

- Arduino 1.0.6.
- X-CTU (Zigbee).

2.1 Ambulance Alert System

Ambulance has attached proximity sensor on its front side; when any vehicle comes within its range, it automatically gives trigger to the Zigbee module which is placed in the ambulance and this module broadcasts this message to the surrounding vehicle, and also a GPS module attached in ambulance broadcasts (latitude and longitude) message to the surrounding vehicle and also sends it to the cloud via Wi-Fi module. The vehicles which are in its path get this message through Zigbee or directly from cloud, and the driver can able to see it on LCD. In normal condition, GPS module sends (latitude and longitude) data into the cloud in 5-min time interval and in critical condition (Zigbee within range) sends data into the cloud in 30-s time interval to compress the data.

2.2 Traffic Congestion Avoidance System

A traffic congestion alert system is controlled by vehicle (driver). When a vehicle passes the traffic congestion area, the driver types that message by using keypad button and broadcasts it to the nearest vehicle which is approaching to that area.

2.3 Methodology

GPS module is connected to the Arduino Uno board for detecting vehicle position, which gives latitude and longitude data shown in Arduino Uno serial monitor screen. Those data are sent to the nearest vehicle by Zigbee module so that driver can know the position of vehicle which is coming towards its direction. If any accident happens, this module can easily detect the position of the vehicle by processing those latitude and longitude data (Fig. 3).

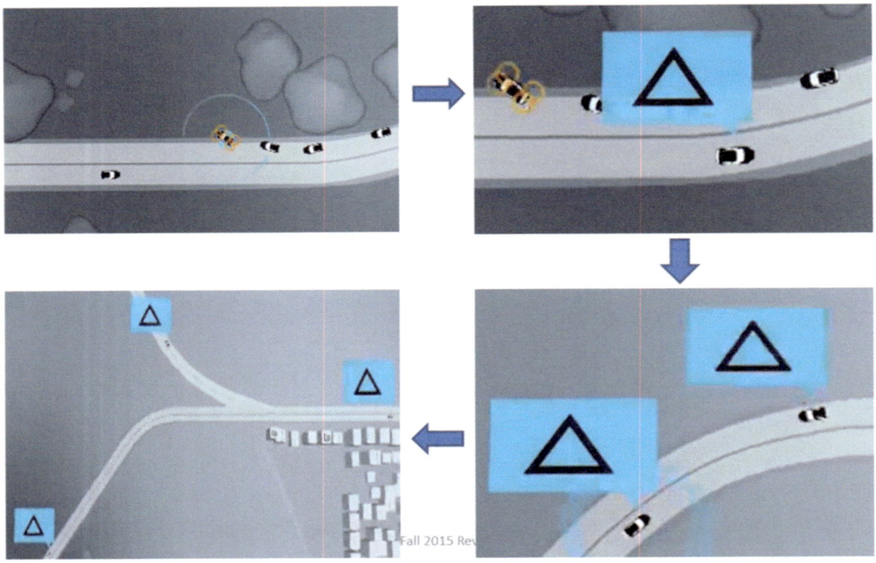

Fig. 3 Vehicle-to-vehicle communication for traffic congestion avoidance approach

2.4 *Flow chart*

See Fig. 4.

3 Result and Discussion

When object comes into proximity sensor's range, it sends a pulse to microcontroller board. Controller board takes care of control part. This digitalized data is transmitted through wireless module (Zigbee) after completing paired device, and that message will be shown on LCD. Two Zigbee modules and two Arduino Uno boards are used here, and Zigbee is programmed in AT mode command. One Zigbee acts as a coordinator whose destination address of high value is 0000 and low value is FFFF. So, it broadcast message to all Zigbee module which are in its range. Other Zigbees act as a router whose destination address of high value is 0000 and low value is 0000. So, data communication is done between Zigbee modules. When router comes into coordinator range, coordinator sends a message ("EMERGENCY AMBULANCE") to the router and displays on LCD (Fig. 5) which is connected to the receiver side Arduino board.

This module (Fig. 6) explained various codes which convey specific information to base station from which appropriate action should be taken. Those codes are 101 = ambulance and 100 = police; in similar manner, those codes for avoidance

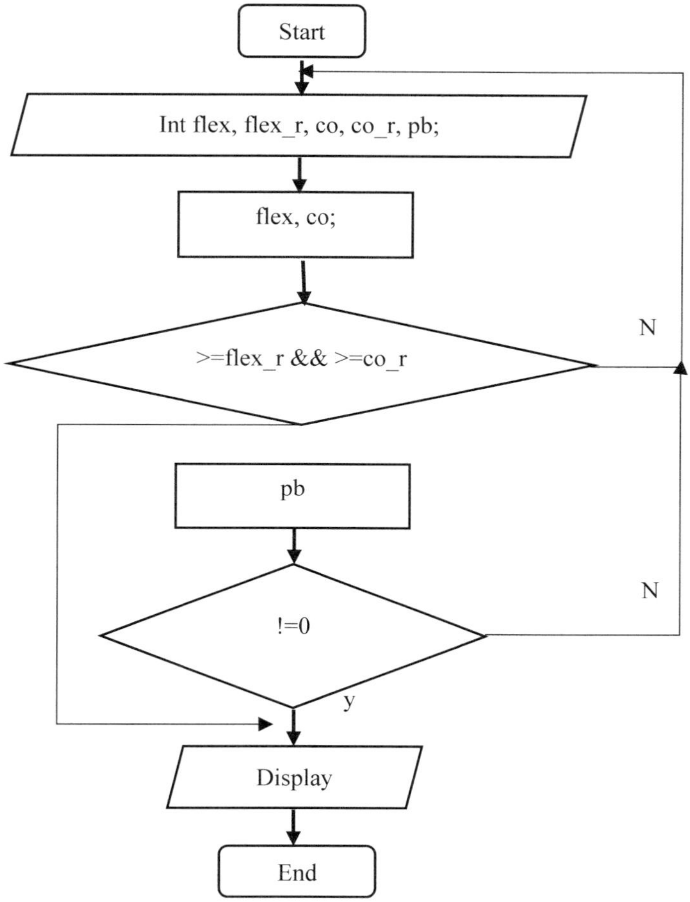

Fig. 4 Flow chart of traffic congestion avoidance. Flex = flex sensor, Co = carbon monoxide sensor, Pb = push button, Flex_r = threshold value of flex sensor, Co_r = threshold value of carbon monoxide sensor

of traffic conjecture are such as 2 = car break down, 3 = traffic due to land siding, 4 = one way; these codes can be sent by the car driver by pressing key as driver observes such scenario. These codes can be connected in two ways to receiver (nearby vehicle, base station) (1) Zigbee (2) Cloud. By this way, this module can avoid traffic congestion.

Fig. 5 Hardware for ambulance alert message

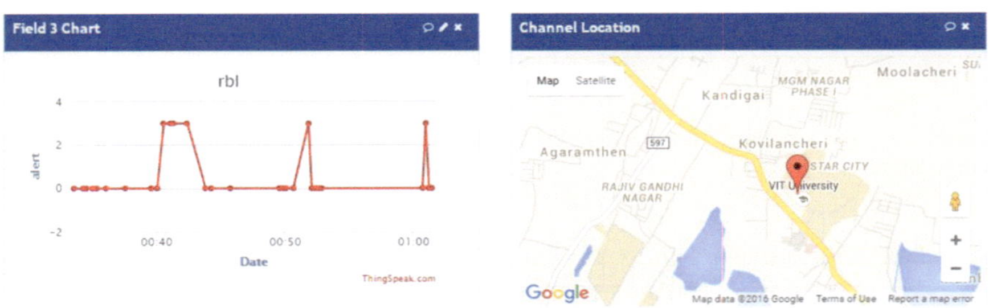

Fig. 6 Alert message sent by driver to cloud (left), respective location of driver (right)

4 Conclusion

This research paper has dealt with pairing of Zigbee with Zigbee and also passes emergency alert to the nearest vehicles and cloud for giving way to the ambulance for emergency purpose. We obtained Latitude and Longitude readings using GPS module (Skytra S1216R8) and successfully transmitted that reading to the surrounding vehicle and cloud data base. This paper has proposed vehicle-to-vehicle communication by using Zigbee module. Basically, this project is low cost and efficient. This module has high speed data transfer rate, and those data are also monitored from base station for taking immediate action in emergency purpose. This paper also proposed to avoid traffic congestion, in that case driver itself sends message to the upcoming vehicle for alert traffic congestion.

References

1. Yu H, Guo M (2010) An efficient freeway traffic information monitoring systems based on wireless sensor networks and floating vehicles. In: Proceedings of the international conference on pervasive computing, signal processing applications, Sept 2010, pp 1065–1068
2. Koh BK-P, Kong P-Y (2006) Performance study on zigbee-based wireless personal area networks for real-time health monitoring. ETRI J 28(4):537–540
3. Leontiadis I, Marfia G, Mack D, Pau G, Mascolo C, Gerla M (2011) On the effectiveness of an opportunistic traffic management system for vehicular networks. IEEE Trans Intell Transp Syst 12(4):1537–1548
4. Dai X, Ferman M, Roesser R (2003) A simulation evaluation of a real-time traffic information system using probe vehicles. In: Proceedings of the IEEE international conference on intelligent transportation systems, pp 475–480
5. Yongchang M, Chowdhury M, Sadek A, Jeihani M (2012) Integrated traffic and communication performance evaluation of an intelligent Vehicle Infrastructure Integration (VII) system for online travel-time prediction. IEEE Trans Intell Transp Syst 13(3):1369–1382
6. Willke TL, Tientrakool P, Maxemchuk NF (2009) A survey of intervehicle communication protocols and their applications. IEEE Commun Surveys Tutorials 11(2):3–20 (2nd Quart.)
7. Wolf MT, Burdick JW (2008) Artificial potential functions for highway driving with collision avoidance. In: Proceedings of the IEEE ICRA, May 2008, pp 3731–3736
8. Garcia-Costa C, Egea-Lopez E, Tomas-Gabarron JB, Garcia-Haro J, Haas ZJ (2012) A stochastic model for chain collisions of vehicles equipped with vehicular communications. IEEE Trans Intell Transp Syst 13(2):503–518
9. Xu Q, Sengupta R, Jiang D (2003) Design and analysis of highway safety communication protocol in 5.9 GHz dedicated short-range communication spectrum. In: Proceedings of the IEEE VTC, vol 57, no 4, pp 2451–55
10. Zhu J, Roy S (2003) MAC for dedicated short range communications in intelligent transport systems. IEEE Commun Mag 41(12)

Smart Mobile Diagnostic Laboratory and Doctor Annunciation System in Ambulances

Nikita Bais, R. Shubha, V. Yamuna and M. Kalyan Chakravarthi

Abstract Background/Objectives: Globally increasing health care issues have been an immense inconvenience to the mankind. Pre-treatment checkup in case of emergencies leads to catastrophic results. So this paper presents a method to carry out checkups in the ambulance on the way to the hospital. **Methods/Statistical analysis**: The two major setbacks which prevents the treatment of a patient at critical situations are the city traffic and the distance from remote area to the hospital. The main drawback of the existing system is that it provides only tele-assistance in ambulance which becomes difficult for the doctor to interpret the situation with the obtained information. **Findings**: This paper provides solution to the existing problem, wherein a diagnostic laboratory would be present inside the ambulance and all the required tests would be taken in prior. The entire system would be designed using FPGA as a stand-alone system with reconfigurable I/O ports using LabVIEW as software. A GUI is created with the patient details, and the data of the monitored parameters along with the vocal sheet of the patient and the image of the wound will be sent to the doctor via this GUI. **Improvements/ Applications**: IoT can be included to update the monitored parameters to the doctor at any instant. And a Google Map can be created to show the exact location so the exact time taken for the ambulance to reach the hospital will be known. A few more sensors can also be added.

Keywords Smart ambulance · Diagnostic laboratory · FPGA · GPS

N. Bais · R. Shubha · V. Yamuna · M. K. Chakravarthi (✉)
School of Electronics Engineering, VIT University, Chennai 600127, Tamil Nadu, India
e-mail: maddikerakalyan@vit.ac.in

N. Bais
e-mail: nikita.bais28@gmail.com

R. Shubha
e-mail: shubharavi20@gmail.com

V. Yamuna
e-mail: veligaramyamuna@gmail.com

D. Thalmann et al. (eds.), *Intelligent Embedded Systems*, Lecture Notes in Electrical Engineering 492, https://doi.org/10.1007/978-981-10-8575-8_17

1 Introduction

With ever-increasing health issues, pre-hospital treatment is very much essential for the survival of a patient. Earlier techniques used various methods to increase the survival rate among which telemedicine plays a major role, a device which managed to get a temporary doctor telematically. This doctor would provide pre-hospital treatment until they reach the hospital. The key information like wound's position is very significant in understanding the criticality of the injury so as to give the patient the apt first-aid treatment. Despite the fatal injury, the patient's condition can become more vulnerable, if his pre-injury health status is not good. For attending these kinds of complaints, it would be smart to collect the health history or pre-injury health record of the patient from his attendant using an audio file. There is a need of an interface where the doctor in the distant hospital is given a leverage to access the health history audio file and wound image data base of the patient. Further sections in this article discuss about the existing technology, proposed methodology, and the experimental validation of the same.

2 Existing Methods

The existing method consists of a communication network which provides the patient an assistance by directing them to the appropriate clinic or the physician who is fully aware of the patient's condition [1]. Mobile health care has gained lot of attention from healthcare professionals, patients, and researchers, especially in case of elderly people. It resulted in advances in expanding healthcare coverage, improving decision making, managing chronic conditions, and providing suitable health care in emergencies [2]. A wearable wireless device embedded with sensors for cardiac patients were designed. The system also has a two-way communication via an interface. Whenever the patient's condition gets critical, alert message will be sent to the doctor [3]. The system uses a mobile application which monitors the vitals of the patient and sends it to the hospital database which can be stored onto a Web site, and this is made available to the doctor at any instant [4]. The system aims at finding the exact location of the vehicle by the combination of GPS and GSM [5]. A portable device was designed which will send the biosignals to the consultation site, and it will telematically get a doctor to treat the patient who will then evaluate the emergency situation and direct accordingly [6]. One of the systems alerts the doctor through e-mail in case of critical situations [7]. Another method proposes a system using sensors, unified communication channel, and GPS [8]. Another work aims at developing a tele-home healthcare system which utilizes wearable devices, wireless communication technologies, and multisensory data fusion methods [9]. Some authors have presented a software agent approach which is based on telemonitoring and uses some sensors along with the alarm to alert [10]. Some research has been carried to provide tele-healthcare exclusively for particular disease like diabetes. These are low-cost systems based on two ends viz. patient's end and a Web end to provide health care [11]. There has been development for miniature biochemical analysis system which uses smart passive microfluidic control system to control parameters such as blood gas concentration and glucose and lactate concentrations [12].

3 Proposed Work

The proposed technique for Smart Mobile Diagnostic Laboratory in ambulance provides the doctor with patient's current health information in prior, thereby the doctor will be able to interpret the patient's current condition and will be able to do the necessary arrangements. Also, the location of the ambulance will be updated to the doctor. Since all the required tests would have been carried out in the ambulance, there would not be any need to perform this in the hospital. Thus, this technique reduces the latency in treatment of a patient which may reduce the probability of death. The parameters to be monitored are ECG using electrodes, body temperature using DS18b20, and pulse rate through SEN11574. There will be an attendant present who will have all the pre-medical records of the patient which will be sent to the doctor through a voice message or an audio file before the patient reaches the hospital. Monitored parameters will be sent to the doctor in an appropriate form. In case of accidents, images of the wound will also be sent to the doctor for him to interpret the patient's current condition via a two-way communicating interface. And the doctor will be periodically updated with the distance of the ambulance with the help of GPS. Figure 1 shows the block diagram for the proposed work. NI CRIO 9075 is the FPGA used which acts as a stand-alone device for data acquisition and processing. NI cRIO-9075 integrated systems combine real-time controller and reconfigurable field-programmable gate array (FPGA)

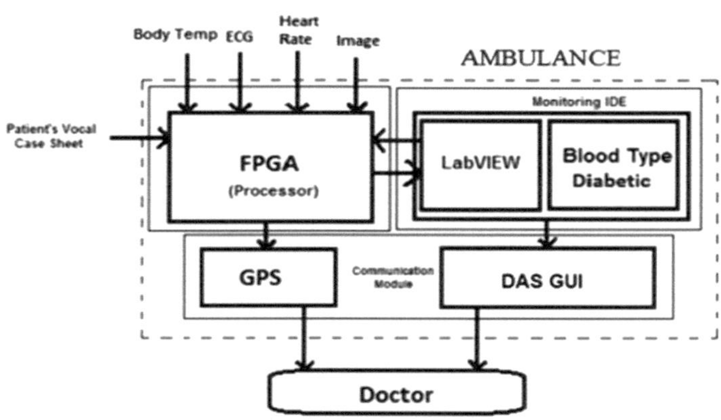

Fig. 1 Block diagram of the functional DAS

chassis for control and monitoring applications. It contains a 400 MHz real-time processor, a four-slot chassis with an embedded, reconfigurable LX45 FPGA chip, and a high-speed USB port. This system features built-in nonvolatile memory and also has fault-tolerant system [13]. And the GPS module is SEA 9414. The SEA 9414 is an extended temperature range GPS module. It is for both stationary and mobile usage. It provides position data for global and precise time base for timing and synchronization purposes [14].

4 Work Done

The parameters are monitored by interfacing ECG, pulse rate, and body temperature sensor with the FPGA. Sensors interfaced with cRIO are shown in Fig. 2. And the following Figs. 3, 4, and 5 show the data acquired through sensors. And to get the location of the ambulance, GPS module is interfaced with the FPGA. This has been shown in Fig. 6 along with the GPS co-ordinates in Fig. 7. A report format which includes the details of patient such as name, place contact number, blood group, insulin level, type of emergency, and ambulance number was created. Some patients might suffer from a long-term disease for which they will have a pre-medical record. For this case, the report contains a pre-medical record section, which if selected as yes will record sound file. The sound file recording is shown in Fig. 8. The sound file will be recorded by attendee in the ambulance which will narrate those pre-medical records which is to be sent to the doctor. Provision for image acquisition in case of accident or burn is also facilitated and is shown in Fig. 9. A GUI created will contain all the information, and this will be sent to the doctor. The GUI interface designed on LabVIEW is shown in Fig. 10.

Fig. 2 Pulse rate and body temperature interfaced with cRIO

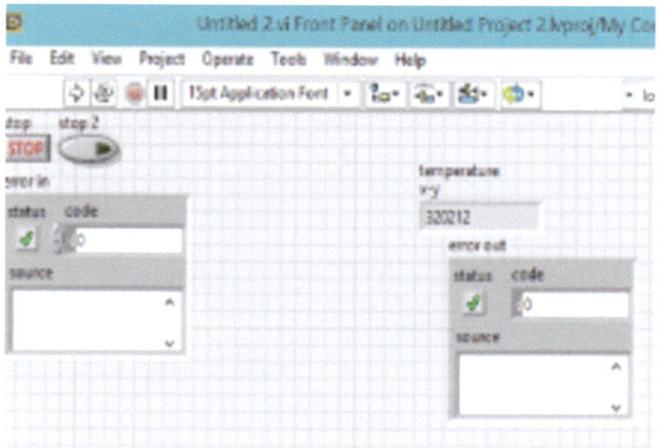

Fig. 3 Body temperature in front panel

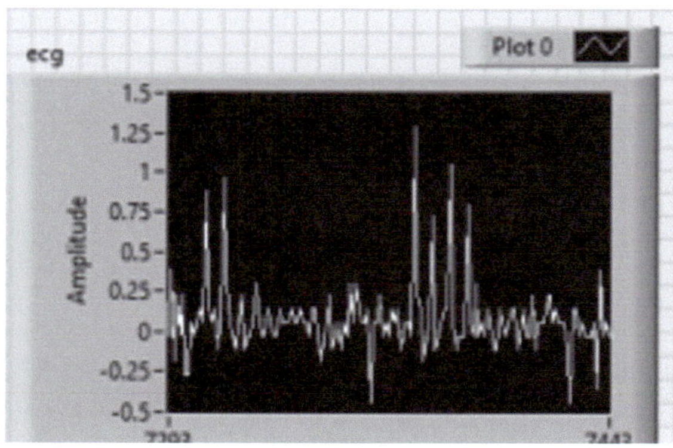

Fig. 4 Raw ECG signal in the front panel

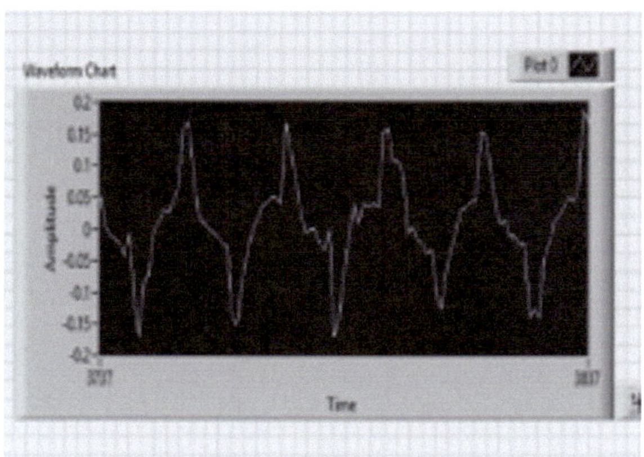

Fig. 5 Pulse rate in the front panel

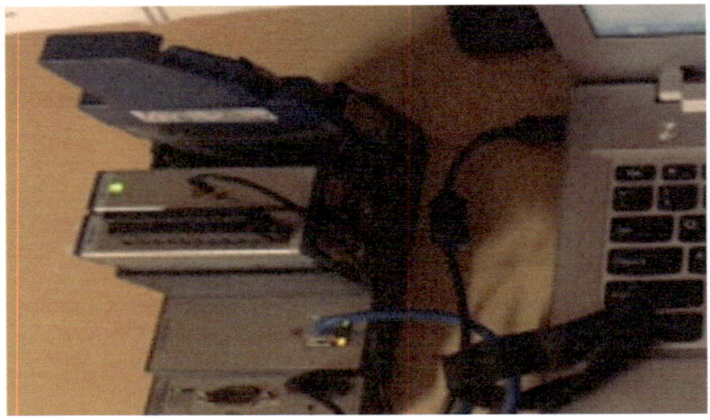

Fig. 6 GPS interfaced with cRIO

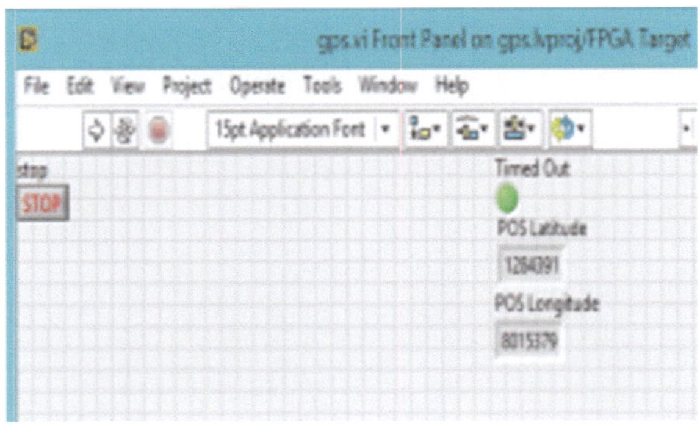

Fig. 7 GPS co-ordinates in the front panel

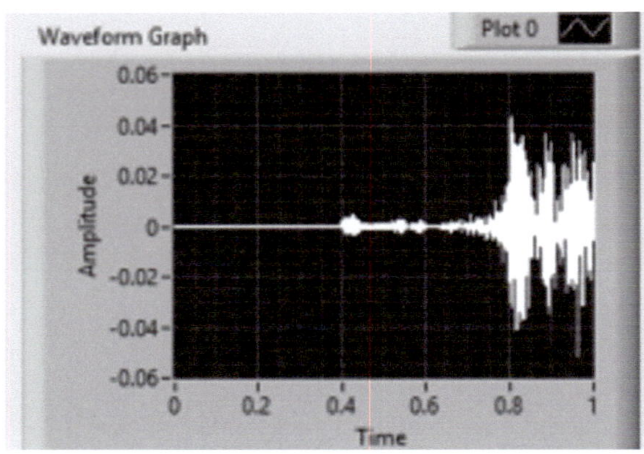

Fig. 8 Sound file acquisition in the front panel

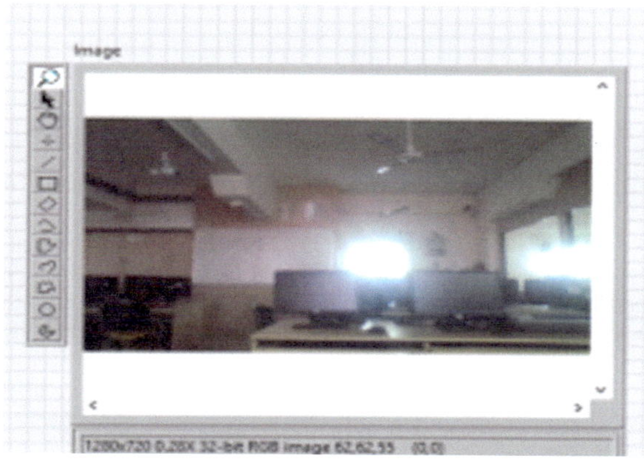

Fig. 9 Image acquisition in the front panel

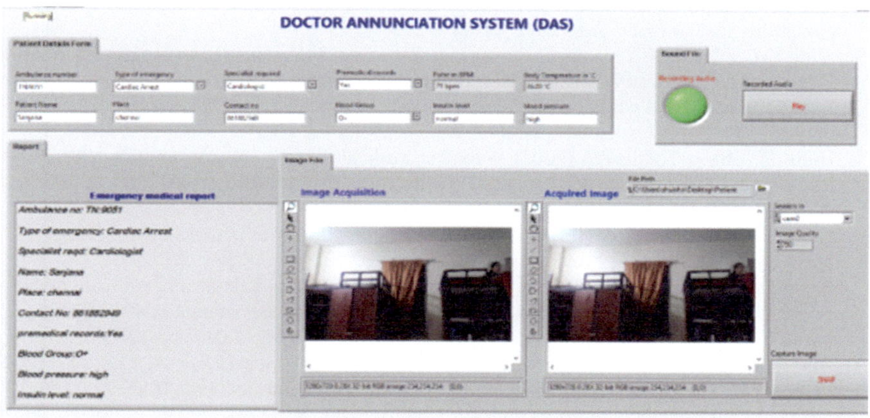

Fig. 10 Doctor annunciation system (DAS) GUI for the Doctor

5 Conclusion

The paper introduces a system, wherein the patient will be diagnosed on the way to the hospital. The parameters to be diagnosed are limited to body temperature, ECG, and pulse rate of the patient. But other vital checkups can be done using available biomedical sensors. The data along with the patient's wound images and his vocal sheet recorded by the attendee will be sent to the doctor via two-way communicating interface. This will be helpful for the doctor to do the arrangements in the hospital in critical cases. The system will be kept in the ambulance and will also increase the life span of a patient by not delaying the treatment.

References

1. Rahbar A (2010) An e-ambulatory healthcare system using mobile network. In: Seventh international conference on information technology: new generations (ITNG), April 2010, pp 1269–1273
2. Varshney U (2014) Mobile health: four emerging themes of research. Decis Support Syst 66:20–35
3. Kakria P, Tripathi NK, Kitipawang P (2015) A real-time health monitoring system for remote cardiac patients using smart phone and wearable sensors. Int J Telemed Appl
4. Majeed Q, Hbail H, Chalechale A (2015) A comprehensive mobile E-healthcare system. In: 7th international conference on information and knowledge technology IKT2015, IEEE, 15 Sept 2015
5. Saaid MF, Kamaludin MA, Ali MM (2014) Vehicle location finder using global position system and global system for mobile. In: IEEE 5th control and system graduate research colloquium, IEEE Aug. 11–12, UiTM, Shah Alam, Malaysia
6. Pavlopoulos S, Kyriacou E, Berler A, Spyros A (1998) A novel emergency telemedicine system based on wireless communication technology ambulance. In: IEEE transaction on information technology in biomedical, Dec 1998, vol 2, No 4
7. Malmathanraj R, Arun M (2012) A distributed e-healthcare system for patient monitoring and diagnosis. In: 2012 IEEE international conference on advanced communication control and computing technologies, Aug 2012, pp 71–77
8. Shih C-C, Chou K-Y, Keh H-C, Cheng Y-C, Yu P-Y, Huang N-C (2013) Building long-distance health care network using minimized portable sensors and active alert system. In: 16th international conference on network-based information systems (NBiS). IEEE, 2013, pp 401–404
9. Hung K, Zhang YT, Tai B (2004) Wearable medical devices for tele-home healthcare. In: Proceedings of 26th annual international conference, IEEE EMBS, San Fransico, 1–5 Sept 2004, pp 5384–5387
10. Rialle V, Lamy JB, Noury N, Bajolle L (2003) Telemonitoring of patients at home: a software agent approach. Comput Methods Programs Biomed 72(3):257–268
11. Benaissa M, Malik B, Kanakis A, Wright NP (2012) Tele-healthcare for diabetes management: a low cost automatic approach. In: 34th Annual international conference of the IEEE EMBS, San Diego, California USA, 28 Aug to 1 Sept 2012, pp 1290–1293
12. Ahn CH, Choi JW, Beaucage G, Nevin JH, Lee JB, Puntambekar A, Lee JY (2004) Disposable smart lab on a chip for point-of-care clinical diagnostics. Proc IEEE 92(1):154173
13. Operating instructions and specifications, CompactRIO, cRIO 9075/9076. Accessed on 22 Feb 2016
14. Application note—software driver migration, SEA 9405/9410/9414, GPS modules for CompactRIO. Accessed on 22 Mar 2016

Magnetic Braking System for Automotives

Arjun Nair and K. Srivatsan

Abstract Background: The existing braking systems suffer from a lot of drawbacks like inferior time-delay response and noise due to mechanical parts, wear and tear due to friction and contact, low performance and bulky size. **Methods/ Statistical analysis**: In this paper, design implementation of an automotive electromagnetic brake has been discussed, so as to replace already existing complicated braking systems with a relatively simple electromagnetic system. In the proposed system, the mechanical parts of the braking systems are completely replaced by the electronic parts, where a magnet is attached to the rim of the tyre of the vehicle and an electromagnet is placed parallel to the face of the rim of the tyre. The motion of the vehicle is controlled by controlling the supply to the electromagnet. As the electromagnet is energised, it acts like a magnet and tries to lock with the magnet on the rim of the tyre. As a result, the tyre stops rotating and the braking action occurs. The working is similar to that of an antilock braking system. **Findings**: A study was conducted on a prototype model analogous to the real-time system, and the plausibility of the work was observed. The problems and the constraints based on the real-time scenario were accounted for and thus optimisation method is also suggested. **Application/Improvements**: The proposed work is most suited for hybrid vehicles and for four-wheel automotive systems. Since only an analogous prototype was implemented and researched upon, the future works involve implementation of real-time system and conduct its analysis.

Keywords Electromagnetic braking · Magnetic locking · Traction
Ignition coil · Antilock braking

A. Nair · K. Srivatsan (✉)
School of Electronics Engineering, VIT University, Chennai, India
e-mail: srivatsan.k@vit.ac.in

A. Nair
e-mail: arjun.nair2014@vit.ac.in

D. Thalmann et al. (eds.), *Intelligent Embedded Systems*, Lecture Notes
in Electrical Engineering 492, https://doi.org/10.1007/978-981-10-8575-8_18

1 Introduction

The braking system is one of the vital parts of an automotive system. There are various types of braking systems. The conventional braking systems include frictional brakes, pumping brakes, hydraulic brakes, servo brakes which are mechanical in nature. These systems suffer from a lot of drawbacks like inferior time-delay response due to mechanical parts, wear and tear due to friction and contact, low performance and bulky size. The modern braking systems include electromechanical brakes such as eddy current brakes and hybrid brakes, which make use of electrical parts like motors and generators for braking action and use electronic circuits for controlling the braking operation. They have better control compared to mechanical braking systems. Although these systems have better control and time response, they lead to considerable power loss and also suffer from wear and tear, and performance issues as in the case of conventional mechanical brakes. For instance, the use of magneto-rheological brake is affected by its limiting torque [1]. Also there are other classes of brakes specially designed for parking and emergency braking applications. Although these brakes are more efficient and reliable compared to the normal mechanical and electromechanical braking system, they cannot be used for normal braking and speed control applications.

Thus, with the advancement in the speed and performance of automobiles increase there is a need for more advanced and powerful braking system which can ensure a better performance, safety and reliability. This paper proposes an electromagnetic contactless braking system using electromagnets and ignition coil.

The proposed braking system overcomes all the above-mentioned drawbacks as it provides better time response, performance and less power loss, owing to its electronic operation. Also the following scheme is used for traction mechanism along with the braking application. Since the following system is contactless type there is no wear and tear, and it is very lightweight and robust [2, 3]. The braking system arrangement consists of magnets placed on the rim of the tyre, and electromagnets are placed parallel to these magnets. When the electromagnet is energised, the magnets on the rim of the tyre lock with the electromagnet, thus antilock braking is initiated [4, 5].

Since the whole system is electronic in nature, it has better control and thus by observing the difference between the speeds of the tyres on the two sides of the vehicle, traction can be obtained precisely by controlling the supply to the electromagnet. Also the independent tyre control feature of the proposed braking system makes it apt for parking and emergency applications.

The scope of this research work is to enhance the braking in an automotive system by implementing a contactless electromagnetic braking system which provides better time response, control, performance and reliability, and minimises the size, weight and noise of the system compared to existing braking systems. Also increase the functionality of the braking system by making it capable of performing emergency braking and traction applications along with normal braking.

2 Related Works

There are currently a number of studies being conducted in the field of electromagnetic braking systems. The literature presents the various works that are currently being researched upon, in parallel to the proposed work.

- Enhancement of torque using a magneto-rheological brake [1] by magnifying the magnetic field strength. Thus, also studying the effects of brake structure and input power on the torque enhancement indicated that the brake achieved a high braking torque.
- Structural analysis and validation of the multi-pole magneto-rheological brake for motorcycle [6]. Comparison of its configurations inner rotor structure and outer rotor structure and verification of the feasibility of the application. The following braking scheme not only provides the necessary braking torque but also satisfies the size constraints.
- Optimal robust control of a contactless brake system using an eddy current for automotive applications by conducting a comparative study between the electromagnetic-based antilock braking system and the hydraulic-based antilock braking system. Brake torque analysis is conducted using an approximate theoretical model which is then modified throughout the experimentation for a reliable result. It is observed that the following model is effective in minimising the braking distance in accordance with the desired slip ratio for the corresponding road conditions [3].
- Application of the existing mathematical techniques for the modelling and controlling of the eddy current braking system [2].
- Improving braking torque generation capacity of an eddy current brake with time-varying magnetic field, by means of frequency control of AC and DC magnetic fields. The frequency control is further optimised via genetic algorithms [5].
- Designing of a magnetic braking system and thus analysing the effect of air gap and the materials of the track on the magnetic flux density [7].
- Designing a magnetic brake on the basis of magnetic drag force and magnetic lift force using image method based on Maxwell's equations [8].

3 Operation

The model primarily consists of electromagnets and magnets on the wheel as actuators, ignition coil as transformer and Arduino board as control circuitry.

The braking action is achieved through the interlocking of the magnets on the wheel with the electromagnet. The more is the supply to the electromagnet, the more the locking between the magnet and the electromagnet (Fig. 1).

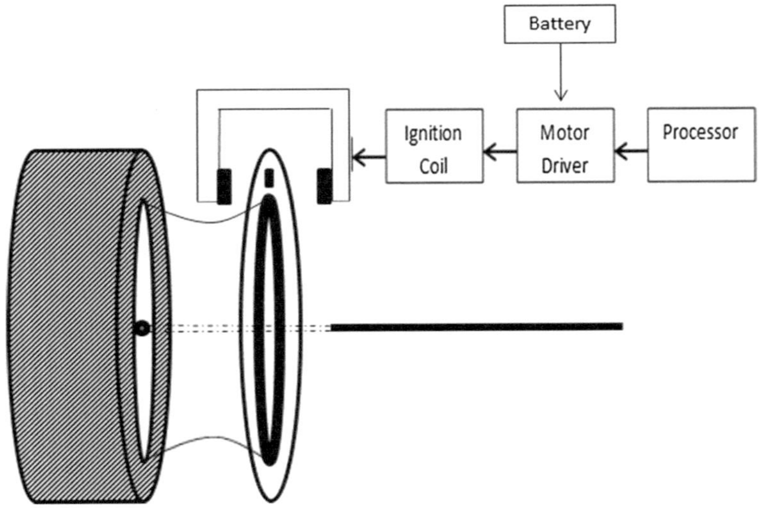

Fig. 1 Schematic of the proposed braking system

Thus, braking is controlled by controlling the supply to the electromagnet via Arduino. The power to the electromagnet is provided through regular car battery. Since for effective braking a large supply is required for the electromagnet, an ignition coil is used to step up the input voltage to electromagnet from the power supply.

3.1 Braking Operation

The battery supplies the power to the various electronic components. The motor driver is used to amplify the voltage output of the processor to the required level for the operation of the components (Fig. 2).

The PWM module is used to gradually build up the input potential to the ignition coil. The ignition coil is a pulse transformer used to generate the power required by electromagnet.

The electromagnet is the actuator used to generate the necessary braking force by creating a strong magnetic field. The distributor is used to control the brakes of the respective tyres independently.

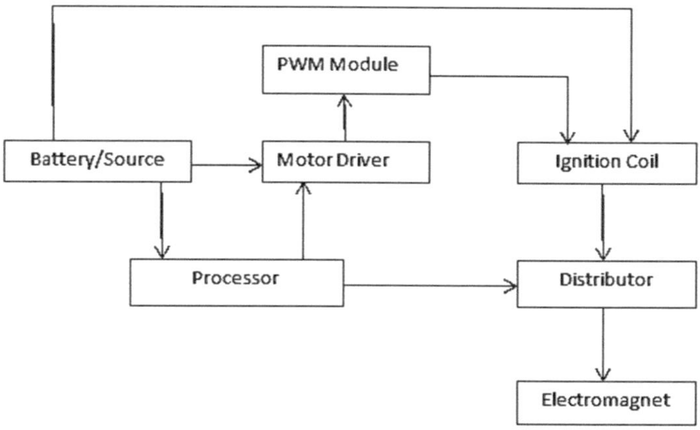

Fig. 2 Block diagram of braking operation

3.2 Traction Operation

The sensors attached to the respective tyres give the value of their respective instantaneous speed. The values so obtained are given as input to the comparator. The comparator is used to calculate the difference in the speed of the two tyres. The output of the comparator is given as input to the processor. If the value of the difference in the speed is above a threshold value then comparator outputs 1, else zero. For a general purpose four-wheel automobile, the maximum permissible difference in speed so as to avoid traction is 8–12 km/h. Hence, the threshold value is set as 8 km/h. Thus on the basis of input from the comparator, the processor activates the brake of the tyre with higher speed. Thus, traction is achieved (Fig. 3).

4 Algorithm

1. Start
2. Define pin for brake switch B.
3. Define pins for all the brakes, that is, backward and forward brakes; B1, B2, F1 and F2.
4. Define pins for input R1, R2, D1 and D2.
5. Read input for the comparator values, indicating the difference in the speed of two tyres; D1, D2.
6. Read the input values from the sensors and calculate the difference values between two tyres

    For back tyres, R1, and for front tyres, R2.

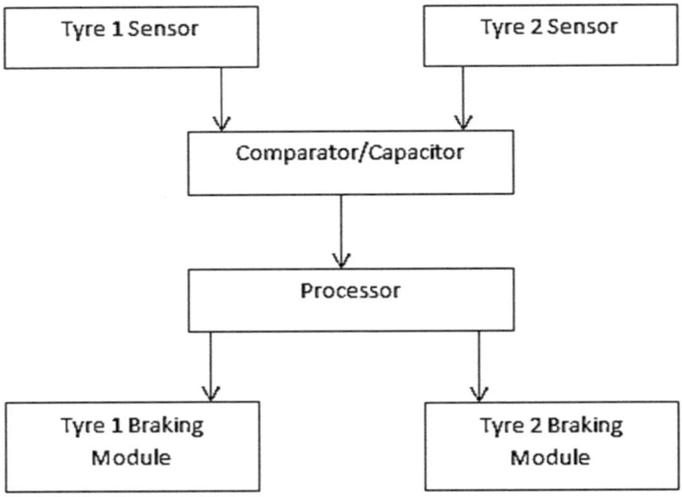

Fig. 3 Block diagram of traction operation

7. Set a threshold value th = 8 (maximum permissible difference in speed so as to avoid traction-8–12 km/h).
8. Read S1 and S2 to store the values of R1 and R2.
9. Read C1 and C2 to store the values of D1 and D2.
10. Read A to store the value of brake switch; B.
11. If B = 0, go to step 13
    Else go to next step.
12. Turn on all the brakes B1, B2, F1 and F2, and gradually increase their potential.
13. If C1 = 1, go to step 14
    Else go to step 15
14. while (|S1| > th)
    If S1 > 0, turn on B1 gradually increase the potential.
    If S1 < 0, turn on B2 and gradually increase the potential.
15. If C2 = 1, go to step 16
    Else go to step 13
16. while (|S2| > th)
    If S2 > 0, turn on F1 gradually increase the potential.
    If S2 < 0, turn on F2 and gradually increase the potential.
17. Stop

5 Experimental Setup

The experimental setup consists of two 300 rpm, 12 V DC motors, two 2.5 N 12 V electromagnets, two opamp (IC741)-based differential amplifiers, two motor drivers and an Arduino board. Here an analogy of a real-time system of a two-sided two tyre model is created, where the motors represent the tyres of the automobiles and electromagnets as the brakes, and Arduino is used as controlling element. The motor driver is used to couple electromagnets with the Arduino.

An artificial real-time scenario of traction is generated by means of running the two motors at same current and different voltages. As a result, the two motors run at different speeds and thus the difference in the potential acts as indicator of difference in speeds between the two motors. The connection of the differential amplifiers is such that the input connections to one of the differential amplifier are complement to the input connections of the other differential amplifier. The Arduino is programmed such that it takes the input values from these two differential amplifiers and accordingly sets the value of traction variable as 1 or 0. If the inputs from both the amplifiers are same, the traction value is set as 0, and if the value of the differential amplifiers is different, the traction value is set as 1. A brake variable is provided for braking application and it has high priority than traction. Whenever the brake value is set as 1, braking application takes place else it is set as 0. Thus, Arduino is programmed by default perform traction, if such event occurs, when the braking variable is set as 0.

A simulation of the prototype model was conducted in LabVIEW in order to demonstrate the working of the proposed work as in reference to the real-time system.

6 Results

The results show that the following prototype model successfully demonstrates the working of the proposed model. In the event of the value of the brake variable being set as 0, whenever the two motors are running at different speeds, the value of differential amplifiers is not same and as a result the traction action is initiated and the electromagnet connected to the motor running at higher speed is energised so that it can run in sync with the slowly running motor.

Whereas the brake variable is set as 1, both the electromagnets are energised simultaneously and apply braking action to both the motors. The following figures show the LabVIEW simulation of the prototype model.

Figures 4 and 5 demonstrate the traction operation, when the brake is not applied, and the speeds of the two motors are not equal as indicative of the difference in the potential supplied to the two motors. When speed of motor1 is greater than speed of motor2, the electromagnetic brake of the motor1 is activated.

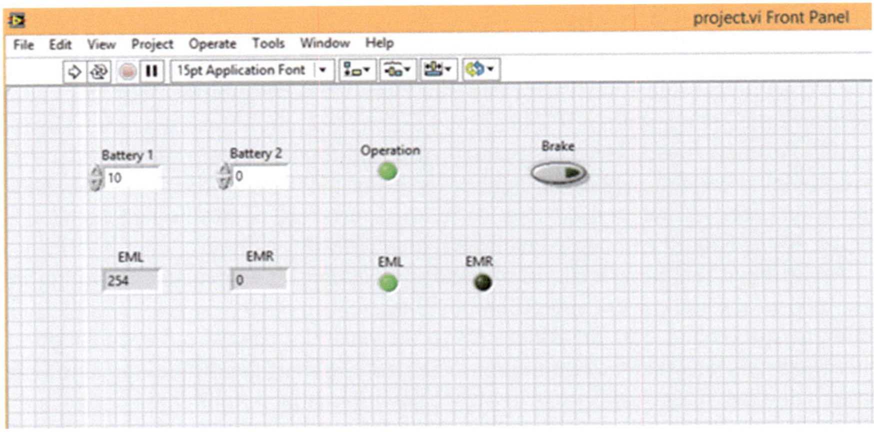

Fig. 4 Speed of motor1 greater than motor2

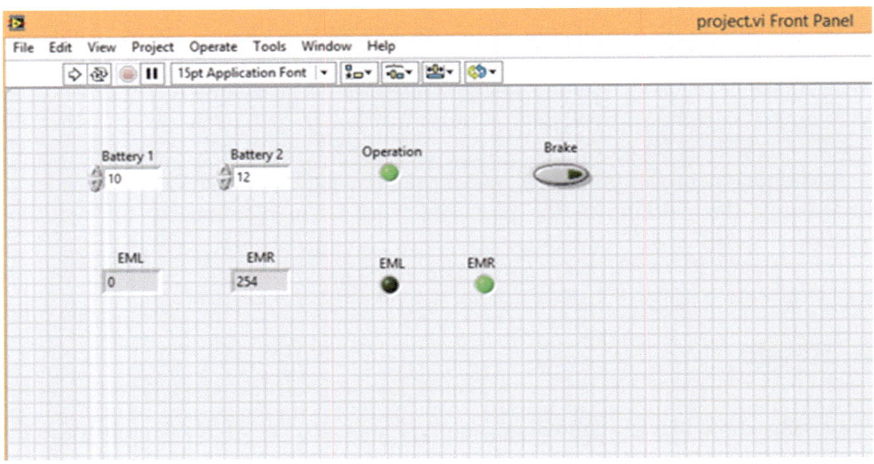

Fig. 5 Speed of motor2 greater than motor1

Similarly, when the speed of motor2 is greater than that of motor1, electromagnetic brake of the motor2 is activated. Figure 6 demonstrates the braking operation. When the brakes are applied, the electromagnetic brakes connected to both the motors are activated.

In the event of braking, the traction operation is prevented. However, the maximum braking action that can be achieved by this low-power electromagnet is only a decrease in speed of 10 rpm.

Thus, by properly designing and using appropriate electromagnet, complete braking action can be achieved.

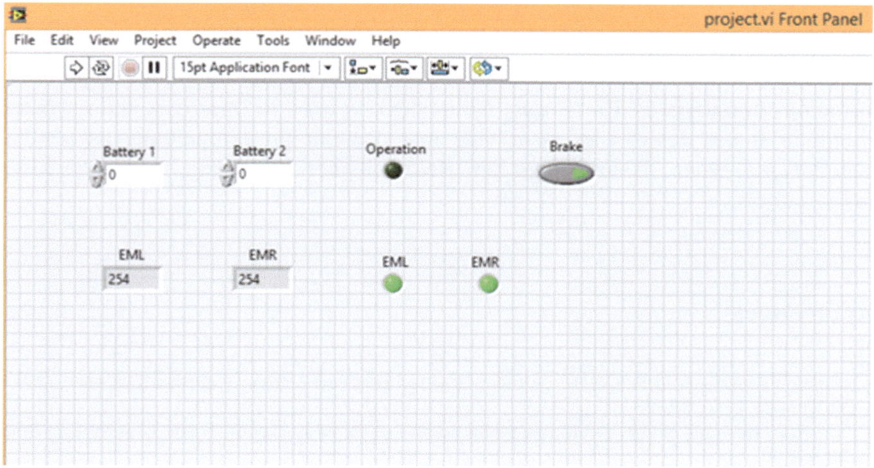

Fig. 6 Brake is applied

7 Application

The application of the proposed scheme extends to nearly all the automobiles. However, it can be most effectively used in four-wheel automobiles. Along with the antilock braking and traction applications, it can be used for emergency braking and parking brake applications.

8 Conclusion

In this paper, an electromagnetic brake design has been introduced as a viable alternative for the currently existing mechanical and electromechanical braking systems. Since all the mechanical parts are completely replaced by electronic components, it has several advantages over currently existing braking systems in terms of better performance, safety, reliability. Since it is completely electronic and contactless type brake, it has no wear and tear, lightweight and robust, and can easily be implemented for software control.

The design procedure consisted of building an analytical model of the proposed system, thus derive the equations for required magnetic field strength, electric potential and power dissipated, respectively [8]. A simulation model was generated in LabVIEW in order to demonstrate the working of the proposed model.

The reliability and the performance of the system are tested using an experimental setup which consisted of DC motors as tyres, electromagnets as the brakes and Arduino as the control unit. The experimental results showed that the braking torque increases with the increase in potential and perform the traction as per the conditions.

However, the maximum reduction in speed achieved when applied full potential to the electromagnet is 10–12 rpm. Thus, there is need to improvise the magnetic circuit in future designs so as to take into account the temperature as well as material properties. Also design an electromagnet pertaining to the required specifications.

References

1. Shiao Y, Nguyen Q-A (2014) Torque enhancement for a new magnetorheological brake. Procedia Eng 76:12–23
2. Simeu E, Georges D (1995) Modeling and control of an Eddy current brake. Control Eng Pract 4:19–26
3. Lee K, Park K (1998) Optimal robust control of a contactless brake system using an eddy current. Mechatronics 9:615–631
4. Karakoc K, Park EJ, Suleman A (2008) Design considerations for an automotive magneto rheological brake. Mechatronics 18:434–447
5. Karakoc K, Park EJ, Suleman A (2012) Improved braking torque generation capacity of an eddy current brake with time varying magnetic fields. Finite Elem Anal Des 59:66–75
6. Shiao Y, Nguyen Q-A (2014) Structural analysis and validation of the multi-pole magnetorheological brake for motorcycles. Procedia Eng 76:24–34
7. Jou M, Shiau J-K, Sun C-C (2006) Design of a magnetic braking system. J Mag Magn Mater 304:234–236
8. Hribar J (2008) Magnetic braking. University of Ljubljana

Raspberry Pi-Based Surveillance System with IoT

Arvin Joseph Kumar Jayakumar and S. Muthulakshmi

Abstract Background: The proposed system describes a surveillance system developed using Raspberry Pi and a camera, which keeps monitoring a certain highly secured region continuously. When the system recognizes a change in motion (human motion) compared to its previous frame, the system starts recording video and stores it primarily in its memory and also in the cloud (for the reason that even if the burglar tries to destroy the system his image/video will be saved in the cloud storage), and the user receives alert mail from the system stating "human motion detected" along with the captured image attached with the alert mail. The system contains database of face patterns of local suspects which is compared with the face pattern of the person triggering the system, and image processing is done in real time to correctly identify the detected face; the system also keeps tracking the face throughout the region even if the person moves out of the frame by a camera mounted over a servo motor. The system turns on a buzzer alarm when the burglar attempts to cause damage to the system. The system allows the user to remotely access the camera to monitor live streaming video output and control the rotation of the camera. **Methods/Statistical analysis**: In this project, different types of surveillance systems which already exist are analysed, and the methods of having a portable surveillance system were developed using Raspberry Pi. Image processing methods for facial identification and face recognition is used. **Findings**: A study based on various image processing techniques is done; it is found that Haar-cascade and linear binary pattern are the suitable algorithm for performing image processing in real time. **Application/Improvements**: For better surveillance, face tracking in introduced, which can track the detected face throughout the region even if the person goes out of the camera frame, and remote accessing with control of the camera through IoT is introduced.

A. J. K. Jayakumar · S. Muthulakshmi (✉)
School of Electronics Engineering, VIT University, Chennai, India
e-mail: muthulakshmi.s@vit.ac.in

A. J. K. Jayakumar
e-mail: arvinkumar92@gmail.com

D. Thalmann et al. (eds.), *Intelligent Embedded Systems*, Lecture Notes in Electrical Engineering 492, https://doi.org/10.1007/978-981-10-8575-8_19

Keywords Raspberry Pi · Camera module · IoT monitor and control
Cloud upload · Face matching · Face tracking

1 Introduction

Indoor surveillance monitoring is the most vital in the field of well-being and security framework. The significance of surveillance is checking of the conduct, exercises or other changing data by and large of individuals with the end goal of impacting, overseeing, coordinating or ensuring them. These sorts of frameworks are generally habituated in homes, office, manufacturing plant or vehicle checking and picture recognizable proof; yet this framework requires a superior centre, since it needs to manage continuous picture handling at a few bodies of evidence which conflicts with a few favourable circumstances of implanted frameworks, for example, low power utilization and ease. The proposed system is exceptionally useful to the administration and law execution to keep up social control, rearrangement and observing of dangers, and counteract/examine criminal exercises.

Video surveillance systems assume exceptionally indispensable part to satisfy the security acknowledges from various perspectives. Thievery and burglary have dependably been an issue for typical inhabitants, particularly for those living in the enormous urban areas. Video surveillance is exceptionally useful to law requirement to explore/avert criminal exercises, for perceiving and checking dangers. The fundamental outline of any surveillance system comprises of investigating the requirements of the general population, inspecting the framework costs in view of the current equipment and innovation, checking decisions then finally making arrangements for the establishment. It is necessary for every home owner to consider adding a home security system, as burglaries, thefts and murders have become routine in big cities [1].

2 Existing Systems

This chapter describes details about the existing surveillance systems. Also, this overview helps the surveillance system project to be a better one. Some researches are somewhat equivalent to the proposed systems, but they have some limitations like it may be cost-effective but not efficient or it may be efficient but there may be more complexity in implementation. Moreover, a similar kind of project or platform, microcontrollers, operating system, sensors, input/output devices and controlling mechanism contributes the knowledge to improve the productivity of the developers. Such kinds of face detection, face recognition and face tracking techniques are explained in the following literature survey.

- Senthilkumar et al. [2]: In this work, the properties of Raspberry Pi and the process of image capturing using Raspberry Pi's own camera module alone are explained; no other feature are included in this work. This is considered to be the basic and preliminary work to start with a security system as the first thing of a security system is to capture images in the embedded device.
- Sanjana Prasad et al. [3] stated that using Raspberry Pi as the main embedded device to perform remote surveillance by creating a VNC server in the system and streaming live video to the Internet Web page, and later, the window is monitored through a Web page from a portable mobile device to ensure authentication and reliability at the receiver side; the system uses ordinary webcam to capture images. This work does not seem to be an efficient surveillance system, as it does not have anything about image processing and control remote controlling, but from this work the remote monitoring of the camera attached to the Raspberry Pi is a work that is to observed and studied [3].

3 Proposed System

The proposed system talks about real-time data acquisition and analyses of the features extracted from the data, which is the image. Raspberry Pi 2 B model board very efficiently processes multiple images frame by frame. The image processing unit gives commands to the Raspberry Pi unit to tilt servo motor that is attached with the system, and the motor drives the camera which is pivoted on it. Primarily the real-time image processing was tested with the Raspberry Pi itself, but later during the implementation, the Raspberry Pi was found to have difficulties in running the real-time face recognition loop of resulting in a delay in processing data; since this system has to be real time, other idea of performing to image processing in the Linux platform core and transferring the face tracking values to the Raspberry Pi's servo motor to perform the camera tilt is carried out.

Otherwise the sensor's process, image uploading into cloud, continuous distance/proximity monitoring using ultrasonic sensor and remote access of the camera live video stream monitoring and controlling of the camera and all are performed in the Raspberry Pi system itself. This system is comes under real-time data acquisition, data processing and controlling system. A new image processing technique is used face detection, face recognition and tracking, which works based on open computer vision (OpenCV). Most of coding part is done with the help of OpenCV library. Python language is used for coding, which is user-friendly and helpful to resolve the error efficiently. OpenCV 2.7.9 library with python is used in this system [4].

4 Design Methodology

The surveillance system follows the given steps to process to perform the image processing operation in following ways:

- Passive Infrared Sensor (PIR) is used for human detection. The system is initiated when it senses human. The camera is triggered, initially starts taking photos and images and saves it in the system and the cloud/mail box. For sending email, "import smtplib" is imported. SMTP is a protocol, which handles sending email and routing email between mail servers.
- The face detection and face recognition techniques are performed using image processing algorithms Haar-cascade and linear binary pattern Histogram in the system by the OpenCV package. The algorithm is modified in such a way that if there are multiple faces available, then the system will not allow the access to control the system [4].
- The system matches the detected human face with a predefined face pattern database, the face is marked in a rectangular frame, and the predefined label on the rectangular frame shows the correctly recognized face or else it labels as "unknown".
- For face tracking, the image position of the moving face is sent from the image processing platform, i.e. the Linux system with OpenCV package to the Raspberry Pi via serial communication.
- The image tracking is attained by obtaining the centre point of the image. The rectangular frame sides are $x1$, $x2$, $y1$, $y2$. The centre x position of the image is calculated by the function $(x1 + x2)/2$; likewise, centre y position is obtained by $(y1 + y2)/2$.
- The centre x is alone required for obtaining the horizontal position, since the camera is set to rotate horizontally. The difference between centre x value and read value for the moving object is the horizontal position. Depending upon the centre x position and the read value from the moving object, the camera which is mounted over the servo motor tilts. It is actually the servo motor which tilts with respect to the horizontal position values obtained from the moving object; the camera mounted on servo motor tilts in accordance. The horizontal position is transmitted to the Raspberry Pi using serial communication.
- Camera is mounted on a servo motor, and the servo motor oscillates in accordance with the position of the moving object (face), thereby tilting the camera. The servo motor converts the angles of the face position to electric signal PWM duty cycle for rotating in accordance with the tilting position.

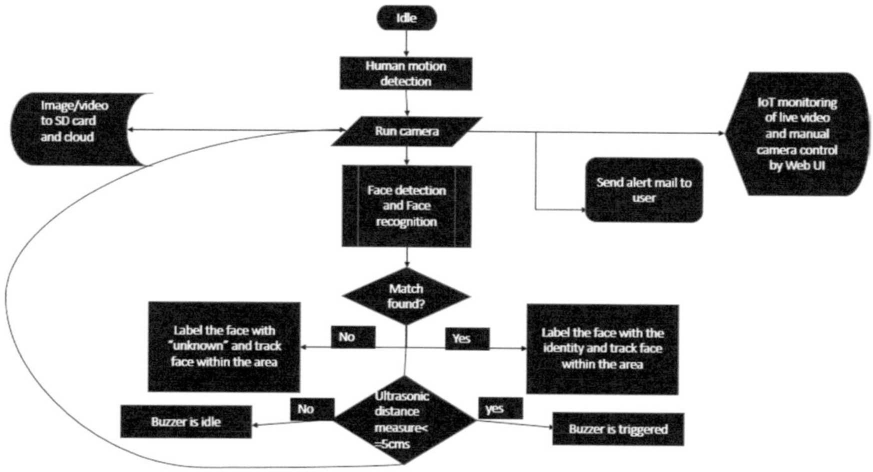

Fig. 1 Flow chart of the working system

- The Raspberry Pi's camera can be remotely logged in through IoT using its IP address to monitor what is going on there. The Raspberry Pi acts as a server when VNC server package is installed in it. The apache server installed in the Raspberry Pi enables the user to create control buttons in a Web page to have a manual control of the servo motor remotely through IoT.
- The ultrasonic sensor is used to measure the proximity of the burglar to the system; if the distance is very less (say 5 cm), the burglar alert alarm is triggered.

The whole design process is depicted in the form of flow chart in Fig. 1.

5 System Algorithm Description

5.1 Face Detection—Haar-Cascade Algorithm

The Haar-like features detection algorithm is a part of OpenCV library. The images captured by the camera are processed and the resulted data for face detection based on Haar-cascade algorithm is acquired. Face detection at its simplest is finding the face in an image.

A "cascade" is a series of "Haar-like features" that are combined to form a classifier. The basic definition of a Haar-like feature can be stated as Haar wavelet which is actually a mathematical function that produces square wave output. When talking about Haar-like features, it actually means that the patterns here are usually rectangular. Every component is a solitary worth acquired by subtracting total of pixels under white rectangle from the aggregate of pixels under dark rectangle, and there is distinctive introduction of it going vertical, Level, and even diagonal [4].

5.2 Face Recognition—Local Binary Pattern Algorithm

Facial representation taking into account local binary pattern (LBP) highlights for individual free outward appearance acknowledgement. LBP features were proposed initially for surface examination also known as texture analysis and as of late have been acquainted with facial image investigation. The most critical properties of LBP features are their resistance as light changes and their computational simplicity [4–13].

5.3 Face Tracking Algorithm

The face tracking algorithm is mentioned as follows.

- The stationary point is going to be 0 with respect to the resolution. The extreme left point will be −340 and the extreme right side will be 340.
- The stationary point of a servo motor is 90°. From the centre point, the face tracking is at four different points in the left direction and also in the right direction.
- For each 22.5° tilt, the angle is converted into PWM duty cycle which the servo motor will understand. The duty cycle acts as electric signal which makes the servo motor rotate.
- The graph below depicted the concept of approach and the face tracking calculation of the eight positions along with the angle at each position, for the corresponding image position and the required duty cycle is plotted in Fig. 2.

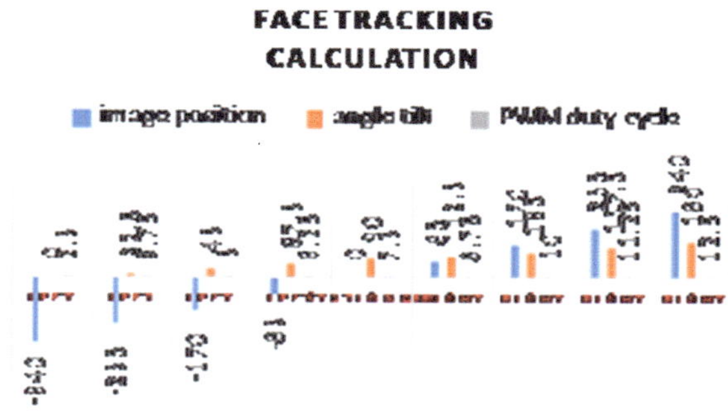

Fig. 2 Face tracking values calculations—angle, direction, duty cycle

Fig. 3 System setup

6 Implementation

The proposed framework/system usage depends on ongoing information obtained which is known as real-time data acquisition and data processing (Fig. 3).

6.1 Face Detection

The OpenCV library of image preparing is utilized for the face detection, as shown in Fig. 5. First, Haar course calculation is utilized for face identification, and the human face is tagged with a green box when the camera is running (Fig. 4).

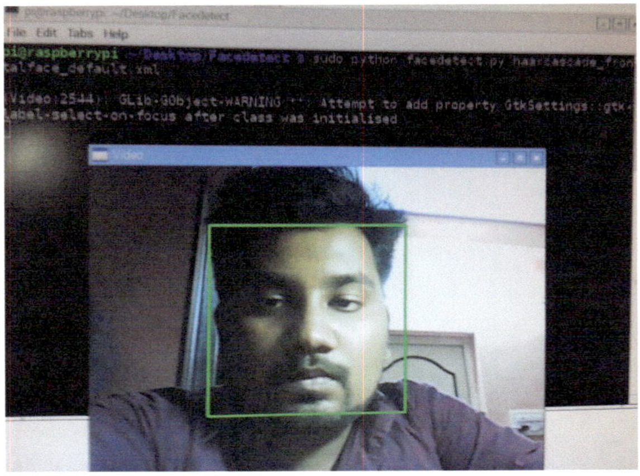

Fig. 4 Human face detection

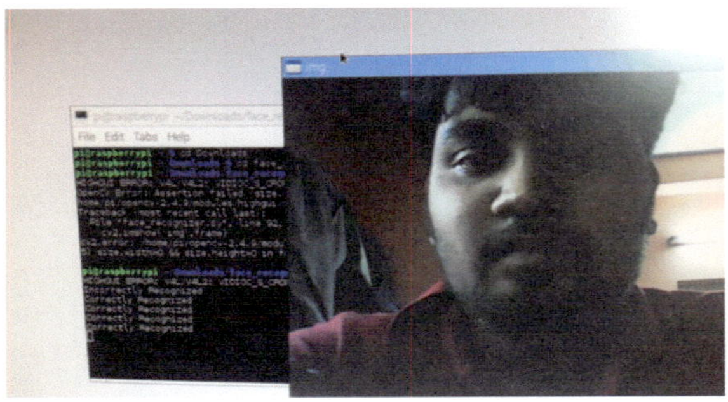

Fig. 5 Face recognition with LBP algorithm

6.2 Face Recognition

Face recognition is performed using linear binary pattern algorithm; the algorithm compares the captured image with the trained face samples saved in the database [6, 8, 9].

When the captured face matches with the trained face, the system recognizes the human face and displays "correctly recognized" message in the terminal window as shown in Fig. 5.

6.3 Face Tracking

The image tracking is attained by obtaining the centre point of the image. The rectangular frame sides are $x1$, $x2$, $y1$, $y2$. The centre x position of the image is calculated by the function $(x1 + x2)/2$, and centre y position is obtained by $(y1 + y2)/2$. The centre x is alone required for obtaining the horizontal position, since the camera pivoted on the servo is set to rotate is a horizontally 180°. The difference between the centre *x* value and read value for the moving object is the horizontal position. The position is displayed in the tracking panel.

6.4 Raspberry Pi Remote Webcam Streaming

The Raspberry Pi's webcam can be accessed remotely, locally and globally depending upon the type of IP. The port number is essential for opening the webcam streaming port online; by default, the webcam port is 8086, and "sudo service motion start" is run in the terminal window to start webcam streaming.

When the IP of the Raspberry Pi IP is given in a browser with webcam port number, remote access to the live streaming camera output is achieved.

6.5 Raspberry Pi Webcam IoT Control

The Raspberry Pi's webcam is provided with IoT control, the GPIO pins are given commands to have control over the servo motor on which the camera is placed, the system lets to have a 180° view angle of the Raspberry Pi's webcam remotely by controlling the IoT-based Web UI consisting of button like "Right", "Left", "Off" as shown in Fig. 6 [10–13].

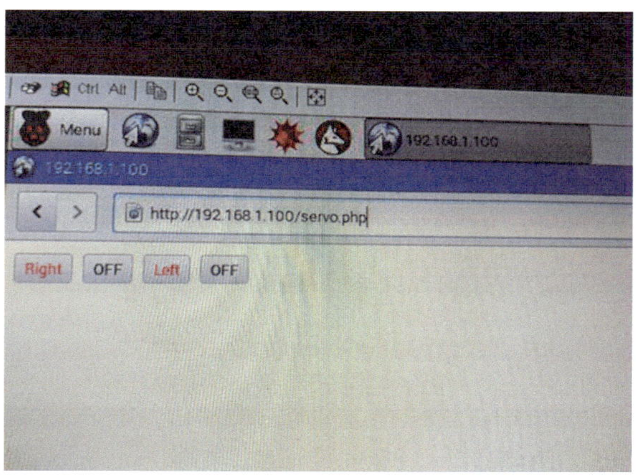

Fig. 6 IoT camera control Web user interface

Fig. 7 Real-time face detection, face recognition and face tracking

7 Results and Discussions

7.1 Image Processing

Figure 7 shows the real-time system getting an input face from a live video running from camera module. In the live running video itself, the different image processing techniques are implemented to detect and recognize the input face.

Within the same panel, the horizontal position of the face is continuously updated; for horizontal position in the range 0–340, the camera rotates right, and it rotates left for the range −340 to 0. When the horizontal position is above the minimal and maximal value of the ranges, the camera gets set to 0 and 180°, respectively.

7.2 Sensors

The PIR sensor triggers the Raspberry Pi camera to capture images to be saved locally inside the system and sends it to user's mailbox.

The ultrasonic sensor is interfaced to continuously measure the distance between the system and the object. If the distance between the system and the object turns to be below 5 cm, the buzzer alarm is initiated.

7.3 IoT Resultant Data

7.3.1 Remote Webcam Accessing

The IoT monitoring, i.e. remote webcam accessing of the Raspberry Pi, is done by the "service motion start". For accessing the webcam remotely, the IP of the Raspberry Pi is typed in the Web browser. Accessing them remotely can be done through locally or globally. For accessing webcam locally, the device from which the webcam is being accessed should be connected in the same network of the Raspberry Pi; this is usually needed when the Raspberry Pi is connected to a dynamic network whose IP keeps changing often at regular intervals. For global access, i.e. accessing the webcam of Raspberry Pi throughout the global, the Raspberry Pi needs to be connected to a static network whose IP is constant throughout the global [5]. The live streaming video is displayed in a Web browser in a 320 × 480-resolution window.

7.3.2 Camera Control Through IoT

The camera of the Raspberry Pi can be remotely controlled through IoT using a Web-based user interface consisting of right, left and off buttons as shown in the Fig. 8. The commands are given through the GPIO pins for controlling the buttons.

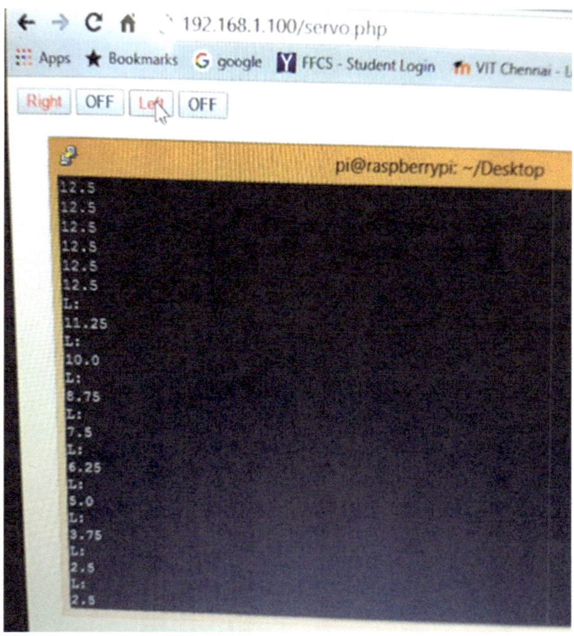

Fig. 8 Rotation from right position to left position

The camera is initially in the stationary position, i.e. 90° (7.5 duty cycle). When the “Right” direction button is pressed, the camera angle increased from 90° to the maximum of 180° (12.5 duty cycle) which is the extreme right position. When the “Left” direction button is pressed, the camera angle decreased from 90° to the minimum of 0° (2.5 duty cycle) which is the extreme left position.

8 Conclusion

The concept of Image processing and remote IoT-based surveillance system not only represents the alternative resource for surveillance. It is actually a concept of making the IoT-based applications familiar to the modern world, as IoT-based products are going to be revolutionary by the year 2020. Image processing-based surveillance system makes the user and the police department to identify the suspect early and properly, minimizing the human effort in investigation time, and the stolen properties can be retrieved soon when the identity of the burglar is already known when he happens to escape the surveillance spot. Image processing is an added advantage to a security system. Few real-time configuration constraints measure like the system takes much time (6 s) to execute the picture processing from the video obtained from the real-time environment as it requires better processor core image processing since the image recognition loop is continuously being run for real-time environment. For this reason, the image processing is performed in a computer and the face position values are transmitted to the Raspberry Pi for tracking the face; eventually face tracking is done by the Raspberry Pi only. But in future, a cluster of Raspberry Pi, i.e. many Raspberry Pi connected together, can be a high-performance core to make this work in a stand-alone system.

This is a real-time data acquisition and controlling system and it provides simple user interface technique to have complete control over the camera in the left and right direction with video streaming. Although the system is a fragile one and can be easily broken, the system has an ultrasonic sensor which notifies by an alarm if someone tries to near the system in attempt of breaking it. The photos of the burglar is saved in the system and sent to the user mail inbox, so that the user can access the captured image even if he is not near the surveillance location.

References

1. Chandana R, Jilani S, Javeed Hussain S (2014) Smart surveillance system using thing speak and Raspberry Pi. Int J Comput Sci Inf Technol (IJCSIT) 4:214–2018
2. Senthilkumar G, Gopalakrishnan K, Sathish Kumar V (2014) Embedded image capturing system using Raspberry Pi. J IEEE Intell Syst 3:213–215
3. Prasad S, Mahalakshmi P, Sunder AJC, Swathi R (2014) Smart surveillance monitoring system using Raspberry Pi and PIR sensor. J IEEE Intell Syst 5:7107–7109

4. Shan C, Gong S, McOwan PW (2005) Robust facial expression recognition using local binary patterns. International conference on image processing (ICIP), Genoa, vol 2, pp 370–373
5. Mei F, Shen X, Chen H, Lu Y (2011) Embedded remote video surveillance system based on ARM. J Control Eng Appl Inform 13(3):51–57
6. Lakshmi Devasena C, Revathí R, Hemalatha M (2011) Video surveillance systems—a survey. Int J Comput Sci (IJCSI) 8(4):1
7. Singh S, Kaur A, Taqdir A (2015) A face recognition technique using local binary pattern method. Int J Adv Res Comput Commun Eng 4(3):165–168
8. Alsiba MH, Manap HB, Abdullah AAB (2015) Enhanced face recognition method performance on android vs windows platform. ARPN J Eng Appl Sci 10(23)
9. Sharma RK et al (2014) Android interface based GSM home security system. In: 2014 international conference on issues and challenges in intelligent computing techniques (ICICT)
10. Bai YW, Shen LS, Li ZH (2013) Design and implementation of an embedded home surveillance system by use of multiple ultrasonic sensors. IEEE Trans Consum Electron 56
11. Raspberry Pi remote webcam streaming. https://www.youtube.com/watch?v=oIUHw0VChEU
12. Wang M, Zhang G, Zhang C, Zhang J, Li C (2013) An IoT-based appliance control system for smart homes. In: 2013 fourth international conference on intelligent control and information processing (ICICIP), June 2013
13. Dumbre K, Ganeshkar S, Dhekne A (2015) Robotic vehicle control using internet via webpage and keyboard. Int J Comput Appl (0975–8887) 114(17)

Development of Roads Pothole Detection System Using Image Processing

Harshad Sawalakhe and Ramchandran Prakash

Abstract Driving a vehicle on the roads, it is required to consider many parameters of roads which include distress like potholes, crack, patches size of road. Road conditions influence the comfort and safety of people who are sitting in the vehicle. Detection of pothole helps for early warning to the driver and assessment to the road maintenance authority. The position and structure of pothole can be shared with road maintenance authority. In this paper, an embedded system is designed to detect pothole from images of the road and sends the GPS location to road maintenance authority crew server and upload it in Google map. This system designed on small single-board computer Raspberry Pi in which it uses the computer vision library for processing the input video which is taken from a camera and identifications of road pothole present.

Keywords Pothole · Computer vision library · Raspberry Pi · Image processing GPS

1 Introduction

Accidents due to defective road conditions are very common all over the world. Driving vehicle without identifying the pothole presents on can be very dangerous. The presence of the pothole on roads results in wastage of fuel, time, wear and tear of tyre, each time when the driver is passing the pothole. It also reduces the life time of the vehicle. Therefore, it is very important to identify distress like pothole on roads and repair it to increase the comfort level and safety of the passengers.

H. Sawalakhe · R. Prakash (✉)
School of Electronics Engineering, Vellore Institute of Technology,
Chennai 600048, Tamil Nadu, India
e-mail: prakash.r@vit.ac.in

H. Sawalakhe
e-mail: harshad.vinodsawalakhe2014@vit.ac.in

D. Thalmann et al. (eds.), *Intelligent Embedded Systems*, Lecture Notes
in Electrical Engineering 492, https://doi.org/10.1007/978-981-10-8575-8_20

Fig. 1 Generalized block diagram of proposed system

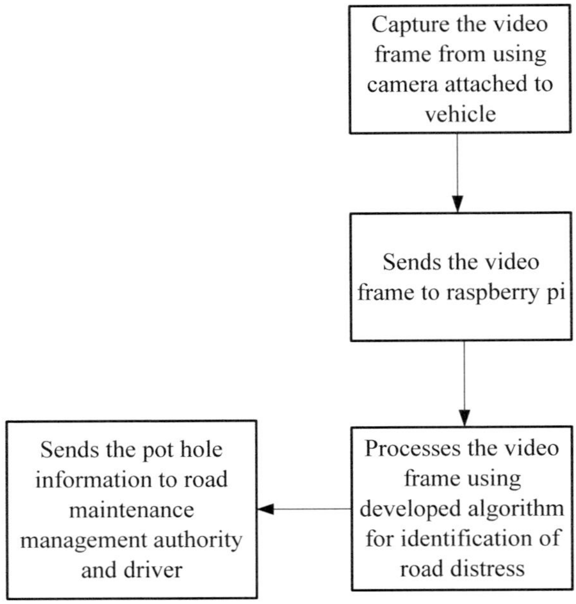

The pothole detection system proposed in this paper uses the stand-alone single-board computer Raspberry Pi for the processing. In this system, Raspberry Pi connected with camera captures the video of the roads and the video frame is processed to determine the presence of pothole and this information is used to warn the driver and to update the database of road authority so that it provides road maintenance authority prior information where it is required to repair the road.

The general block diagram of the proposed system is shown in Fig. 1.

2 Related Work

The assessment of road surface distress in a road maintenance management system is very important task for the road developing and maintenance strategy. About 40–50 years before, the detection of road distress like potholes, cracks, or patches was performed by manually by observing it by human eyes. Then after that, automatic systems were developed which uses vibration sensors for detection of road distress. The image processing-based system is developed later for detection of road distress.

For selecting a proper feasible method for road pothole detection, following literature survey is done. Mainly, there are three types of methods developed for the detections of pothole distress on the roads. These are a method using vibration sensor, 3D scanner-based method, or 3D vision-based method as depicted in Fig. 2. The method proposed in [1] is applied on Indian highways. In this method, first all

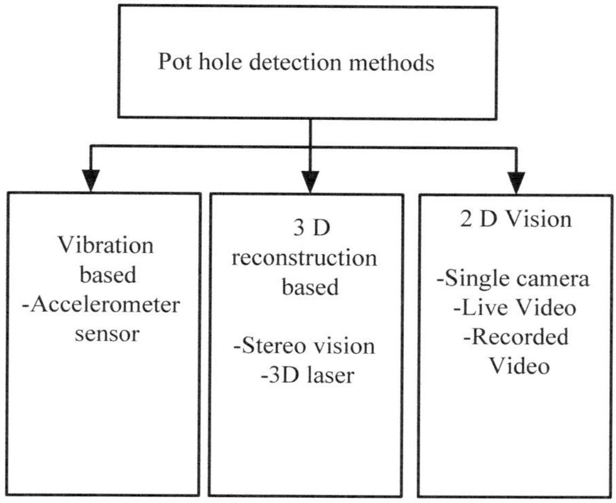

Fig. 2 Different methods of pothole detection system

frames of raw video are divided into two groups. Video frames which contain distress like potholes, crack, or patches are in one group of database and other are frames which does not contain any types of such distress. After this, the frames, which contain distress, are processed with an algorithm called Critical Distress Detection Measurement and Classification (CDDMC). In this, they calculated the standard deviation, circularity, width, area, and elongation properties of pothole, and after analyzing these different properties of pothole, it decided whether it is pothole, crack, patch, or non-critical distress. In method [2], the system is designed using vibration sensors (accelerometer). The system consists of an accelerometer, which is generally placed in the front part of the vehicle; each time when the vehicle passed from potholes, it detects gives the reading for identification of potholes. The problem with pothole detection using a vibration sensor is that many times it provides an inaccurate reading of pothole. The response of vibration-based sensor system is also very low.

A 3D laser scanner reconstruction for detection of pothole is proposed in [3]. 3D laser scanner uses a technology in which laser beams are transmitted and the reflected back laser pulses are used to identify the object. This is an expensive technique for detection of pothole and cost of maintenances of this system is more. Also, the algorithm required for this system is very complicated.

Pothole detection algorithm using two cameras, stereo vision camera is developed in [4] which gets 3D measurement of pothole geometric features. The problem with this system is that the cost of such system makes it expensive; these systems require highly complicated algorithm and powerful processors. This system does not provide a very accurate result of pothole and misses many times because the foreground and background image data considered as same many times because variation in that is very little for detection using stereo visions camera [5].

3 Methodology for Pothole Detection

In this paper, a portable embedded system is proposed which is effectively useful for detection of the roads, pothole, cracks and identifying their positions. For, this video is captured using a camera which is mounted on the vehicle and connected with Raspberry Pi board. Detection of distress from input video from camera given to Raspberry Pi-2 is performed by using OpenCV image processing libraries. This complete software system is built in raspberry on top of the wheezy operating system.

For this, we conduct experiments on the roads of Chennai, which consists of bituminous surface. Whenever the system detects the road pothole, road crack, or road patch, then GPS connected with Raspberry Pi sends the coordinate position, area, and other information about potholes, cracks, or patches to the road maintenance crew. The number of potholes in each image is also calculated. After, these images can be classified into different types which includes image contains potholes, image contains cracks, image contain patches and images without distress.

4 Algorithm for Distress Detection

The algorithm to be ported in the proposed Raspberry Pi system is developed based on CDDMC [1] as shown in Fig. 3. Mainly, this algorithm developed for detection of any type of distress on the road consists of five sequential steps: (1) enhancement of the image, (2) segmentation of the image, (3) extraction of the visual properties from image, (4) detection and classification using decision logic, and (5) analysis of the extracting and detecting images.

1. Enhancement of the image: By using the image enhancement, the brightness and contrast of the image is adjusted for improving the quality of the image.

   (a) Bit selection:
   In case of computer image processing, bit is useful for the storage of data. Each individual pixel of the image stored 24 bits of data (i.e., blue, green, and red channel; each channel stored 8 bits of data). For this, we are converting image from 24 bits into a 8-bit image by selecting the blue channel of the image. By selecting only the blue channel of the image instead of 24 bit of image, it becomes just 8 bit of image as the now image contains less data, processing on that image become more easy, fast and provides fast computations further.

   (b) Median Filtering:
   Filtering is used for removing of unwanted noise presents in the image. Median filtering used for image processing in noise removal because median filter very effectively removes the salt and pepper noise which in this case is like small particle and dusts. In case of the median filter, it calculates median of all the pixels under the kernel window, then central pixel is replaced by the calculated median value.

Fig. 3 Flowchart of design system

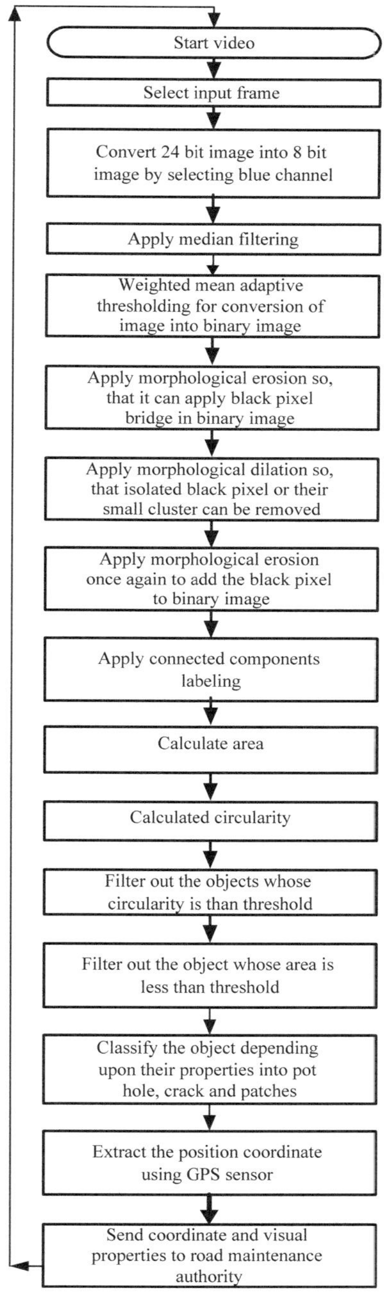

2. Segmentation of image:
   In segmentation of image, digital image is divided into multiple parts so that the representation of images becomes more easy and useful.
   (a) Weighted mean-based adaptive thresholding: Thresholding in image processing is useful for segmenting an image into two parts: foreground part and background part. The image which has a pixel intensity value greater than threshold consider as foreground and pixel having less value than the threshold are considered as background. In case of conventional thresholding, it uses a global threshold value for all the pixels. In case of adaptive thresholding, threshold for each pixel is changed over the image dynamically. Adoptive thresholding is more cultivated than simple thresholding because adaptive thresholding can hold more changes of lightening conditions because in global threshold uses a fixed threshold for all pixels, but in adaptive threshold, considers the individual threshold for each pixel depending upon the range of values in its neighborhood pixel.
3. Extracting the visual properties from images:
   After this, image segmentation visual properties of images are taken into account for further processing. We are extracting three visual properties from image, and these are image texture, shape factor, and dimensions:
   (a) Image texture:
   Distress present in road, i.e., pothole, crack, and patches, contains more variations of contrast comparing with the surrounding road without distress. Still, the variations of the contrast in different distress are different, e.g., variation of contrast in pothole difference in contrast variation in crack or patch.
   The morphological operation applies structure elements to an input image and generates an output image which reduces noise, isolation of the individual images and joining separate elements in the images. The two morphological operations available are: morphological erosion and dilation
   (i) Morphological erosion: The idea of the morphological erosion is like soil erosion. Morphological erosion tries to distinct the foreground object and background objects and tries to make a foreground object white. For this, kernel is used to isolate. We are using a 3 $\times$ 3 matrix kernel in this case first after the morphological operation. If the pixels under kernel in image are 1, then it is considered as 1 otherwise it is considered as 0. Therefore, it reduces the thickness or size of the foreground image by removing white nose and detaches two elements.
   Morphological dilation is totally opposite of morphological erosion; in case of morphological dilation, pixel element is considered as '1' if at least any one element under the kernel is '1'. It is also used for the removal of white noise. But it It shrinks the object when the noise is removed. The required object area is increased. It is also useful in case of joining the objects broken parts.

(b) Identifying the structure of distress:
The distress pothole and pitch consist of more circular compare with distress crack which is more elongated. So, by identifying the shape factor in terms of circularity provides a better understanding about the distress type.
(c) Dimension of distress: Area of the crack is less than pothole or patch. Therefore, it provides a useful identifying type of distress.

4. Detection and classification using decision logic:
In the detection and classification, logic uses if else conditions for deciding distress type necessary and also ignore the unnecessary part of image so only required portion acquired in this part.
5. Analysis of the extracting and detecting images: In this part, analysis of extracting and detecting image is done.

The algorithm explained above is applied for detection of distress on roads. The result of the applied algorithm is shown in the figures.

5 Result

An experiment is conducted by using the input road images which contain pothole shown in Fig. 4a. The image binary conversion of the original image is shown in Fig. 4b. First morphological erosion operation applied on binary image is shown in Fig. 4c, and morphological dilation after the morphological erosion operation is shown in Fig. 4d. Second morphological erosion operation applied after the morphological dilation is shown in Fig. 4e, the drawing after the second morphological erosion operations is shown in Fig. 4f, and the details of the visual properties of the pothole captured are shown in Fig. 4g. From the final result, it can be observed that the pothole can exactly detect from the input road image.

6 Conclusion

In this project use, full road distress detection system using image processing is developed. The effective algorithm for this is developed and tested. Advantages and disadvantages of the proposed are discussed below.

Advantages of system:

1. Pothole present on the road can be effectively identified.
2. Road maintenance authority gets information and prior knowledge regarding the places where they required for repairs to the road.
3. Cost of identification of pothole is reduced compared with other systems for detections of pothole.

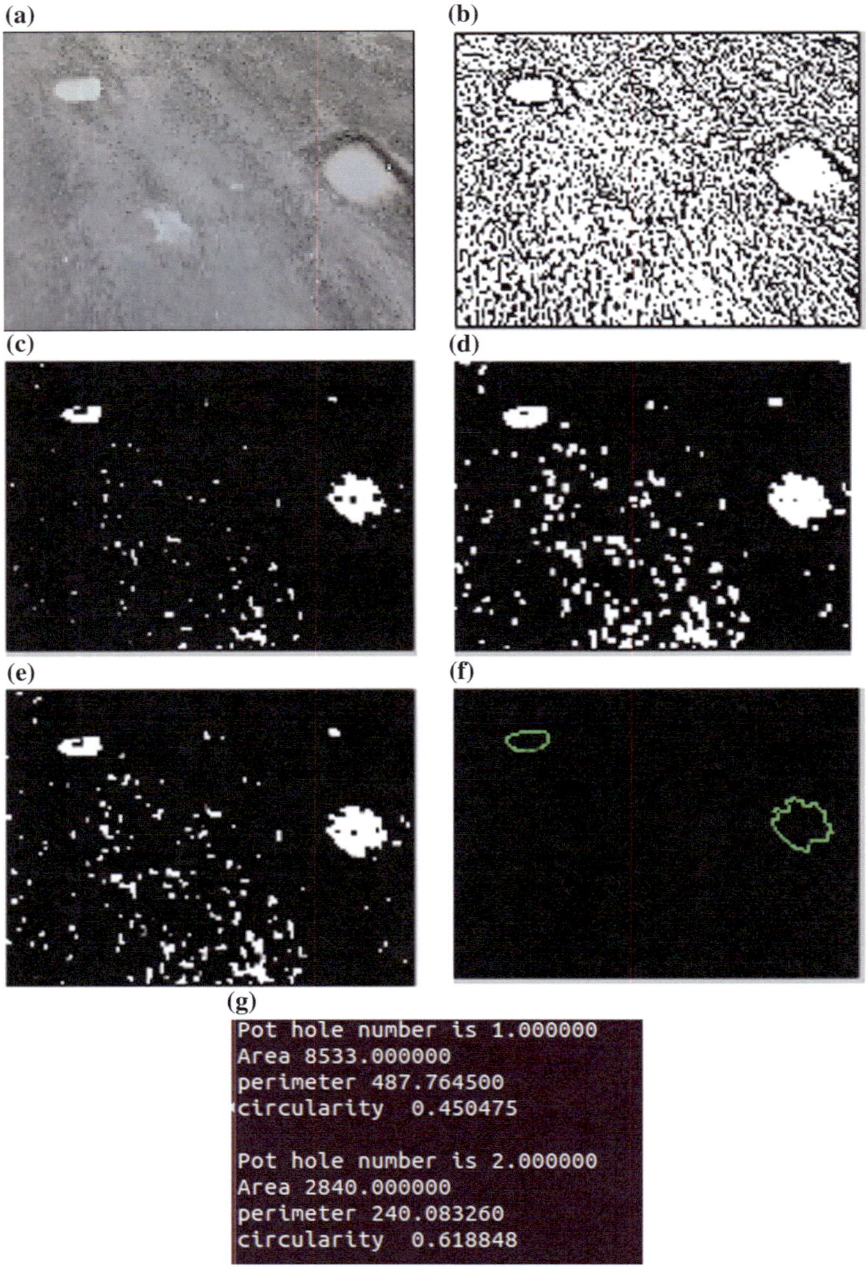

Fig. 4 **a** Original image of pothole. **b** Binary conversion of original image. **c** Morphological erosion of binary image. **d** Morphological dilation after morphological erosion. **e** Morphological erosion again after morphological dilation. **f** Drawing the distress of road object. **g** Details of the pothole visual properties of the image

Disadvantage of system:

1. If the vehicle is at very high speed, accuracy of pothole detection may be less.
2. In the rainy season and night, the identification of the pothole on the roads becomes a difficult task.

References

1. Huidrom L, Das LK, Sud SK (2013) Methods method for automated assessment of potholes, cracks and patches from road surface video clips. In: 2nd conference of transportation research group of India (2nd CTRG). Procedia 2013, pp 312–320
2. Rode SS, Vijay S, Ghoyal P (2009) Pothole detection and warning system: infrastructure support and system design. In: ICECT '09 proceedings of the 2009 international conference on electronic computer technology, pp 286–290
3. Chang KT, Chang JR, Liu JK (2005) Detection of pavement distress using 3D laser scanning technology. In: Proceeding of the ASCE international conference on computing in civil engineering, pp 1–11
4. Zhang Z, Ai X (2009) An efficient algorithm for pothole detection using stereo vision. In: IEEE international conference on acoustic, speech and signal processing (ICASSP), pp 565–568
5. Punjabi H, Nanwani R, Vaswani A, Jotwani R, Kunte A (2014) Intelligent pot hole detection system. Int J Emerg Technol Adv Eng 4(7):590–595

Automated Interoperability Testing of Optical Network Terminals for VoIP Call Features Using Robot Framework

Kavya Ajith, Kalaiselvan Ramalingam and Muddukrishna Dandu

Abstract **Background**: Coexistence of different technologies, protocols and frameworks for Voice over IP results in many interoperability issues. The Gigabyte-capable Passive Optical Network (GPON) that uses fiber-based point-to-multipoint networking escalates the possibilities of VoIP. The GPON-based VoIP consists of Optical Line Terminal (OLT) which is present at the service provider or server side and many Optical Network Terminal (ONT) at the customer end. The inception of GPON technology into telecommunication, unfolded the possibility of triple play (transfer of voice, data and video). **Methods/ Statistical Analysis**: This paper proposes an automated testing mechanism of Optical Network Terminals for the SIP-based voice call features over IP. Out of the triple play service (data, voice and video) provided by the Optical Network Terminal, this work concentrates only on the voice-related functionality. The proposed system uses the Robot Framework for automating the SIP-based Voice over IP test scenarios. The paper also covers implementation of Asterisk-based Voice over IP Client-Server setup on a Raspberry Pi to understand the background functionality. **Findings**: The automated system captures the software faults of the Optical Network Terminal beforehand and ensures proper interaction with other network elements without any interoperability issues. In addition to the ease in using Robot Framework, this method reduces the human interventions in testing process. **Application/Improvements**: Automated interoperability testing provides a better coverage for the functionality of the ONT considering the manual testing. Our future work is extending this current work in all data and voice and thus creating a complete testing mechanism for the ONT system.

K. Ajith (✉)
School of Electronics Engineering, VIT University, Chennai, Tamil Nadu, India
e-mail: kavuajith@gmail.com

K. Ramalingam · M. Dandu
Nokia India Limited, Chennai, India
e-mail: kalaiselvan.ramalingam@nokia.com

M. Dandu
e-mail: muddukrishna.dandu@nokia.com

D. Thalmann et al. (eds.), *Intelligent Embedded Systems*, Lecture Notes in Electrical Engineering 492, https://doi.org/10.1007/978-981-10-8575-8_21

Keywords Optical network terminal · SIP · VoIP · Robot Framework
GPON · Raspberry Pi · Asterisk · PBX

1 Introduction

Presently, Internet or Web is not merely a platform for information sharing; it acts as a convergence zone of numerous services. As worldwide reach is a salient feature of Internet, more and more providers are utilizing this platform for serving their ever-increasing customer base. Voice over IP (VoIP) is one of the spellbinding innovations to transport voice correspondence over the existing Internet like IP network. This innovation gives the ability of making telephone calls by sending networks information in bundles instead of the customary circuit switching of the Public Switched Telephone Network (PSTN).

VoIP is a class of hardware equipment and custom-made software that empowers usage of Internet as the transmission medium. VoIP has many advantages of which most captivating one is its incredibly low call costs as it uses Internet as its medium of transmission. The ease in extending the service and mobility provided by VoIP is very high compared to other traditional networks. This leads VoIP is drawing in more clients from the traditional communication area, which eventually resulting in more organizations working towards the improvement and use of the VoIP frameworks. It is expected that the VoIP will totally replace the PSTN framework in near future.

Numerous VoIP protocols have been proposed since the inception of VoIP. H.323 is one among the first VoIP protocols which was proposed in 1996 by International Telecommunication Union—Telecom Standardisation Sector (ITU-T) and has been redesigned a several times [1]. It recommends a standard for multimedia correspondence over Local Area Networks (LAN), Wide Area Network (WAN) and Web. Another VoIP standard proposed by the Internet Engineering Task Force (IETF) is the Session Initiation Protocol (SIP) [2]. It incorporates call setup and multimedia correspondence over the network. SIP has become most encouraging protocol due to its straightforwardness, adaptability and its modular character.

The Gigabyte-capable Passive Optical Network (GPON) that uses fibre-based point-to-multipoint networking escalates the possibilities of VoIP [3]. The GPON based VoIP consist of many Optical Network Terminal (ONT) at the customer end, Optical Line Terminal (OLT) which is present at the service provider or server side, other network components such as Switches, Remote Authentication Dial-In User Service (RADIUS), Broadband Remote Access Server (BRAS) etc. and finally Internet. Figure 1 illustrates basic elements in an optical network. The genesis of GPON technology into telecommunication, unfolded the possibility of triple play (transfer of voice, data and video).

The SIP protocol is used to signal and control multimedia transfer session establishes a session in GPON-based VoIP service. Signalling flows are between

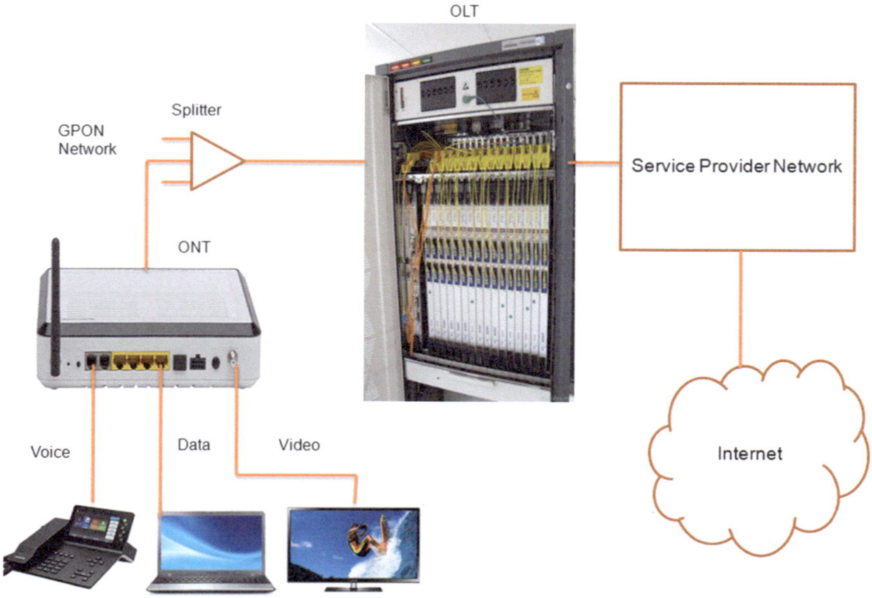

Fig. 1 Optical network elements

the ONT and the soft-switch/server which may consist of a SIP call server and SIP registrar. The actual voice traffic is carried out using real-time transport protocol (RTP) which flows through the network independent of the signalling carried out.

Even though SIP protocol is widely used for VoIP session initiation and management, the conventional PSTN and H.323 which is considered as VoIP "legacy" Protocols are still in use around the globe. Though the new technologies replace old ones with time, it is critical for the clients to ensure their capital that has as of now been invested in existing resources are protected. The coexistence of different frameworks and technologies raises the issue of achieving interoperability.

In this paper, we use the Robot Framework, a keyword-driven generic test automation framework, to carry out the VoIP-related testing on ONTs. This ensures the interoperability of the ONTs with the service provider side and other devices in network so that the proper signalling is carried out and a session is established in reference to different call features such as 3-way calling, Call Hold, Music on Hold, Call Transfer, Call Forwarding, Conference.

Robot Framework as shown in Fig. 2 has a modular architecture which provides independent data, script and services. Test data are defined as files and these data together with written test cases form a test suite. Parsing test cases and completing the system service calls is carried out by Robot Framework. Built-in libraries are collectively called Test Libraries. Any additional equipment or software used come under test tools [4, 5]. Wireshark, the Network Protocol Analyser, is a test tool.

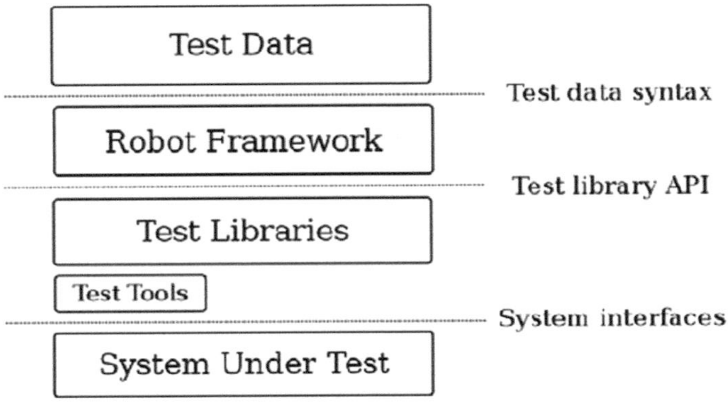

Fig. 2 Robot framework

2 Related Works

Every VoIP implementation is based on certain standards [6] that ensure both data link and physical layer interoperability. This portrays the system-level prerequisites to facilitate multi-vendor interoperability at service level. The major goal of these standards is to provide flexibility for plug and play interoperability in an ever-growing complex communication network.

A passive approach of interoperability testing is illustrated in different works [7–9]. Passive method operates over real network without disturbing the normal operation of the system. For the ease in handling the network elements are partitioned into different one-to-one systems and such complex systems are further divided to form different functional units.

There are many interoperability between two unique protocols has been the topic of discussion in many documents. A Signalling Gateway to interpret the signalling between happening SIP and H.323 protocol is proposed in this work [10]. For the H.323 and PSTN networks to interwork a Gateway, which incorporates a Signal Gateway, Media Gateway Controller and a Media Gateway are recommended to handle transport, control and signalling by the ITU-T. Interworking issues in SIP and PSTN protocol with Gateway is discussed in this work [11]. But all these approach was addressing issues between any two protocols.

A VoIP Web service base to facilitate the interoperability among diverse VoIP conventions is proposed in this work [12]. With this foundation different VoIP, protocol's signal handling is covered up through the relating Web service abstraction. This facilitates a common Web interface for all VoIP Protocols to provide solution for Interoperability.

With a specific goal to enhance the quality and the proficiency of software testing, automation test is utilized in this work [13]. This paper examines the utilization of testing apparatuses in test mechanization, with the Robot Automation Test Analysis Framework.

SIP implementation in software requires significantly more time for processing when compare to a hardware implementation of SIP. A reconfigurable architecture is proposed by work [14] for the SIP Protocol on Field Programmable Gate Arrays (FPGA). In this system, SIP processing and generation is handled by hardware section and software section interconnects the hardware implementation with other devices in network.

An experimental setup of VoIP for using radio voice from hams using open source technology was proposed in work [15]. This work was done as part of radio amateurism, a hobby, and was proved very useful in disaster relief and emergency situations. The system developed has capability update data on Amateur Position Reporting System (APRS). This facilitates user activity monitoring for VoIP service providers and mobility for the users.

Asterisk is used as the base for the hardware implementation of VoIP client-server system on Raspberry Pi. Researches on the suitability of using the Asterisk PBX server to give VoIP correspondence capacities an considerable Mean Opinion Score (MOS) quality to a vast number of clients is carried out in paper [16]. Exploratory results have demonstrated that the Asterisk PBX can adequately handle more than 160 simultaneous voice calls with a blocking likelihood of not exactly 5% while giving voice calls normal MOS above 4.

3 Methodology

The work carried out can be classified into two distinct parts, former being the Robot Framework-based automation and interoperability testing of real Optical Network Terminal for VoIP call features and later being the open source framework Asterisk-based VoIP client-server implementation and testing various call features and performance.

Software testing can be considered as a process that starts with the early decision-making stage of a project and lasts till the system gets implemented in a customer field, in most cases extending further. A product delivered carries the image of the organization into the market and extensive testing is done before any device is ready to market.

When it comes to embedded software testing, much more care should be given at both hardware and software end and their interaction. ONT being an embedded system requires proper hardware or emulators that can provide the operation of a new feature being implemented. For all the testing procedure, actual hardware is used. Figure 3 provides a flow of testing carried out by the software development team. Every small feature is unit tested to find all the possible bugs at the earlier stage when the cost of fixing it is less.

Once the software feature is delivered to the testing team rigorous testing activities as shown in Fig. 4, are carried out as per the test plan developed during the initial stage of Software Life Cycle. The software build's requirements and

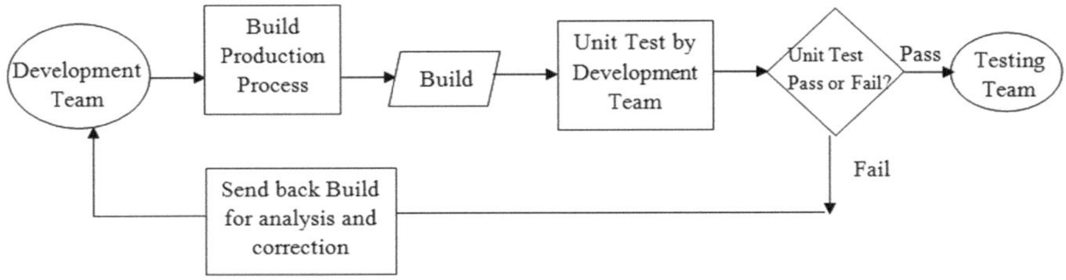

Fig. 3 Development team test process

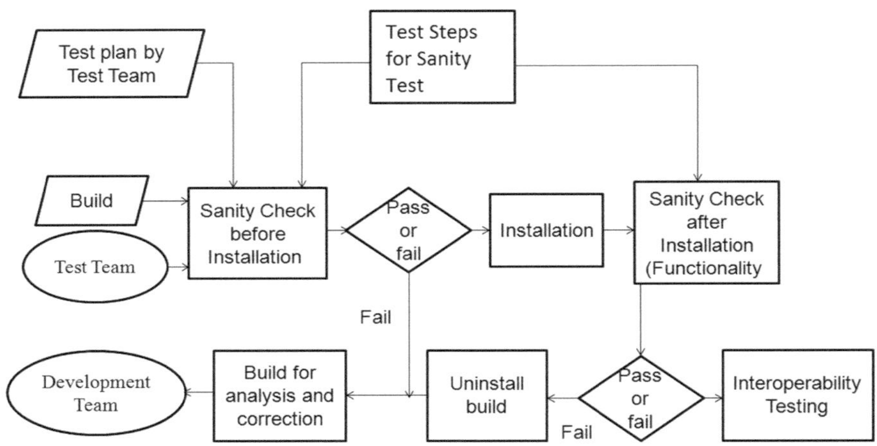

Fig. 4 Testing team test process

functionality of feature is ensured by this sanity and feature testing. To ensure the test procedure is fault free, a Sanity check is done on the actual test steps.

The next step in testing process is the interoperability testing, where the software is loaded on to a ONT that is connected to all other network elements that is actually present in the customer environment. This process is very tedious as it involves multiple combinations of configurations on each of these devices, all possible software versions on each element, multi-vendor scenarios and associated traffic in the network. A manual test process for this would delay the market time of the product or feature considerably. An automated test procedure based on Robot Framework is used for the automation process.

The hardware implementation in the work consists of a Raspberry Pi which acts as a SIP-based VoIP server/SIP Registrar and a Private Branch Exchange (PBX). This part of the work is done using the open source Asterisk framework and a Web-based Graphical User Interface (GUI) which is open source, Free PBX that is used to control the Asterisk. The client section consists of two soft phones installed in Windows-based personal computers and configured as clients with the PBX.

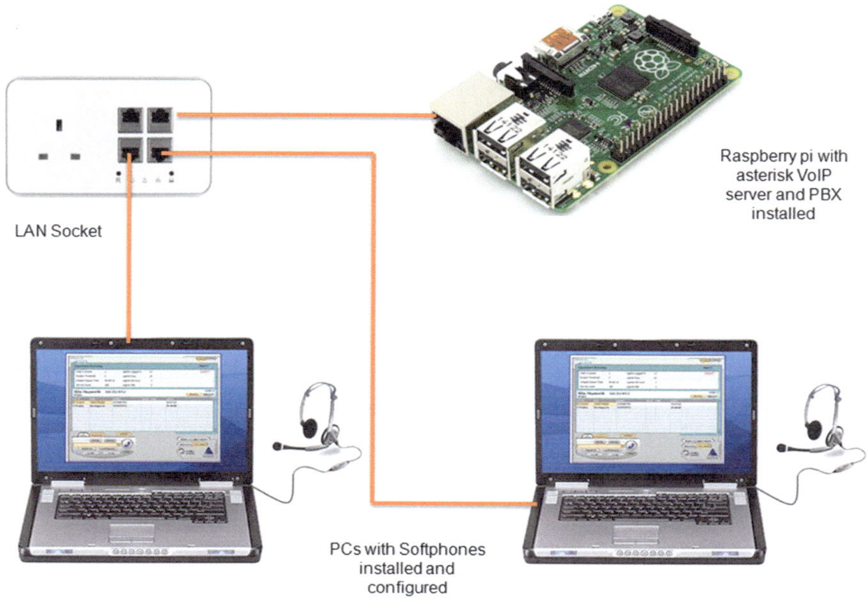

Fig. 5 Hardware implementation

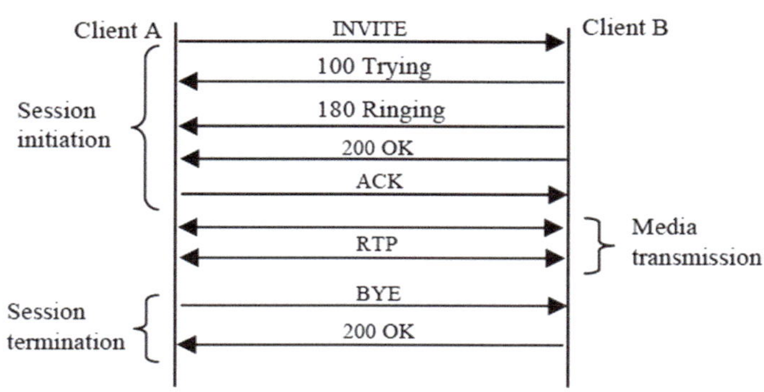

Fig. 6 Basic SIP call

The hardware implementation illustrated in Fig. 5 shows two clients and a server connected to Internet through LAN socket. Once the configuration part is complete, a call from Client A to Client B can be successfully handled by the server and we can evaluate the SIP signalling carried out between the entities by using any network analyser. A basic SIP call as shown in Fig. 6 between A and B is taken as example throughout the paper.

4 Experimental Setup

The setup for automated interoperability testing of different VoIP call features consists of all network elements like ONT, OLT, Switches, and SIP Server/Registrar as demonstrated in Fig. 7. Each ONT has two Plain Old Telephone Service (POTS) slots where we can connect a normal or a SIP phone. In case of manual testing, actual phones are connected. To automate the procedure, the POTS line is connected to a Synway voice card installed in Windows environment which emulate the on hook and off hook based on the test script. OLT, Switch and SIP server form the service provider's end.

The Robot Framework and the test Scripts are running in a Linux-based system which further connects either to a port in OLT or Switch where the traffic is mirrored. Once all setup is ready, the test case is triggered from the Linux PC. The system which is in same network as of the Windows PC will trigger Client A to go off hook and dial Client B. Table 1 shows the settings for the test, and Table 2 shows the variables used.

The Robot Framework takes the variables defined in the variables file mentioned and performs the actions as per the script. The variable script varies as per the ONT variants to capture the hardware- and software-specific details. Complete flow of the basic call test is shown in Fig. 8

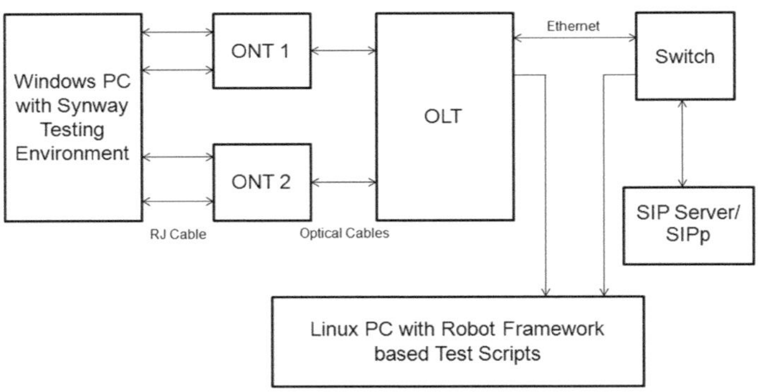

Fig. 7 Testing setup in robot framework

Table 1 Settings for testing

Setting	Value
Documentation	Test
Source	/repo/kajith/robot
Test case	Basic call between A and B
Test setup	Wireshark and voice card session open
Test teardown	Wireshark and voice card session close

Table 2 Test variables

Variables	Value
${user1}	A with num = num_A
${user2}	B with num = num_B
${dialnum}	A Dial num_B
${DtmfStr_A_to_B}	String to check RTP path

5 Results and Discussions

Out of the many advantages of Robot Framework generation of test report is one that deserves special mentioning. After the successful completion of running test cases, robot framework generates three output files. These files are report.html which gives a brief description about the status of the test cases run, log.html that saves the complete log of the test execution and output.xml which saves the details in a computer-readable format. Here, Fig. 9 shows the test case run for the basic call. Figures 10 and 11 illustrate the Command Line Execution Results and the Test Execution log, respectively.

The test case also opens a Wireshark session to trace the actual SIP signalling between the two parties. This captures the message session starting from INVITE method till BYE as depicted in Basic SIP Call. Once the test case is completed, the Wireshark session is terminated and the message flow is saved. Figure 12 shows the Wireshark capture for the basic call test case. All message exchange between Client A and B are captured here.

Once the test cases are run, we are made available with all the details adequate for analysing the system. If any interoperability issue occurs in the ONT network, a immediate action can be triggered to rectify it before product reaching the market. Usually faults find during the interoperability testing is errors in message parsers, timing delays due to software issues, fault in RTP transport and so on.

In this work, we have limited the scope to evaluate the SIP-based VoIP call features facilitated by the ONT. This can be extended such that the entire features can be subjected to automated testing. Thus, a complete overview of the device performance can be obtained using the automated interoperability testing of ONT and other network elements.

Fig. 8 Flowchart for basic call test case

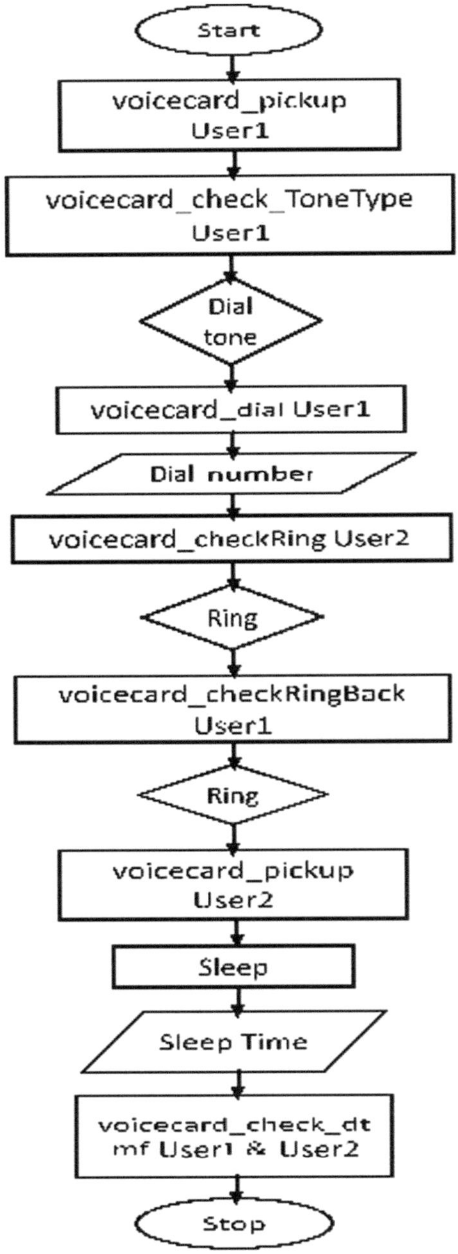

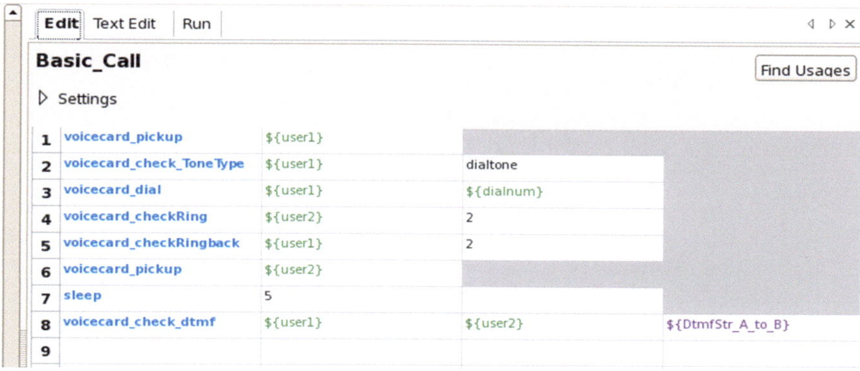

Fig. 9 Basic call test case

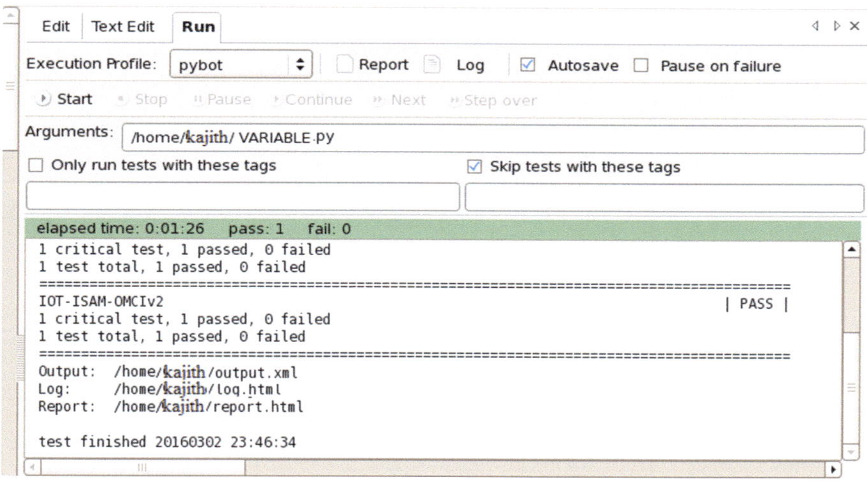

Fig. 10 Command line execution result

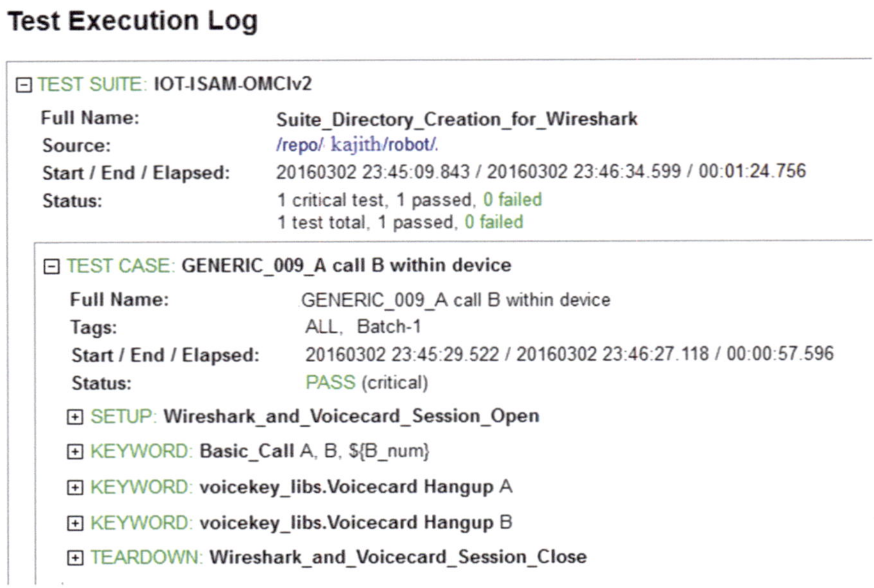

Fig. 11 Test execution log

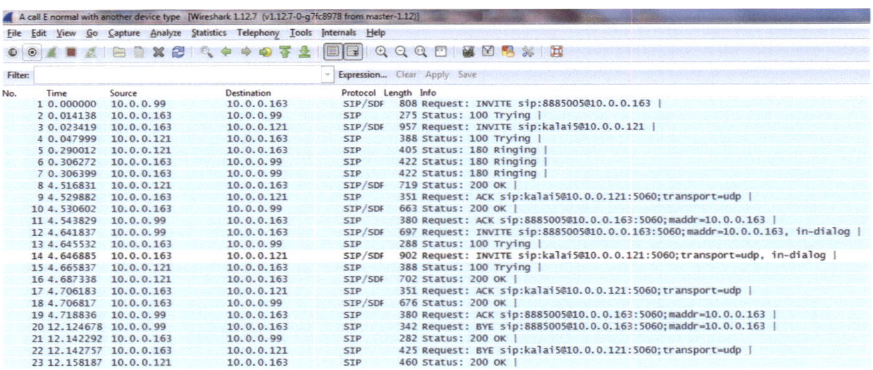

No.	Time	Source	Destination	Protocol	Length	Info
1	0.000000	10.0.0.99	10.0.0.163	SIP/SDF	808	Request: INVITE sip:8885005@10.0.0.163 |
2	0.014138	10.0.0.163	10.0.0.99	SIP	275	Status: 100 Trying |
3	0.023419	10.0.0.163	10.0.0.121	SIP/SDF	957	Request: INVITE sip:kalai5@10.0.0.121 |
4	0.047999	10.0.0.121	10.0.0.163	SIP	388	Status: 100 Trying |
5	0.290012	10.0.0.121	10.0.0.163	SIP	405	Status: 180 Ringing |
6	0.306272	10.0.0.163	10.0.0.99	SIP	422	Status: 180 Ringing |
7	0.306399	10.0.0.163	10.0.0.99	SIP	422	Status: 180 Ringing |
8	4.516831	10.0.0.121	10.0.0.163	SIP/SDF	719	Status: 200 OK |
9	4.529882	10.0.0.163	10.0.0.121	SIP	351	Request: ACK sip:kalai5@10.0.0.121:5060;transport=udp |
10	4.530602	10.0.0.163	10.0.0.99	SIP/SDF	663	Status: 200 OK |
11	4.543829	10.0.0.99	10.0.0.163	SIP	380	Request: ACK sip:8885005@10.0.0.163:5060;maddr=10.0.0.163 |
12	4.641837	10.0.0.99	10.0.0.163	SIP/SDF	697	Request: INVITE sip:8885005@10.0.0.163:5060;maddr=10.0.0.163, in-dialog |
13	4.645532	10.0.0.163	10.0.0.99	SIP	288	Status: 100 Trying |
14	4.646885	10.0.0.163	10.0.0.121	SIP/SDF	902	Request: INVITE sip:kalai5@10.0.0.121:5060;transport=udp, in-dialog |
15	4.665837	10.0.0.121	10.0.0.163	SIP	388	Status: 100 Trying |
16	4.687338	10.0.0.121	10.0.0.163	SIP/SDF	702	Status: 200 OK |
17	4.706183	10.0.0.163	10.0.0.121	SIP	351	Request: ACK sip:kalai5@10.0.0.121:5060;transport=udp |
18	4.706817	10.0.0.163	10.0.0.99	SIP/SDF	676	Status: 200 OK |
19	4.718836	10.0.0.99	10.0.0.163	SIP	380	Request: ACK sip:8885005@10.0.0.163:5060;maddr=10.0.0.163 |
20	12.124678	10.0.0.99	10.0.0.163	SIP	342	Request: BYE sip:8885005@10.0.0.163:5060;maddr=10.0.0.163 |
21	12.142292	10.0.0.163	10.0.0.99	SIP	282	Status: 200 OK |
22	12.142757	10.0.0.163	10.0.0.121	SIP	425	Request: BYE sip:kalai5@10.0.0.121:5060;transport=udp |
23	12.158187	10.0.0.121	10.0.0.163	SIP	460	Status: 200 OK |

Fig. 12 Wireshark capture for basic call

6 Conclusion

This work effectively gives a solution to identify and control the interoperability issues faced in SIP-based VoIP calls facilitated by the ONT before it actually being deployed in customer network. Automation of the test scenarios reduces the human intervention in the testing process. Use of Robot Framework in addition to the ease in preparing test cases provides a complete analysis report which is easy for the users to understand. The testing mechanism proposed can be extended to cover all features testing in an ONT and thereby reduce the tedious effort of testing.

Acknowledgements We would like to thank the Nokia for giving us such a wonderful opportunity to carry out this research work in their company premises by providing us the requisite resources and infrastructure for carrying out the research.

References

1. H.323: Packet-based multimedia communications system, ITU-T Recommendation, Nov 2000
2. Rosenberg J, Schulzrinne H, Camarillo G, Johnston A, Peterson J, Sparks R, Handley M, Schooler E (2002) SIP: session initiation protocol. RFC 3261, IETF, June 2002
3. Selmanovic F, Skaljo E (2010) GPON in telecommunication network. In: International Congress on Ultra Modern Telecommunications and Control Systems and Workshops (ICUMT), Oct 2010
4. Na Q, Huaichang D (2015) Extension and application based on robot testing framework. In: 6th IEEE international conference on software engineering and service science (ICSESS), Sep 2015
5. Nokia Siemens Networks. Robot framework user guide. 2008–2009. pp 6–74
6. IEEE standard for service interoperability in ethernet passive optical networks (SIEPON)", IEEE Communications Society, IEEE Std 1904.1, Sep 2013
7. Guo Y, Huang X, Di M, Ma Y (2014) Research on modeling and development process for passive interoperability testing. In: IEEE 3rd international conference on cloud computing and intelligence systems (CCIS), Nov 2014
8. Lin H (2007) Research on theory and method of protocol interoperability testing[D]. University of Science and Technology of China
9. Chen D (2004) Passive testing theory and its application in protocol fault detection[D]. Tsinghua University
10. Singh K, Schulzrinne H (2000) Interworking between SIP/SDP and H.323. In: Proceedings of the 1st IP-telephony Workshop (IPTel'2000), April 2000
11. Vemuri H, Peterson J (2002) Seesion initiation protocol for telephones (SIP-T): context and architecture. RFC 3372, IETF, Sep 2002
12. Hillenbrand M, Zhang G (2004) A web service based framework for voice over IP. In: Proceeding of the 30th EUROMICRO conference, Aug 2004
13. Jian-Ping L, Juan-Juan L, Dong-Long W (2012) Application analysis of automated testing framework based on robot. In: Third international conference on networking and distributed computing, Oct 2012
14. Peterkin R, Abou-Gabal M, El-Hassan F, Ionescu D (2009) Hardware implementation of session initiation protocol servers and clients. IEEE symposium on computers and communications ISCC, July 2009
15. Hajdarevic K, Konjicija S, Subasi A (2014) Svxlink VOIP implementation using raspberry pi in education and disaster relief situations. International symposium on telecommunications (BIHTEL), Oct 2014
16. Costa LR, Nunes LSN, Bordim JL, Nakano K (2015) Asterisk PBX capacity evaluation. IEEE international parallel and distributed processing symposium workshops, May 2015

Design and Implementation of Smart Helmet Using Low Power MSP430 Platform

Yogya Indupuru, K. Venkatasubramanian and V. Umamaheswari

Abstract An accident is an unexpected, unintended, and unusual external action that occurs at particular place and time which leads to a marked effect on human life. Most of the countries have rules for motorcycle users to wear a helmet for their safety purpose where the rate of two-wheeler accidents increasing day by day. Smart helmet automatically checks whether a person has undergone any accident and sends messages to the emergency numbers. This system finds the impact with which the head hits the ground when an accident occurs. It makes use of a low power MSP430 processor which receives the analog value from the piezoelectric disk sensor based on the impact of helmet and compares with critical value. Once the critical value is obtained, GSM transmits a message to emergency numbers. As an additional feature, there is a temperature sensor which senses temperature in helmet and checks the temperature at every instant based on the variations of temperature in helmet control cooling system when temperature is high.

Keywords MSP430 · GPS · GSM · Safeguard system · Cooling system DC fan

1 Introduction

In present-day scenario, many accidents occur in our everyday life. There are about 1.3 million accidents occurring per year in world. In India, about 4 lakh accidents and 1.5 lakh fatal accidents occurring every year, out of them 26% two-wheeler accidents are occurring [1]. The reasons for the accidents may due to no fitness of

Y. Indupuru · K. Venkatasubramanian (✉) · V. Umamaheswari
School of Electronics Engineering, VIT University, Chennai 600127, Tamil Nadu, India
e-mail: venkatasubramanian.k@vit.ac.in

Y. Indupuru
e-mail: yogya.i24@gmail.com

V. Umamaheswari
e-mail: umamaheswari.t1@vit.ac.in

D. Thalmann et al. (eds.), *Intelligent Embedded Systems*, Lecture Notes in Electrical Engineering 492, https://doi.org/10.1007/978-981-10-8575-8_22

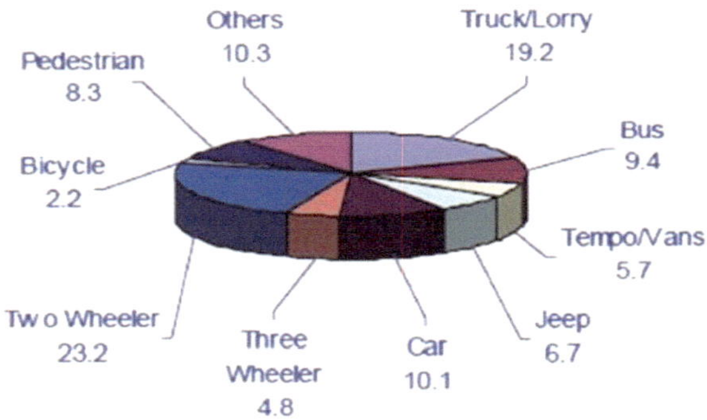

Fig. 1 Road accidents death by different means of transport

the bike, drunken driver, and rash driving etc. Most of the accidents lead to major head injuries, and people may die due to these injuries.

- In the year of 2011, the total number of accidents occurred in rural areas (53.5%) is more compared to the urban areas (46.5%).
- Rural areas had more fatalities (63.4%) compared to urban areas (36.6%).
- During 2012, about 35.2% of deaths were recorded due to road traffic accidents.
- 23.2% of the victims of road accidents were due to "two-wheelers" (Fig. 1).

Helmet is a protective shield that ensures the safety of two-wheeler rider's head. It reduces the impact on rider's head when accident occurs. But people avoid the use of helmets because of several reasons; few of them are excess heat produced inside, not able to receive calls, not able to hear the other vehicles horn sounds, and limited visibility. Nowadays, everyone is bothered about their comfort compared to safety.

The "smart helmet design and implementation using MSP430" provide a way to safeguard bike riders and help to provide immediate help in case of emergency. It also provides an effective cooling mechanism for person's convenience [2–7]. Objective is to provide safety measures to rider based on the impact on a person's head when an accident occurred.

This work proposes a solution for the heat generated inside the helmet and ensures the safety of rider and provision to enhance the vehicle tracking using GPS which will also be integrated to zero in the position. Once the critical section is reached, we also have an additional feature of sending a message to couple of emergency numbers, where the message contains the distance, help message, and the location.

2 Literature Survey

Dipti Bhandari and Dharshit Solanki proposed a work which was published in Academia on "i-biker" where it provides safeguard for biker by using smart intelligent system that ensures the biker has worn helmet, identifies an alcoholic consumer and provides prompt medical attention after met with an accident. This alcoholic consumer detection avoids the drunken driving accidents [8].

Amitava Das and Priti Das presented a journal in International Journal for Advances in Science Engineering and Technology, ISSN: 2321-9009 on "Smart helmet for Indian bikers," that ensure that the rider cannot start the bike without wearing helmet, and this uses simple cable replacement for wirelessly switching on a bike which is attached to the helmet, so that the bike would not start without both the keys. It ensures that the person is wearing a helmet or not using these keys [9].

Mohd Khairul Afiq Mohd Rasli, Nina Korlina Madzhi, and Juliana Johari presented a paper in International Conference on Electrical, Electronics, and System Engineering in the year of 2013, "Smart Helmet with Sensors Accident Prevention," provides a safety measure during the accidents occur and an indication to rider is given when the speed is exceeded. Here sensors used to measure the speed of the vehicle and once the speed is above the limit, it indicates the rider about the speed exceeded. There is also an accident detection sensors to identify the accident occurred and provide safety measures.

Ping Li, Ramy Meziane, Martin J.-D. Otis, Hassan Ezzaidi, and Philippe Cardou presented a paper which was published in IEEE in the year of 2014 on "A Smart Safety Helmet using EEG and IMU sensors for worker fatigue detection" which provided results for smart safety helmet implementation using EEG sensor and makes sure that the workers are not injured in the mines when accidents occur. Here the EEG sensors are used to identify any accidents that occurred in the mines and if any accidents occur, then it sends an alarm intimating the person that a rider has undergone accident and requires help.

3 Methodology

3.1 System Design Model

The work proposed in this paper is a solution for the heat generated inside the helmet and ensures the safety of rider and provision to enhance the vehicle tracking using GPS which will also locate the position. This system provides an immediate help in case of emergency by sending a message to a couple of emergency numbers.

This project consists of two actions occurring in parallel. The first action is the safeguarding helmet which based on the impact on the helmet, and the tracking system provides necessary action in case of emergency. The second action is the heat controlling system which is used to maintain low temperature inside the helmet (Fig. 2).

3.1.1 Safeguard System for Helmet

A piezoelectric sensor is used to measure the impact on which the head hits the ground. Once the sensor reading is greater than the threshold value, the controller receives the GPS location and sends an emergency message to a couple of numbers saved in the GSM (Fig. 3).

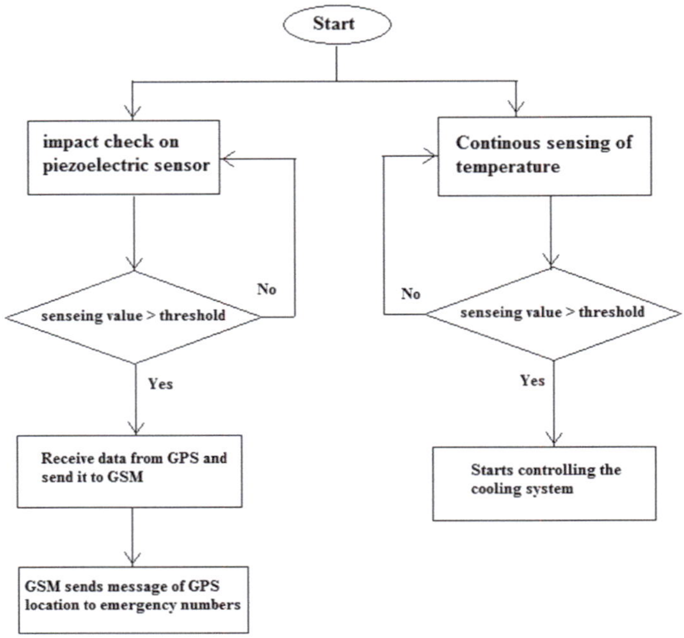

Fig. 2 Flow diagram for the system design

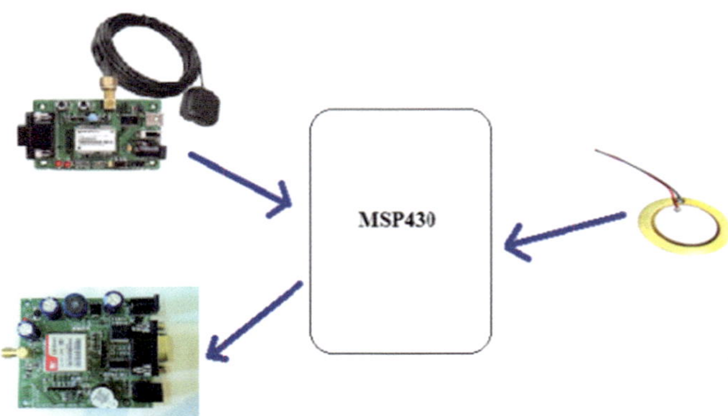

Fig. 3 Block diagram for safeguard system

3.1.2 Cooling System

The cooling system helps in controlling heat inside the helmet to maintain low temperature inside it. Once the temperature value is above threshold value, it automatically turns the cooling system (Fig. 4).

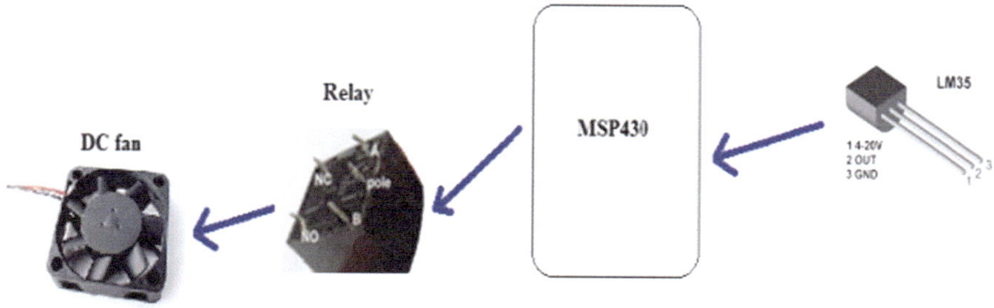

Fig. 4 Block diagram for cooling system

3.2 *Software Description*

Software requirements are the most important requirement. For every processor, we need to control the external hardware by controlling the processor by using software programming. Here we are using MSP430 Texas Instrument launch pad, and here energia software is used for programming.

3.3 *Hardware Description*

3.3.1 MSP430

MSP430G2553 is a low-cost experimental launch pad which is used in this for low power usage.

3.3.2 UART Serial Communication

Pins P1.1 and P1.2 are the UART pins for serial communication. Where P1.1 is the receiver pin and P1.2 is the transmitter pin. In order to enable the serial communication on hardware UART in MSP430, we need to change the jumper position in the J3 header by rotating them 90° from the position (Fig. 5).

3.3.3 DC Fan

A cooling fan is used to create airflow. The DC fan consists of nine rotating vanes or blades which are set to rotate when voltage is given as input and tends to spread airflow.

Fig. 5 MSP430G2553 launch pad with J3 jumper connection for serial communication

Fig. 6 A DC brushless cooling fan

The DC fan is a portable tiny fan which has the dimensions of length is 2.5×2.5 cm and height is 0.7 cm. This is a brushless DC cooling fan with an operating voltage of 5–12 V (Fig. 6).

4 Hardware Implementation

The basic block diagram describes the procedure for implementation of smart helmet safety accruing system, and the cooling system can be set in the helmet and used for regular usage [10]. Figure 1 shows us the basic mechanism between all the components and MSP430 controlling action (Fig. 7).

Fig. 7 Block diagram

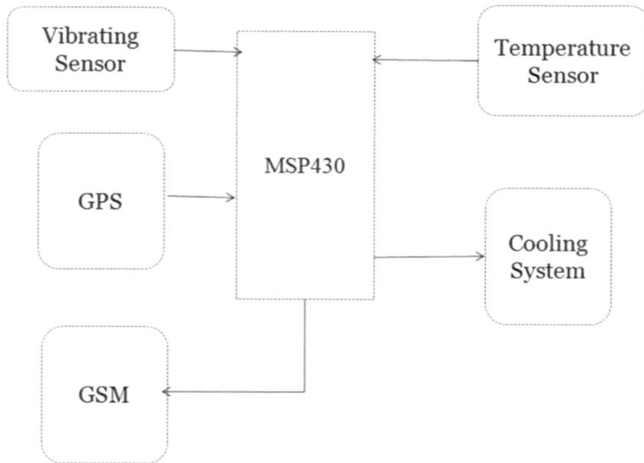

Fig. 8 Piezoelectric disk sensors and MSP430 interfacing

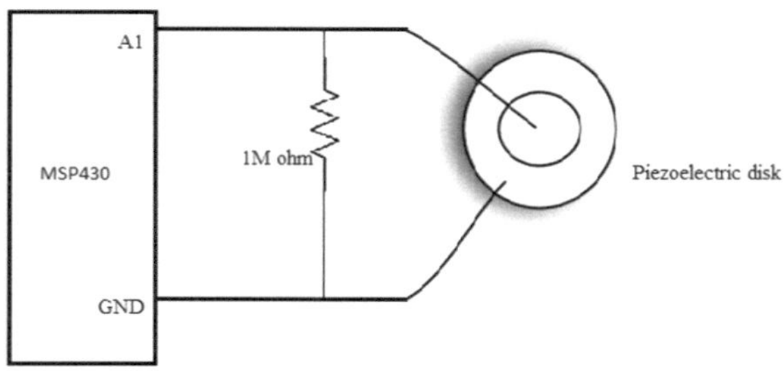

4.1 Interface Piezoelectric Sensor with MSP430

Interface piezoelectric sensor with MSP430 is for identifying the threshold value. Interfacing of piezoelectric sensor is done in three stages:

- Without fixing sensor to the helmet:
  Here the sensor is externally connected to the controller, and it is hit by a hard material and tested at different environments, and the results are noted.
- Fixing the sensor to a hallow helmet:
  Here the sensor is fixed inside the helmet and hit very hard and tested.
- Fixing the sensor to a helmet and fixed that helmet to human head:
  Here the sensor is fixed inside the helmet. A person wearing the helmet is hit with hard metal and tested.

Comparing these experimental results, the threshold value is fixed (Fig. 8).

4.2 Interface GSM with MSP430

Interface GSM module with MSP430 is used to understand the programming of GSM and analyzes its working (Fig. 9).

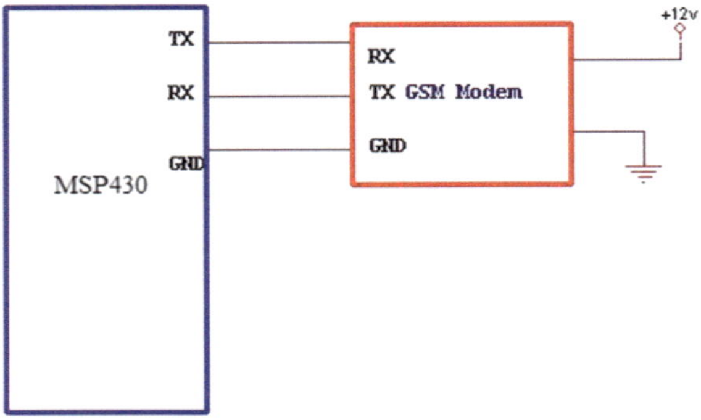

Fig. 9 GSM SIM900A module interfacing with MSP430

In order to enable the serial communication on hardware, UART in the transmitter pin(2) of GSM is connected to the UART receiver pin(A2) of MSP430, and receiver pin(3) of GSM is connected to the UART transmitter pin(A1) of MSP430. GND of GSM is connected to GND of MSP430.

4.3 Interface GPS with MSP430

GPS module is connected to MSP430 and understanding the programming of GPS is tested between GSM and UART of msp430 microcontroller. Here the data related to the location of device is received, and GPS receiver module was tested successful (Fig. 10).

In order to enable the serial communication on hardware, UART in MSP430 transmitter pin of GPS is connected to the UART receiver pin of MSP430, and receiver pin of GPS is connected to the UART transmitter pin of MSP430. GND of GPS is connected to GND of MSP430. The 3.3 V supply is given from the MSP430 pin.

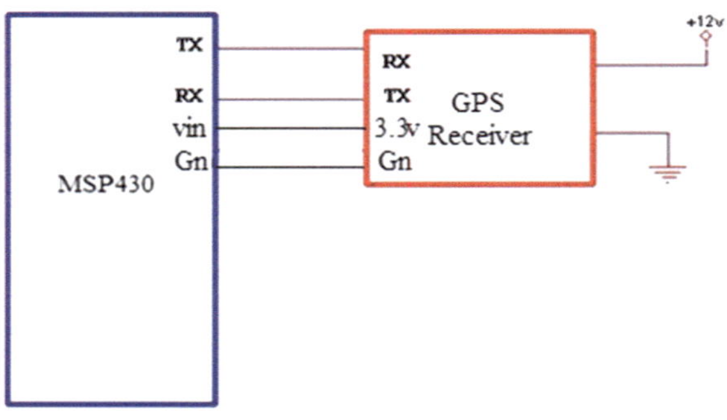

Fig. 10 GPS module interfacing with MSP430

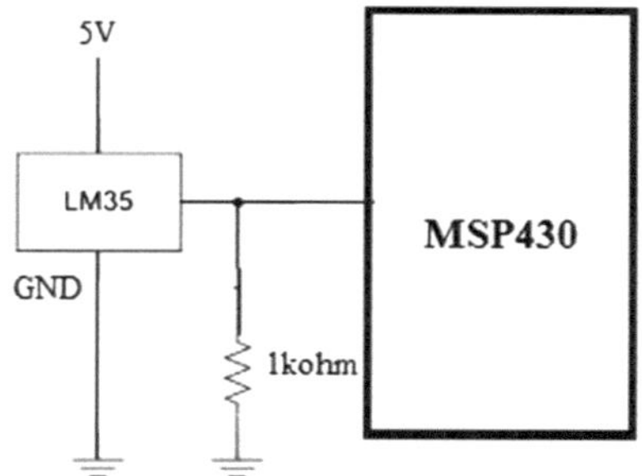

Fig. 11 Circuit connection between LM35 and MSP430

4.4 LM35

The LM35 is an analog sensor to calibrate temperature of environment using a precision integrated circuit.

LM35 has 3 pins, first pin is connected to the voltage supply of 5 V, second pin is connected as analog input value to the MSP430 to read the temperature value, and third pin is given as ground. The connection for the temperature sensor is shown in Fig. 11.

4.5 DC Fan Driving Circuit

See Fig. 12.

To drive the DC fan with a driven voltage of 5 V, we need to develop a circuit which use a 12 V battery and an op-amp amplifier circuit is used to increase the

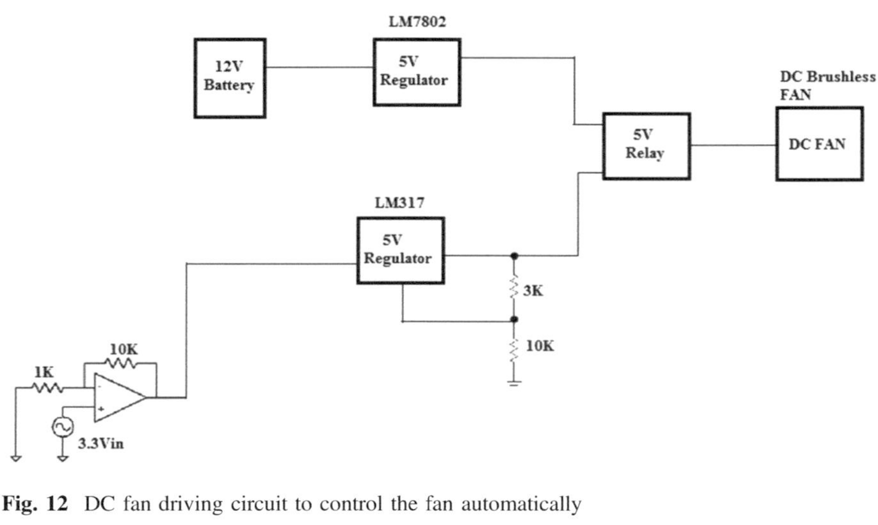

Fig. 12 DC fan driving circuit to control the fan automatically

voltage value, and the voltage is feed to regulator which maintains the constant output voltage of 5 V. This 5 V is feed to relay which starts to drive the fan automatically.

5 Results and Discussions

See Fig. 13.

For testing the sensor, here a 1 MΩ resistance is connected in parallel to sensor and analog pin is connected to the sensor to calibrate the values. Here the first output is the sensor value without fixing inside helmet. The other output is when the sensor is fixed to helmet. Based on the comparing values of two output, the threshold value ranges up to 500 in case of accident occurred impact is high.

GSM module is used to connect to MSP430 and understanding the programming of GSM is tested. Connections between GSM and UART of msp430 are implemented. Here sending a message to the mobile phone is tested.

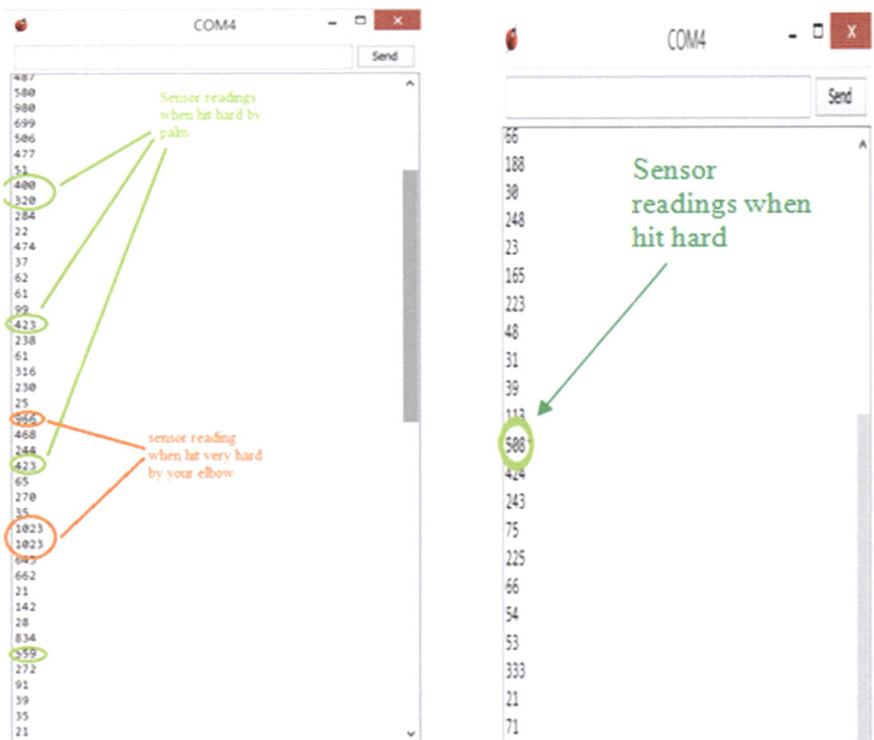

Fig. 13 Output values of sensor

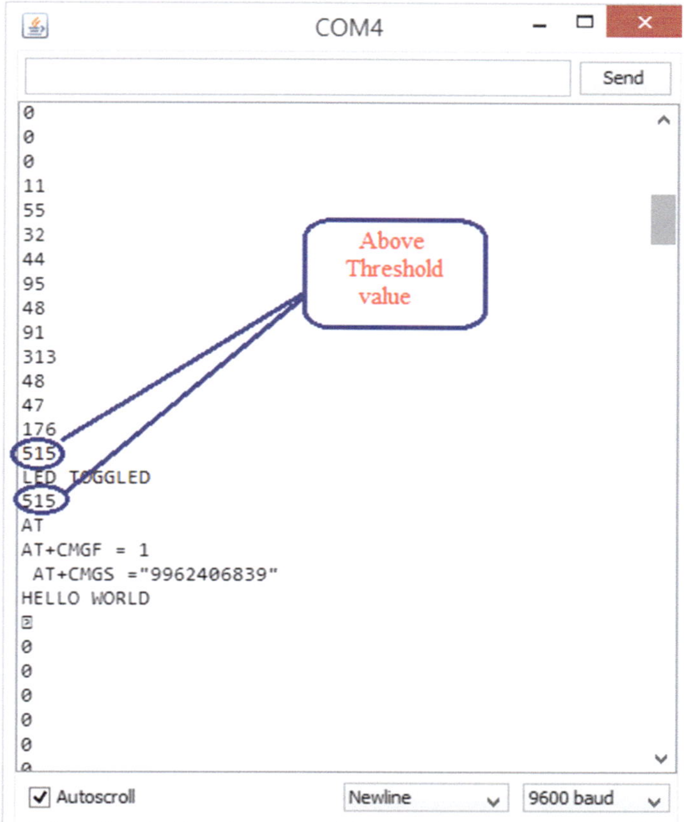

Fig. 14 GSM and piezoelectric sensor interfacing with MSP430 output

Interfacing GSM and piezoelectric sensor together are performed based on the sensor readings where the GSM sends message. Here we fixed a threshold value as 500. If the sensor reading is greater than the threshold value, then it activates GSM and sends a message to the reference number (Fig. 14).

Once GPS is interfaced with the microcontroller, it starts receiving the sequence of information through the serial UART port. This data is received as GPGGA, GPGSA, GPGSV, and so on, followed by numerical data as shown below. Here this data received by the receiver is NMEA data. Using this data, we need to find the geographical points as latitude and longitude (Fig. 15).

Based on the threshold value of the temperature, the fan is controlled. If the temperature is high, then it automatically starts to control the cooling system. This is an output for controlling the cooling system by using circuit that provides 5 V supply (Fig. 16).

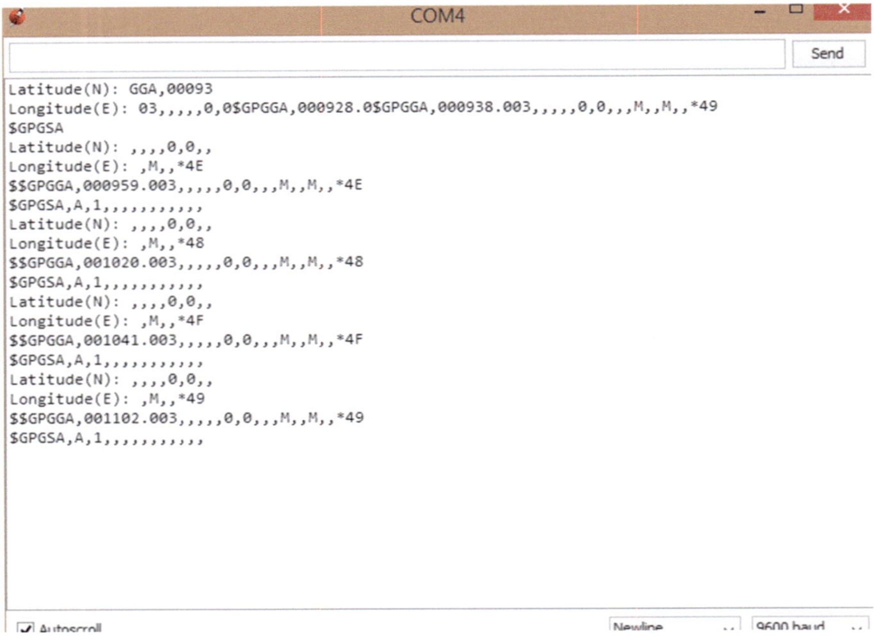

Fig. 15 GPS interfacing with MSP430 output

Fig. 16 Circuit to control the cooling system based on MSP430 output

6 Conclusions

This project is based on the analysis of implemented circuits and testing the threshold value of piezoelectric sensor is fixed, where it indicates a very high impact on the helmet. Based on the threshold, the controlling of GSM and GPS is done. Once threshold is high, GPS geographical coordinates should receive and this is sent to GSM.

On the other hand, the temperature sensor continuously checks the condition and sends to microcontroller; here based on the threshold, the microcontroller starts controlling the DC fan using a relay and regulator circuit.

By using this process, we can assure safety to the helmet users and provide the comfort. It is easy to handle, and it is an automated system. This makes the user comfortable. The DC fan used here is mini-sized and portable. We can use an array of DC fans to improve the performance. This reduces the usage of power as we make use of low power devices. It has an effective low cost.

Acknowledgements I would like to thank my project guide Prof. K.Venkatasubramanian for sharing expertise, sincere, and valuable guidance and encouragement extended to me. I would also thank the anonymous reviews for their comments for improving the paper and I also extend my gratitude to VIT University, Chennai, for their support.

References

1. Cho MH, Lee CH (2010) A low-power real-time operating system for ARC (actual remote control) wearable device. IEEE Trans Consum Electron 56(3)
2. Vijayan S, Govind VT, Mathews M, Surendran S, ME MS (2014) Alcohol detection using smart helmet system, vol 8. Issue 1. ISSN: 0976-1353
3. Chverton J (2011) Helmet presence classification with motorcycle detection and tracking. IET Intell Transp Syst 6 (3)
4. Rasli MKAM, Madzhi NK, Johari J (2013) Smart helmet with sensors for accident prevention. In: International conference on electrical, electronics and system engineering
5. Manjesh N, Sudarshan Raju CH. Safety measures for Two wheelers by smart helmet and four wheelers by vehicular communication. In: National conference on developments, advances and trends in engineering sciences and Int J Eng Res Appl (IJERA). ISSN: 2248-9622
6. http://www.jotr.in/article.asp?issn=09757341;year=2013;volume=6;issue=1;spage=1;epage=6;aulast=Ruikar
7. Thakre K, Waskar P, Sawant P (2015) Smart helmet. Int J Adv Res Comput Sci Softw Eng 5 (2)
8. Bhandari D, Solanki D. i-bike—a smarter way to safeguard bikers. www.Academia.edu
9. Das A, Das P (2014) Smart helmet for Indian bikers. Int J Adv Sci Eng Technol 2(4)
10. http://maxembedded.com/2013/12/ioportoperationsinmsp430
11. http://extremeelectronics.co.in/datasheets/gsm_shield_sim900a.pdf

Vision Intelligence System for Power Management Using Human Activity Detection System

Sukanya B. Pasupuleti and Prakash Ramachandran

Abstract Nowadays, scarcity of power has been increased at much higher rate due to improper power management. In this paper, the concept of power management is demonstrated with a Webcam and OpenCV tool. OpenCV is used for the implementation of face-detection algorithm. The system is a real-time embedded system product, which is employed with a camera fixed in a room, which detects the faces of the people present in the room using OpenCV software which is a computer vision tool; as soon as the faces of human are detected in the room, the system takes the necessary steps and accordingly controls (turn ON/OFF) the AC/DC loads (like fan and lights) of the room. Thus, in this way, the utilization of the power can be maintained in an efficient way.

Keywords Face detection · Image processing · Power management
Haar-Cascade classifiers · OpenCV

1 Introduction

Power crisis is the biggest issue nowadays because of which it may result in the fall of the economic growth rate of any country. In today's era, natural resources like coal, from which power is generated, are depleting at the faster pace. Power management plays a vital role in the economic progress of any country; thus, an immediate need raised during past few years to take appropriate steps so that it leads to proper power management, which indirectly contributes in reduction of wastage of power generated for electricity. Thus, misuse of electricity is the biggest

S. B. Pasupuleti (✉) · P. Ramachandran
School of Electronics Engineering, VIT University, Chennai 600127, India
e-mail: sukanya.bhavaniprasadi2014@vit.ac.in

P. Ramachandran
e-mail: prakash.r@vit.ac.in

D. Thalmann et al. (eds.), *Intelligent Embedded Systems*, Lecture Notes in Electrical Engineering 492, https://doi.org/10.1007/978-981-10-8575-8_23

factor in problem, which can be witnessed mainly in public buildings like school, colleges, companies, conference halls, office cabins. Most of the time, people forget to turn off lights and fans when they leave their rooms, which contributes to wastage of electricity, which in turn adds to power crisis issue. To overcome this power issue, we need to find an efficient and automated method to conserve electricity else in future we need to face a major shortage of electricity.

This paper proposes an idea to implement an automated real-time embedded system which detects the face of the persons those who are present in the room using Webcam. As the Webcam is used everywhere nowadays for surveillance purpose, thus implementing the controlling of appliances using face detection using Webcam will be worth. If face is detected in the room, then the fans and lights of the room gets turned ON, and if no face is detected in the room, then automatically the fans or lights will remain turned OFF, as earlier when the room is empty, implying the zero face detection, or else, if people were present there earlier and left the room, system will detect zero face detection and make the fans and lights to get turned off as soon as people leave the surveillance premises. Thus, this method can be efficient in minimizing the power wastage due to human negligence to some extent and can be implemented effectively over particular surveillance area. Implementation of such system can minimize the wastage of electricity to the large extent which ultimately results in conservation of electricity and its proper utilization. Cost-effective product can be employed for such power management strategies.

2 Objective

The objective of this project is to implement a real-time embedded system, using an image processing technique for face detection through Webcam, so that energy consumption by the electrical loads can be managed. Such system can be employed in those surveillance areas where the maintenance of energy conservation is needed. This system must perform the image processing part to detect the faces while streaming the surveillance through Webcam and should be able to control the electrical appliances on the acknowledgment of face detection if any person is present in the surveillance area. The system should be cost effective and portable so that it can be fixed at any desired area to monitor the surveillance for face detection [1].

Every room is employed with a relay or switch for controlling the electrical loads like fan and light of the room. Someone must take effort to check the room manually, to make sure whether the fans and lights are switched off when the room is vacant. However, manual checking is not feasible due to following reasons:

1. Difficulty in checking manually, to find which room is empty.
2. Employing the dedicated staff for this manual checkup is not feasible.

Thus, this results in

(i) Power wastage.
(ii) Painstaking task to turn off the fans and lights whenever the room is empty.
(iii) Higher bills may result because of human negligence.

Thus, to overcome all the above difficulties, we need the automated human activity detection system for power management.

3 Proposed Work

The proposed concept in this paper is to design an automated system which can detect the human face through image processing and also control the electrical loads simultaneously along with the face detection so as to minimize the wastage of power/electricity [2–5].

Figure 1 shows the block diagram of the System Architecture Design of the proposed system.

In Fig. 1, system is needed to be installed in the desired surveillance area like room, office, fitted along with the Webcam to the system. The Webcam will capture the live stream of the room and give it as input to the computer to process it using computer vision tool like OpenCV. The OpenCV software will enable us to detect the human presence in the room using face-detection module. If no face is found, then via GPIO pins on Raspberry Pi board, a control logic is sent to the relay which then sends the signal to turn off the lights and fan of the room.

The above human activity detection system consists of two parts:

1. Image processing
2. Controlling electrical loads

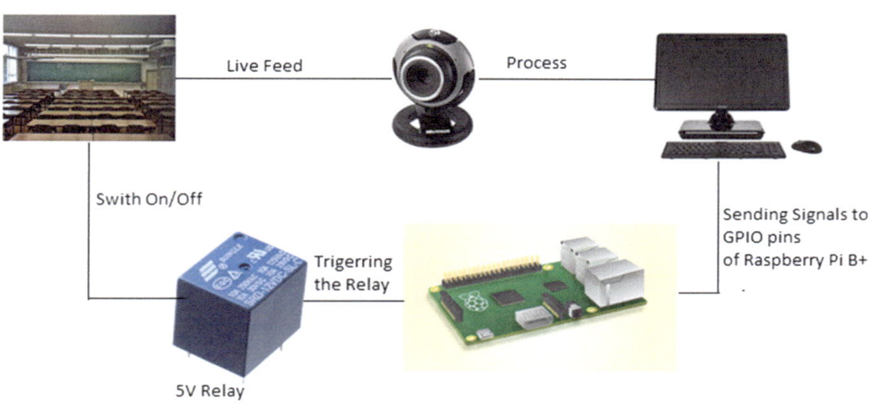

Fig. 1 System architecture design of human activity detection system

3.1 Image Processing

Image processing in this system is done using OpenCV software. OpenCV software is a computer vision tool, which is widely used for object detection. Here we are using OpenCV software for face detection using Haar-Cascade classifier. A Webcam is connected to the system for the live streaming of the video [6–9].

3.2 Controlling Electrical Loads

This part mainly consists of the hardware part of the system, which consists of the AC/DC loads connected to relay, whereas the relay is connected to the main single computing board of the system which is a Raspberry Pi board. Raspberry Pi board is small credit-sized minicomputer system. Relays are interfaced with Raspberry Pi board for controlling the relay automatically through GPIO pins of Raspberry Pi board.

4 Methodology

Implementation of human activity detection of system has four modules: Fig. 2.

1. Camera initialization.
2. Image processing for face detection.
3. Accessing the GPIO pins for hardware controlling.
4. Hardware interfacing.

4.1 Camera Initialization

An USB Webcam is connected through USB port of Raspberry Pi b+ board for live streaming. As soon as Webcam is connected to Raspberry Pi b+ board, it would not start working if we run the program that means camera is not initialized. So as to initialize the Webcam, it is needed to install drivers for it. So it is needed to install V4 l (video for Linux) driver to initialize the camera through program code. Once the Webcam is initialized, it can stream the video effectively and give the frames captured as input for image processing to the OpenCV software.

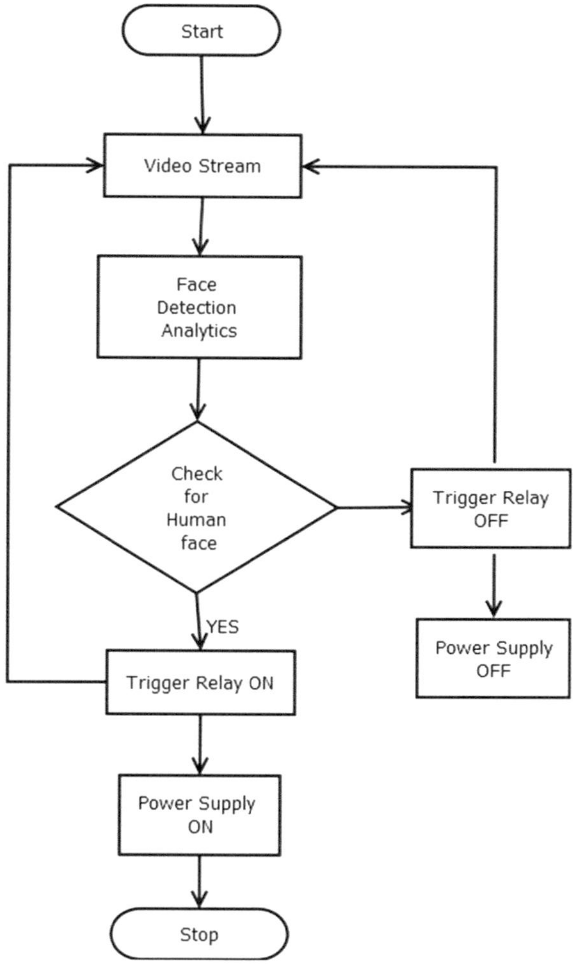

Fig. 2 Generalized flow diagram of the system

4.2 *Image Processing for Face Detection*

As soon as the frames are obtained through Webcam in the form of images, it can be given as an input to the OpenCV software for image processing. Face-detection analytics will be performed in OpenCV software following the steps shown in Fig. 3 which queries each frame to detect whether the face is available or not. If the face is available, it will read those frames first, then the frames are resized; once they are resized, those frames are converted to grayscale image from color image. As soon as the frames are converted in gray scale, they undergo for image processing into Haar-Cascade classifier to detect the face and to draw a rectangle around the recognized facial portion in the image. The Haar-Cascade classifier will divide the frames into two parts. One part is having all positives samples which have features of human face, which tells the system what is to be detected and the other set of negative samples which includes non-facial features of human, which tells the system what is not to be detected. So both the samples are compared with image in frame to detect where the face is detected and where it is not. Once the face is detected, it will resize the window to its original dimensions and the face detection analytics part gets completed.

4.3 Accessing the GPIO Pins for Hardware Controlling

As the face is detected through Haar-Cascade classifier, it should draw a rectangle over facial region of human through paint run. When the face is detected, processor will send the logic 1 on one of the GPIO pin, but we would not get the output unless the GPIO pins are inactive in system since there is no preinstallation for accessing the GPIO pins on Raspberry Pi B+. So as to access the GPIO pins of Raspberry Pi B+, we need a library named RPi.GPIO, which handles interfacing with the GPIO pins. Once we include this library in our program, we can access the GPIO pins. Also to access the GPIO pins of Raspberry Pi, one should operate the system in super-user mode or should have administrative access to the system.

4.4 Hardware Interfacing

Hardware interfacing will include the controlling of the AC/DC loads through GPIO pins via relay. As far as DC loads are considered, it can be connected to GPIO pins through jumper wires. So as soon as the GPIO pin will get signal or logic 1 on it, it will trigger the relay and turn on the DC loads. If it gets logic 0 on it, loads will be triggered off by the relay. So if we want to connect AC loads through it, we need to add relay to the circuit which acts as an electrically controllable switch. Relays allow isolation between two separate sections of system; here, one is GPIO ports and the other one are loads (Fig. 4).

The interfacing of relay with Raspberry Pi board through GPIO pins is shown in Fig. 5. Relay has five pins which are labeled as COM, N/C, and N/O which denotes common, normally closed, and normally open, respectively.

The experimental setup of the system is as shown in Fig. 6. The experimental setup of the system consists of the following equipment:

1. Webcam 2.0MP
2. Raspberry Pi B+ board
3. Display (TV/PC/laptop)
4. HDMI cable/ethernet cable

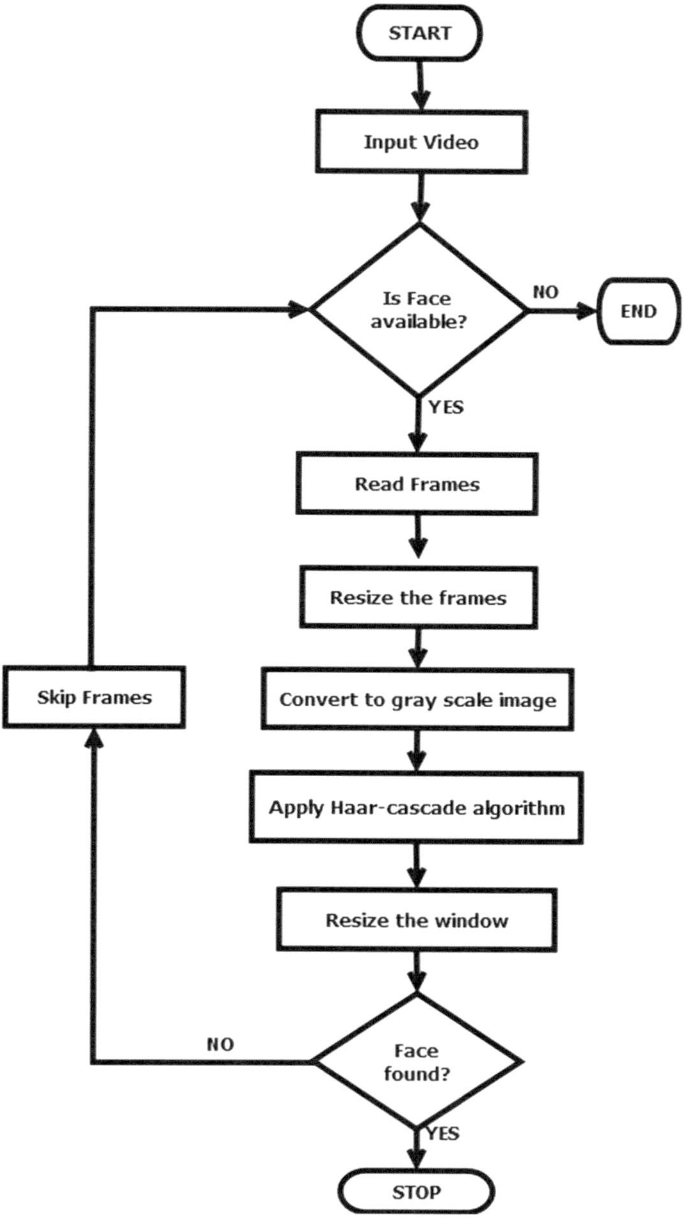

Fig. 3 Image processing steps of the system

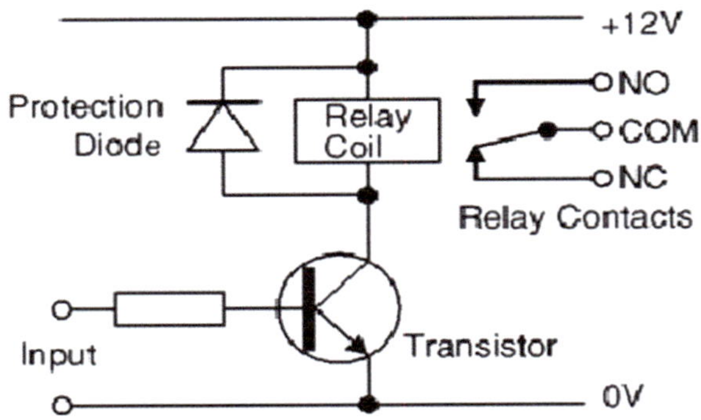

Fig. 4 Circuit diagram of relay

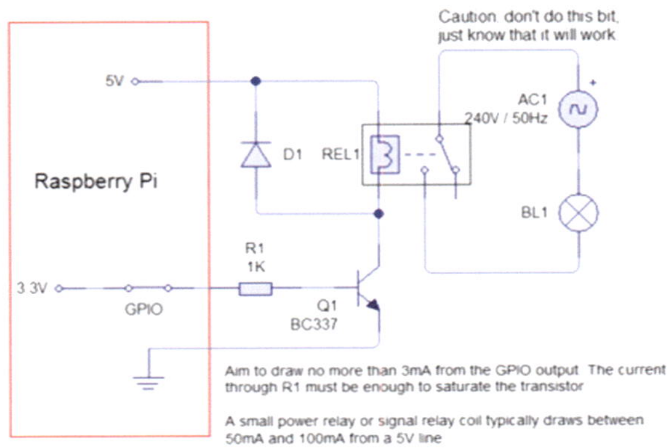

Fig. 5 Interfacing of relay with Raspberry Pi B+ board

5. USB cable for power supply
6. 5 V relay
7. DC loads like DC motor, zero bulb, buzzer

This setup can be boxed into a system and can be fixed wherever we need to install it. The Webcam can be placed at any corner of the room wherever required so as to possible the faces of people present in the room. Here we have placed at the top of the display. Loads such as buzzer, DC motor, and led bulb are coupled to relay, whereas relay is interfaced with GPIO pins of Raspberry Pi B+ board. The display is connected to Raspberry Pi board through its HDMI port for which we have used HDMI to VGA converter cable. USB cable is connected to power up Raspberry Pi B+ board. The USB camera is connected to one of the USB ports of RPi, and also, keyboard and mouse are also connected to USB port of RPi. An SD card installed with Raspbian OS is used which is inserted to SD card slot of the RPi board. If we wish to connect Internet to RPi board, we can use ethernet cable for providing LAN connection. Even, if we need to access RPi through remote connection we can i.e. connect RPi using laptop, though we connect it remotely using Ethernet cable. Thus, this is a setup for the human activity detection system for

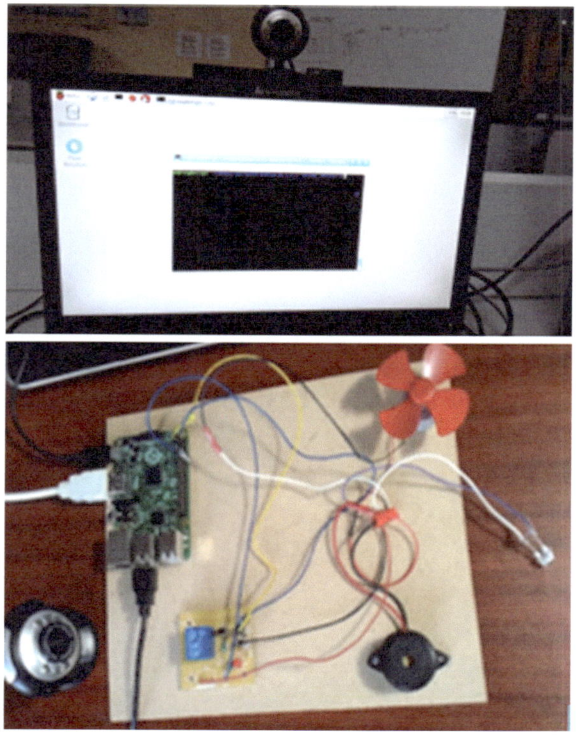

Fig. 6 Experimental setup of the proposed system

detecting faces of the people present in the room and then control the AC or DC electrical loads accordingly.

5 Result

The results obtained for the human activity detection system which we implemented is as follows:

5.1 Face Detection Analysis

OpenCV software is installed in the SD card which is used in Raspberry Pi B+ board. OpenCV software and the pre-trained Haar classifier available in it are used for image processing with Raspberry Pi board which results in the face detection in real time system. These face detection output will be used for controlling the DC devices we have connected in the setup shown in Fig. 7. Once the face is detected,

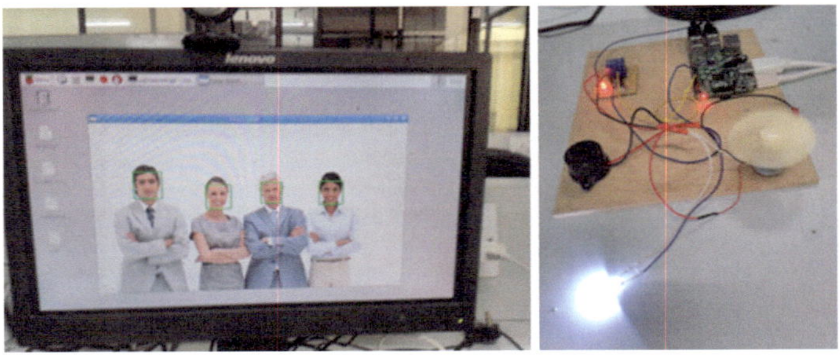

Fig. 7 Controlling the DC loads by triggering the relay on face detection

Fig. 8 Empty room while streaming through Webcam

Logic 1 passes on one of the output mode GPIO pins which trigger the relay, and thus on getting triggered, relay will turn ON all the devices connected to it as the face is detected in the image. Thus, we get the following result as shown in Fig. 7.

In the next experiment, face detection is done in a real-time environment where the Webcam starts streaming the video of the surveillance area or the room to be captured. While streaming the video, it displays the status of the room in real time, which is empty as shown in the Fig. 8 and as well as alongside the process of face detection is also keep going on. The frames captured during streaming are converted to grayscale images, and the Haar-Cascade classifiers are applied for face detection. As soon as the face is detected in the real-time environment itself, it will start drawing rectangle around the face as shown in Figs. 9 and 10.

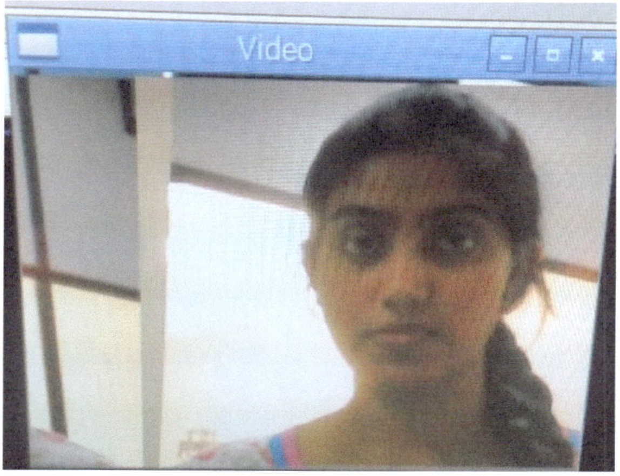

Fig. 9 Capture the person present in the room, through the Webcam while streaming the surveillance in the real-time environment, but no face is detected

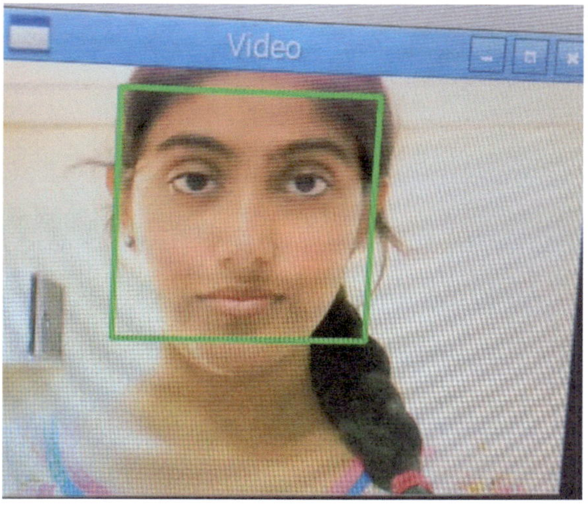

Fig. 10 Detecting the face and drawing the rectangle on the face of the person present in the room in real-time environment

5.2 *Room Occupancy Display*

The system is able to display the room status of the surveillance by proving the information about no. of faces detected in the room which indirectly indicates the no. of persons present in the room.

Condition 1 When no face is detected

When the room is empty, and no person is present, the system will detect the zero face detection, thus indicating no person is present in the room; thus, it displays that "Found 0 faces!" as shown in Fig. 11.

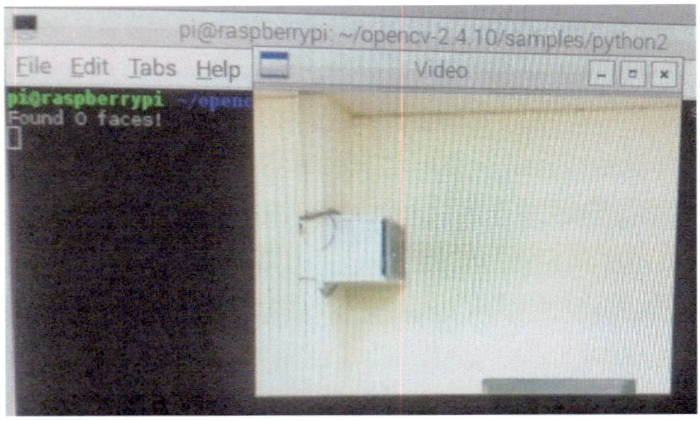

Fig. 11 Zero face detection indicating room status "Empty"

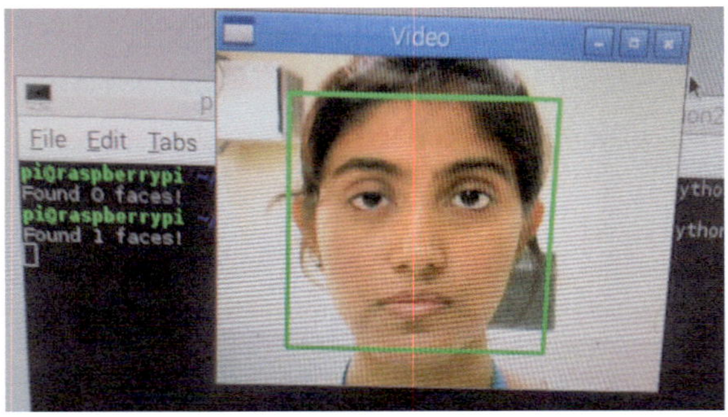

Fig. 12 Single face detection indicating the room status "Occupied"

Condition 2 When face is detected

When the room is occupied and person is present, the system will detect the face detection, thus indicating person is present in the room; thus, it displays that "Found 1 faces!" as shown in Fig. 12.

Condition 3 When face is present but no face is detected

When the room is occupied and person is present, the system will be unable to detect the face, thus indicating insufficient light in the surveillance are to find the face. Thus, even if the person is present in the room, it displays that "Found 0 faces!" as shown in Fig. 13.

5.3 Controlling Loads on Face Detection

Once the face is detected, the immediate task of the system is to control the DC loads by triggering the 5 V relay through GPIO access from Raspberry Pi to control the loads.

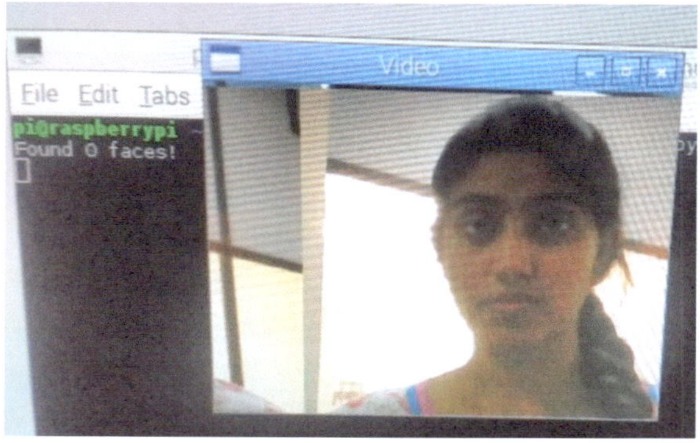

Fig. 13 Zero face detection even after the face is captured in the image, indicating the room status "Empty"

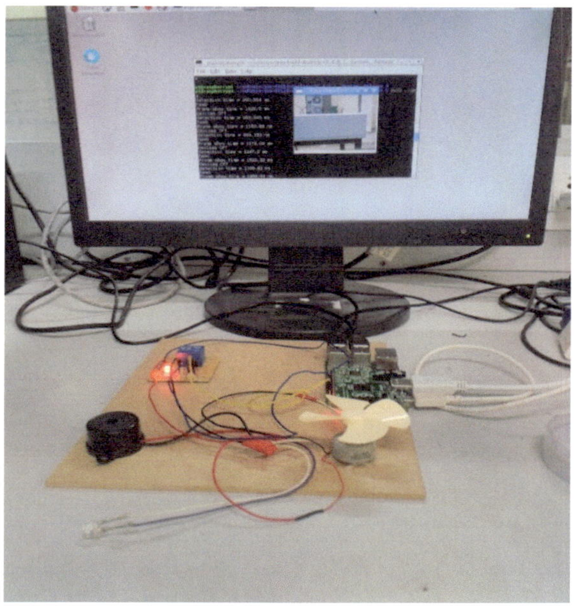

Fig. 14 DC loads are in OFF state when there is zero face detection

Condition 1 When the room is "Empty"

The initial condition of the DC loads is OFF state when room is empty. Zero face detection is occurred, indicating no person is present in the room; thus, relay will not be triggered on and the DC loads will remain in OFF state as shown in Fig. 14.

Condition 2 When the room is "Occupied"

When the face is detected, the DC loads should switch from OFF state to ON state indicating the room is occupied. Single face detection is occurred indicating one

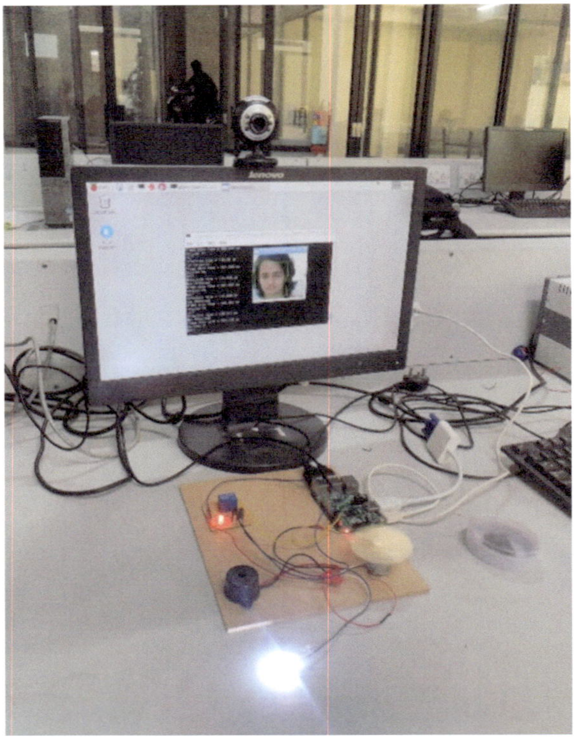

Fig. 15 DC loads are in ON state when there is single face detection

Table 1 Brief summary of results

S. No.	Room condition	Room status empty/occupied	Face detection YES/NO	DC loads ON/OFF state
1	No person is present	Empty	NO	OFF
2	People are present	Occupied	YES	ON
3	People are present, but room is not properly illuminated	Occupied	NO	OFF

person is present in the room; thus, relay will be triggered ON and the DC loads will be in ON state as shown in Fig. 15.

Thus, below table will give brief outlook about the discussion on results and analysis of the status of DC electrical loads according to the number of face detection occurred in real-time environment in Table 1.

Analysis on the capturing area of the surveillance through Webcam:

The distance of camera and the object (here face) to be detected can be estimated as mentioned in Table 2.

Table 2 Distance analysis for face detection using Webcam in OpenCV tool

Range	Low	Medium	High	Extreme
Magnification of Webcam	1×–3×	3×–10×	10x–30x	>30x
Distance	<3 m	3–10 m	10–100 m	>100 m
Impact on face detection	Detecting the face	Detecting the face	Detecting the face with some error	Not detecting the face

The table gives us information about how efficiently and with higher accuracy the face detection can be achieved by choosing proper Webcam magnification, its focal length, and based on this, at what distance it is working with minimum errors to detect the face of the person.

When the person is present in the range of Webcam which is less, i.e., the distance between the Webcam and the object (face) is below 3 m and the Webcam is set to magnification length of 1×–3×, face can be detected with highest accuracy. When the range is medium, i.e., between 3 and 10 m face detection is accurate, whereas the distance keeps on increasing, the accuracy of face detection goes on decreasing. When the distance is between 10 m and 100 m, face detection is achieved with some error. It also detects some non-facial features and also some ghost images appear while detection, which leads to improper face detection. Again when the distance between the Webcam and the face is increased i.e. greater than 100 m, no face will be detected, It means that the person is present out of the range of capturing area of Webcam. Thus, in OpenCV, one can detect the face of the person within or the below the 100 m range of the surveillance area.

6 Conclusion

This paper presents the design and implementation of the proposed human activity detection system using face-detection method which is a probable solution for maintaining the power usage at the desired surveillance area. The proposed solution can be the probable solution for the problem of power management. This system has succeeded over the shortcomings of previous proposed other technologies for maintaining power usage. Thus, we conclude that this automated human activity detection system will be the feasible approach to the problem of power management.

Acknowledgements We would like to thank the anonymous reviewers for their comments in improving the paper, and also, we extend our gratitude to VIT University, Chennai, for their support.

References

1. Ahmad F (2012) Managing electricity consumption using image processing as a tool. 2012 IEEE
2. Tathe SV, Narote SP (2013) Real-time human detection and tracking. 2013 IEEE
3. Mrazovac B, Bjelica MZ, Kukolj D (2012) System design for passive human detection using principal components of the signal strength space. 2012 IEEE
4. A proposal of user friendly alive human detection robot to tackle crisis situation.INB: 2012 12th international conference on control, automation and system (ICROS)
5. Garripoli C, Mercuri M, Karsmakers P, Soh PJ, Crupi G, Vandenbosch GA, Pace C, Leroux P, Schreurs D (2015) Embedded DSP-based telehealth radar system for remote in-door fall detection. IEEE J 1
6. http://developers.face.com/new-detector-demo
7. Deshmukh A, Wadaskar H, Zade L, Dhakate N, Karmore P (2013) Webcam based intelligent surveillance system. Res Inventy: Int J Eng Sci 2(8):38–42
8. Sribhagat Varma N, Taduri G, Bhagirath Sai N (2013) Automatic electrical appliances control based on image processing. Int J Eng Res Technol 2(9)
9. Fully-automated control of lighting and security system of a Room, EE389 Electronic Design Lab Project Report, EE Dept, IIT Bombay, Nov 2009

Embedded System for Classification of Upper Limb Movement During Action Using EEG

Navya Tummala, K. Venkatasubramanian and V. Umamaheswari

Abstract The role of the upper limb in human life is not limited to physical or functional movements but is accompanied with purposes like gestures, communication, and sensation. Many people lose their limb during an accident in the work place or complexity arising out of diseases or infections arising after operations due to which they are unable to perform the daily actions using hands. A healthy brain receives the signals from its body and generates responses to it. Here, a method is devised to detect which part of the upper limb is about to move, and hence, the desired output could be used to control external devices like a robotic arm and for the purpose of neural prostheses.

Keywords Electroencephalogram · Phase synchrony · MATLAB
BeagleBone Black · Simulink

1 Introduction

Millions of people in the world are amputated every year due to diseases, industrial injuries, traffic accidents, and accidental injuries. When a person loses a limb, he or she faces staggering emotional and financial lifestyle changes. He or she then requires a prosthetic device and service which could become a life-long support. The replacement of a missing body part by an artificial body is called as prosthesis. The need for an artificial limb may be due to loss of upper limb during accidents or a disease.

N. Tummala (✉) · K. Venkatasubramanian · V. Umamaheswari
School of Electronics Engineering, VIT University, Chennai, India
e-mail: tummalanavya17@gmail.com

K. Venkatasubramanian
e-mail: venkatasuramanian.k@vit.ac.in

V. Umamaheswari
e-mail: umamaheswari.t1@vit.ac.in

D. Thalmann et al. (eds.), *Intelligent Embedded Systems*, Lecture Notes
in Electrical Engineering 492, https://doi.org/10.1007/978-981-10-8575-8_24

The upper limb can become dysfunctional if there is a heart-stroke or paralysis for a person. In a country like US, approximately 1.6 million are being operated due to the loss of the upper limb each year. Due to accidents by a lawn mower, nearly 9000 children are being operated for the upper limb. Even some children are born with the defect from birth, and hence, they need the prosthetic device from young age itself. People working in the army are the most affected as accidents occurrence is common in battlefield. There are more than 40,000 people who are being operated for the upper limb.

Different movements of the motor imagery of the same limb are very challenging and difficult as the imagined movements have their dimensional representation on the motor cortex area of the brain. It has been proved that the actions are linked to the thinking process. The observing, imagining, and preparing about an action are bonded to the execution of the same action. The region of the brain that is responsible for executing an action is basically activated just by looking at the action and observing it. An evidence also proved that several regions of the brain can get activated while an action is being generated or just by observing the action of other people.

2 Literature Survey

Brain–computer interface (BCI) is a system which provides the connection between the computer and the brain. It transforms the neural activity of the brain to command that can be used to drive external applications. Thus, a BCI is usually a pathway from the brain signals to the real-time applications [1, 2].

The studies show that the recordings of neural activities from the brain give information about the hand movements which are voluntary in nature. Different methods have been researched and implemented to make this as a real-time BCI application. Various algorithms have been used for the classification purpose [3, 4].

In one of the papers, a wavelet-common spatial pattern algorithm was introduced to extract the features and classify them. Here, EEG signals from seven healthy subjects were captured. The subjects were asked to perform right-hand movement in four different directions voluntarily. The data collected had a cutoff frequency of 0.05 Hz, and it was sampled at 125 Hz. The subject had to hold a robot and perform the right-hand movements.

The robot then recorded the required details like position, speed, and other parameters. The center point is the start point, and the other circles on the circumference are the target points. The recorded data was low-pass filtered at 96 Hz.

3 Background

The electroencephalogram (EEG) is a plot of neural activities from the brain which is a time series data of stimulated potentials [5–11]. To record the data from the neural activity of the brain, the data is collected by placing the electrodes on the scalp and a graph of voltage versus time is plotted. The amplitude of the EEG signal is represented by the voltage. Generally, the voltage of an EEG signal varies between 10 and 100 µV, whereas in adults, it varies from 10 and 50 µV. For the classification purpose, the frequency range required is from 0.1 to 100 Hz. The frequency range is divided into several frequency components as below:

- Delta rhythm (0.5–4 Hz): This band has a frequency of 4 Hz or below. This is known to be highest in amplitude and has the slowest waves.
- Theta rhythm (4–8 Hz): This band has a frequency of 4–8 Hz and is usually classified as a slow activity.
- Alpha rhythm (8–13 Hz): This band has a frequency of 8–13 Hz and is majorly seen in normal relaxed adults.
- Beta rhythm (13–30 Hz): This band has a frequency of 13–30 Hz, and this type of rhythm appears when the activity is fast.

4 Methodology

The basic step for the brain–computer interface is the acquisition of the brain signals. These include various brain signals such as the electroencephalogram which are obtained by placing the electrodes on the scalp of the brain. Other signals include the electrooculography and magnetoencephalogram. This deals with the magnetic signals from the brain. The most widely used method of all these is the EEG which is a noninvasive and a cost-effective method. But it is more sensitive to noise and other artifacts. Signal acquisition and signal processing is shown in Fig. 1.

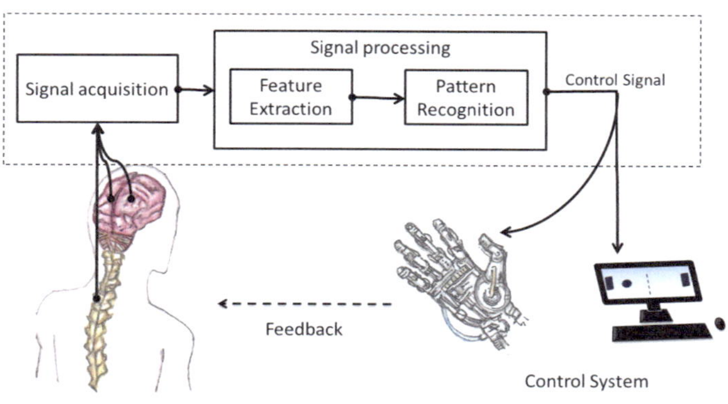

Fig. 1 Signal acquisition and signal processing

4.1 Hardware Specification

BeagleBone Black:

A high-speed microprocessor capable of running an operating system (OS) in the design was highly desirable for signal processing in software. An embedded processor for this requirement was highly recommended. BeagleBone Black meets all the required specifications. The dimensions of the BeagleBone Black are $3.4' \times 2.15'$. It is a single-board computer, and it has the ports like HDMI, USB, and Ethernet. It has the ARM Cortex-A8 of 1 GHz. The RAM capacity is 512 MB, and it also has a flash memory of 4 GB with a support of micro-SD card. The external interface pins are 92 I/O, and it is capable of interacting with devices that would normally require a deterministic microcontroller, due to its two independent on-chip Programmable Real-time Unit Subsystems (PRUSSs). Thus, the Beaglebone is a low-power open-source hardware which is produced by Texas Instruments. The main processor core runs at 700 MHz with a 256 M RAM. All these features make it a powerful machine. It can run Linux, Python, SSH, etc.

EEG Electrodes:

The electrodes consist of a small metal disk which could be made of stainless steel, gold, silver, or tin which has to be covered with a coat of silver chloride. These could be placed at different positions of the scalp. These positions are already specified by the 10/20 electrode system described below. The electrode position on the scalp is labeled with a letter and a number. The letter corresponds to the area of the brain under it. For the frontal lobe, it is specified as F, and for the temporal lobe, it is specified as T. The numbers correspond to the right side or left side of the brain. The even numbers correspond to right side, and the odd numbers correspond to the left side of the brain.

5 Implementation

The basic step of the project is to acquire signals from the brain. Here, the BeagleBone Black board is used to acquire signals from the scalp of the brain. The Simulink is used as the signal processing toolbox for pre-processing and classifying the signals.

The four positions on the scalp, i.e., FC5, FC6, C3, and C4 positions are used to capture the signals and is shown in Fig. 2. These are all in the motor cortex area and hence are selected for experiment purpose as the motor cortex is mainly responsible for voluntary movements in the body. Then, the whole dataset is saved to the workspace in MATLAB. In MATLAB, the pre-processing of the signals is done. To study the signals obtained, the parameters are calculated like energy, variance, skewness, and kurtosis.

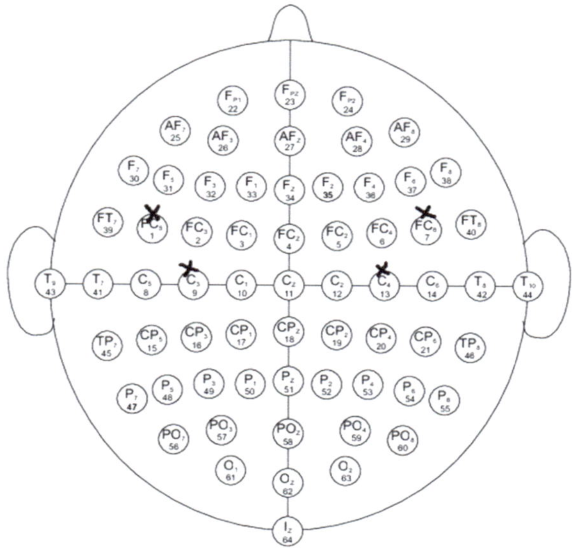

Fig. 2 Positions on the scalp to capture signals

Then, the output obtained from this processing is used to control the external devices such as in neuroprosthesis (Figs. 3 and 4).

Before building the circuit, configuration parameters have to be set according to the BeagleBone Black. Circuit was built using the library function. Sampling rate and frame rates have to be specified manually. After the completion of circuit, it is built and dumped into hardware (Fig. 5).

The experimental setup with the subject is as shown in Fig. 6.

The energy plots for different signals have been plotted but nothing could be concluded from the graphs.

The offline database for the EEG was collected from the physionet.org Web site. The details of the database are explained below. Various parameters have to be calculated to study the database and come to certain conclusions (Fig. 7).

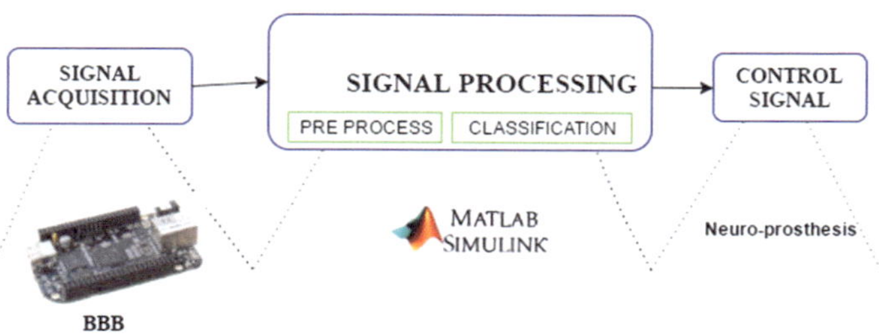

Fig. 3 Signal acquisition from BBB and processing in MATLAB

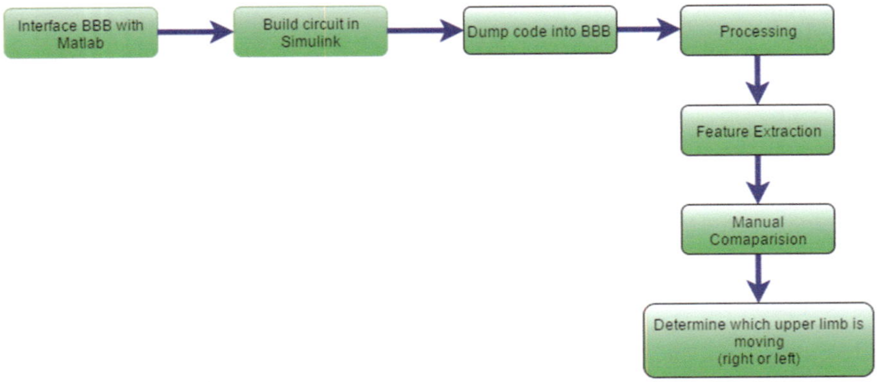

Fig. 4 Hardware block diagram

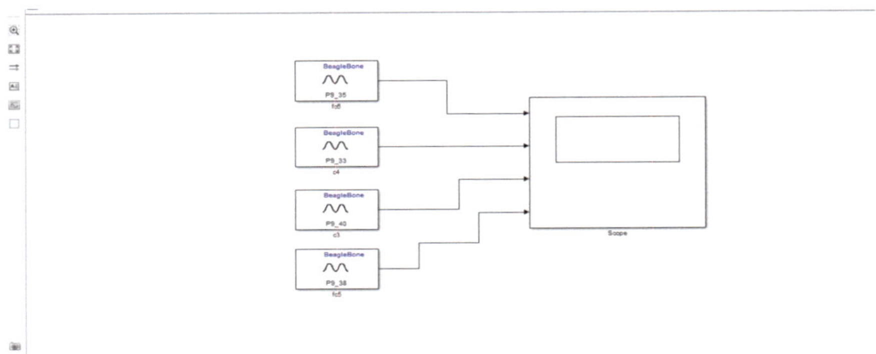

Fig. 5 Circuit for capturing signals from BBB

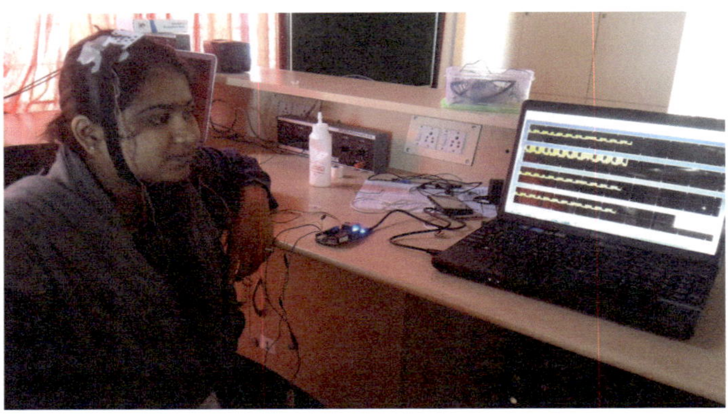

Fig. 6 Subject performing experiment

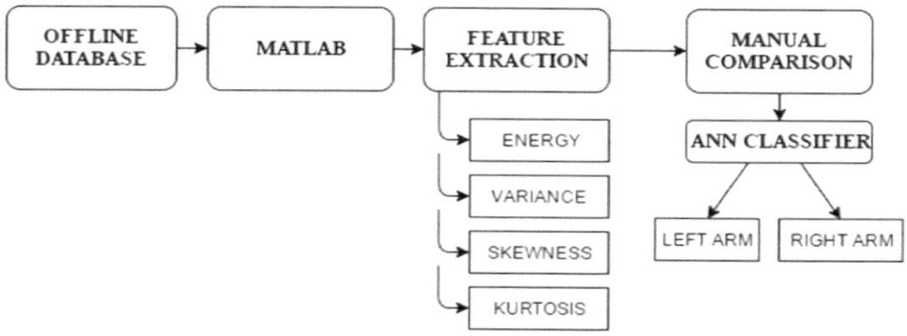

Fig. 7 Parameters from offline database

The developers of BCI2000 instrumentation system created the EEG dataset and contributed it to Physionet. There are more than 1500 EEG records in the dataset obtained from 109 subjects. The dataset is available in the Web site http://www.physionet.org/pn4/eegmmidb/.

The duration is around 1 or 2 min per record. The subjects were asked to perform various imagery tasks. The electrode cap of 64 electrodes was used to conduct the experiment. Fourteen experimental trails were done by the subjects.

The primary considerations for developing any prediction model for detection and classification are the type of features to be extracted from the EEG input signal. These features are then analyzed to find out the most relevant and effective features required. The other features are discarded. The following features were calculated by processing the EEG database:

(a) **Energy**: One of the important feature extractor is the energy parameter. It is the average signal amplitude of the EEG signal.

$$E(n) = \frac{1}{N}\sum_{n=1}^{N} x(n)^2$$

(b) **Variance**: It is the measure of signal spread.

$$\sigma^2 = \sum_{i=1}^{n} \frac{(x_i - \mu)^2}{N}$$

(c) **Skewness**: It is used to measure the asymmetry of signal values around its mean relative to a normal distribution.

$$\text{skew} = E\left[\frac{(x-\mu)^3}{\sigma^3}\right]$$

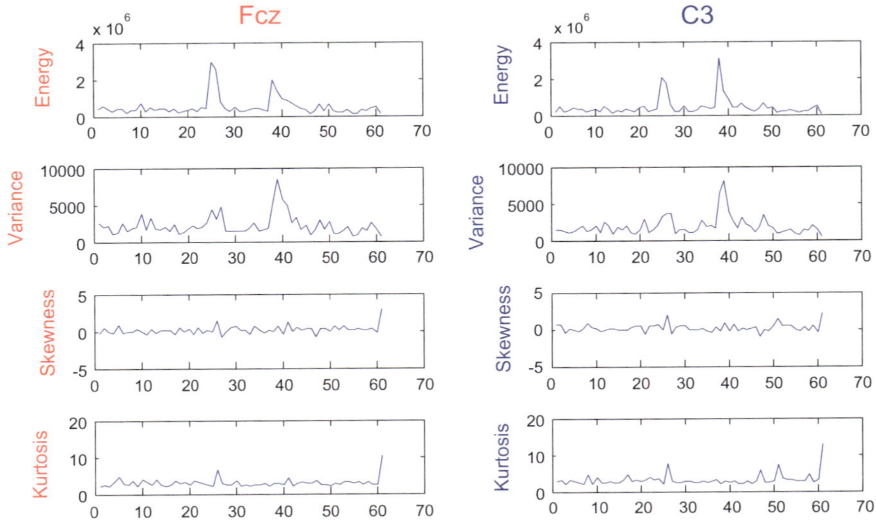

Fig. 8 Plots of Fcz and C3 positions

(d) **Kurtosis**: It is used to measure of the degree of peak of a distribution of signal values relative to a normal distribution.

$$\text{kurt} = \sum \frac{(x-\mu)^4}{\sigma^4} - 3$$

The parameters calculated are plotted as in Fig. 8.

The observations that could be concluded manually from the above graphs are as follows:

- It can be observed that during the time interval 25–30 frame period and 35–40 frame period, the energy plot has peaks which indicates that an action using the upper limb has been done during this period (either left or right).
- As the motor cortex is responsible for voluntary movements, the electrodes FCz, C3, and C4 can be observed for the movement of upper limb.
- It is also observed that the variance is maximum for the left trails, whereas it is minimum for the right trails.
- For right-hand movements, the electrode C4 has to be observed and the variance is low compared to the C3 electrode.
- The kurtosis plot is used to determine the imagined movement by the person. It can be observed that in the C3 plot, in the 30–40 frame period, the peaks change from more to less, whereas in the C4 plot, the peak changes from less to more. This indicates that the person is imagining movement in the right hand.

After capturing the EEG signals in real time, the data was processed and the parameter energy was calculated. These are as shown below. But nothing could be concluded from the graphs as the person might be thinking about various things and could have not been concentrating on the required experiment. A method called phase synchrony has been devised where the phase difference between two signals has been calculated.

Phase locking value (PLV) is a measure of synchronization in the time domain. This concept is applied for analysis of EEG signals recorded during performing a motor imagery task. For pre-processing, a band-pass filter is used for each signal between 0.5 and 30 Hz.

The phase locking value (PLV) is to be computed for the possible pairs of electrodes. In this case, the possible combinations are:

- Fc5-C4
- Fc5-Fc6
- C3-C4
- C3-Fc6

This concept is under process and thus could be used to classify the left- and right-hand movements.

6 Conclusion

This project included various experiments to be conducted to study the EEG signals from the brain related to the upper limb movements. Various parameters were calculated to study the signals acquired. But a definite conclusion could not be devised.

So a method called phase synchrony is being processed which determines which side of the upper limb is moving, i.e., right or left. The phase locking value is calculated for a pair of signals. So if there are four electrodes placed on the scalp, then there would be four different PLV values. After calculating these values, a method is followed which determines which side of the upper limb is moved. Thus, this project could be used for neuroprosthesis.

Acknowledgements I am extremely thankful and indebted to my project guide Prof. K. Venkatasubramanian for sharing expertise, sincere and valuable guidance and encouragement extended to me.

I would also like to thank the anonymous reviewers for their comments in improving the paper, and I also extend my gratitude to VIT University, Chennai, for their support.

References

1. Robinson N, Guan C, Vinod AP, Ang KK, Tee KP (2013) Multi-class EEG classification of voluntary hand movement directions. J Neural Eng
2. Barański R, Kozupa A (2014) Hand grip-EMG muscle response. Acoust Biomed Eng
3. Sun S, Zhou J (2014) A review of adaptive feature extraction and classification methods for EEG-based brain-computer interfaces. In: IEEE transactions on biomedical engineering
4. Planelles D, Hortal E, Costa Á, Úbeda A, Iáñez E, Azorín JM (2014) Evaluating classifiers to detect arm movement intention from EEG signals. Sensors
5. Wang Y, Makeig S (2009) Predicting intended movement direction using EEG from human posterior parietal cortex (PPC). J Neural Eng
6. Tan HG, Shee CY, Kong KH, Guan C, Ang WT (2011) EEG controlled neuromuscular electrical stimulation of the upper limb for stroke patients. Front Mech Eng
7. Blokland Y, Spyrou L, Thijssen D, Eijsvogels T, Colier W, Floor-Westerdijk M, Vlek R, Bruhn J, Farquhar J (2014) Combined EEG-fNIRS decoding of motor attempt and imagery for brain switch control: an offline study in patients with Tetraplegia. In: IEEE transactions on neural systems and rehabilitation engineering, vol 22, no. 2, March 2014
8. Kähkönen S, Komssi S, Wilenius J, Ilmoniemi RJ (2005) Prefrontal TMS produces smaller EEG responses than motor-cortex TMS: implications for rTMS treatment in depression
9. Leeb R, Scherer R (2004) Navigation in virtual environments through motor imagery. Institute of Computer graphics and vision
10. Hassan M, Seddik AF (2008) Classification of the imagination of the left and right hand movements using EEG
11. Loboda A, Margineanu A, Rotariu G, Lazar AM (2014) Discrimination of EEG-based motor imagery tasks by means of a simple phase information method. Int J Adv Res Artif Intell 3 (10)

Intelligent Digital Signage System Based on Gender Identification

Riya Elizabeth Abraham and M. Robert Kennedy

Abstract This paper proposes an intelligent transformation of digital signage systems by making it more audience interactive. The increase in flexibility and enhancement of digital signage display system can be done by providing optimized information and appearance attractive multimedia content through the signage system. This emphasizes more on the advertisement industry especially in public spaces like hotels, airports. The system has been designed to broadcast the advertisement on the signage display system based on the demographic features like gender of the observer. Real-time computer vision algorithms are applied to provide an observer-specific advertisement broadcast on the display system.

Keywords FERET database · Demographic features · Content management Prediction model

1 Introduction

The digital signage display is an important and efficient medium for providing information and hence acts as an effective interface for providing dynamic and attractive multimedia content to the customers. The significance of signage system in the advertisement industry increases its demand to be more observer-interactive and information optimized. Dynamic broadcast of the advertisement in real time allows for full context and audience adaptation. To completely exploit the high potential of the digital signage displays, the displayed content must be made non-generic and more interesting for observers. The major challenge is to make digital signage a more effective information interface. To make the audience-adaptive signage displays, the

R. E. Abraham (✉)
School of Electronics Engineering, VIT University, Chennai, India
e-mail: riyameliza@gmail.com

M. Robert Kennedy (✉)
Qmax Systems India (P) Ltd, Chennai, India
e-mail: robert@qmax.com

D. Thalmann et al. (eds.), *Intelligent Embedded Systems*, Lecture Notes in Electrical Engineering 492, https://doi.org/10.1007/978-981-10-8575-8_25

contents need to be viewed on the basis of their interests. Analysing the demographic features of the observers before the broadcast and displaying the content accordingly can provide targeted information to them.

The proposed system interacts with the demographic feature like gender of the observer and displays corresponding advertisements to the viewer based on their interests. Automatic human gender recognition is the main aim followed by simultaneous advertisement display based on the identified gender from the database. Facial image-based gender recognition is done based on the different facial features in order to recognize whether a given facial image is male or female resulting in a two-class classification problem.

The facial features of the observer are determined and extracted by using several computer vision methods. These features are preprocessed and converted into a model which is then used to classify the gender of the person. Once the gender is determined, the corresponding favorable ads get displayed to the signage system based on the gender. Though more complex hardware enhancements can be made available for better and more accurate results, this system mainly aims at a simple and cost-effective implementation. Also it is a more rational method since the displays with built-in monocular cameras are accessible widely. An interactive digital signage system is designed on the basis of efficient and real-time capable computer vision techniques. The block diagram of the proposed system is shown in Fig.1.

Engaging the users with more interactive content by displaying the set of advertisements in which they are interested based on their gender can reduce the interaction blindness [1], where there is no interaction with the displays because of

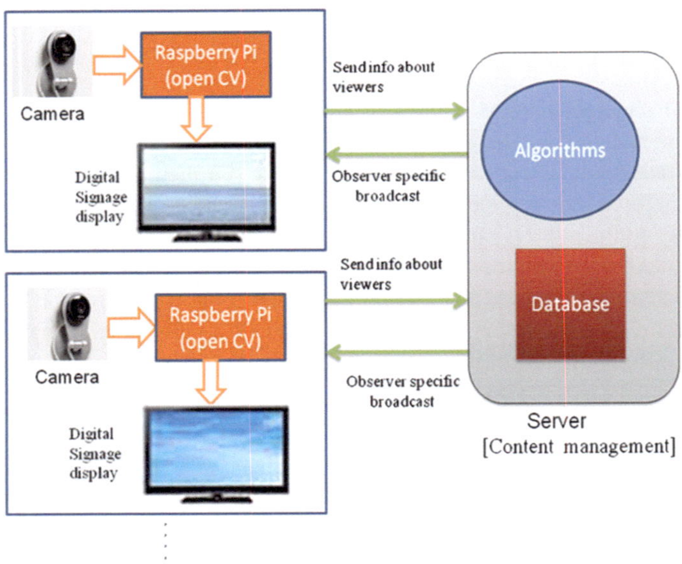

Fig. 1 Block diagram of the proposed system

poor interaction process. Automated facial gender recognition has become an interesting and challenging research recently. It can be applied to many potential application fields like biometrics, demographic data collection which demands more computations. Several novel feature extraction techniques are being used for the identification of gender like eye and eyebrow region extraction [2], geometric-based methods like distance between eyes, face length and width, and appearance-based method where the whole image is considered rather than local features corresponding to different parts of the face [3].

The gender classification is used to increase the user satisfaction by providing individualized services [4]. The input test images are analyzed and processed for testing the accuracy of the computer vision algorithms. The images contain variations in lighting and facial expression, pose angles, and aging effects [5] which can create issues in the prediction of gender. Proper training of images and efficient computation can make the prediction in a better way. In real life, varying illumination is the major challenge for gender classification. Illumination normalization needs to be performed in order to increase the recognition performance illumination. Both gender recognition algorithm training and testing require huge enough color image database [6].

Once the gender prediction is done, broadcasting the appropriate advertisement displays can be done using database operations. The database sends the query to display the particular advertisements to attract the observer, hence helping the advertisement industry to be smooth going. The best-identified classification approaches are adopted using the OpenCV for the real-time application [7]. The Front-end basic flow diagram is shown in Fig. 2.

2 Experimental Setup

The primary aim of the work is to implement the observer-specific display of advertisements based on their gender. The automatic gender recognition is done by using OpenCV algorithms in Python language in Raspberry Pi. The real-time implementation of the gender prediction was done using OpenCV library. A computer display enhanced with a Logitech Webcam camera is used as broadcasting unit prototype. The real time hardware setup is shown in Fig. 3.

2.1 Raspberry Pi

For making the system cost effective and reasonable, Raspberry Pi is used as the main hardware. The entire system is controlled by the Raspberry Pi. The Raspbian OS is used as the platform on which open computer vision is installed and the computations are made. It acts as the interface between both ends where the gender prediction technique is done and the advertisements are displayed in Fig. 4.

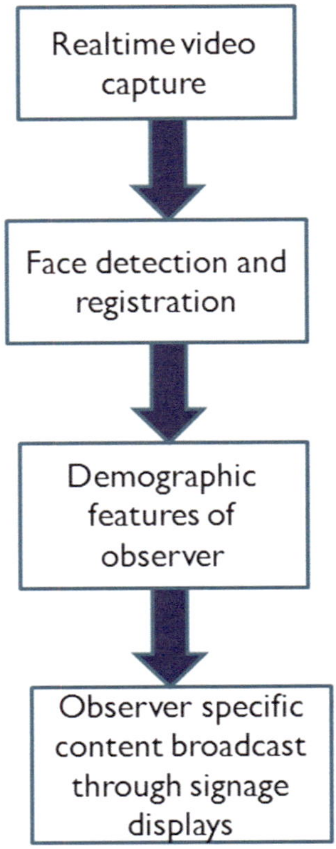

Fig. 2 Front-end basic flow diagram

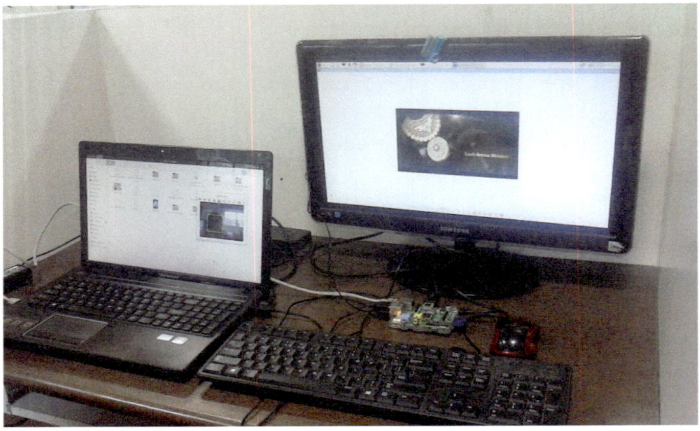

Fig. 3 Real-time hardware setup

Fig. 4 Raspberry Pi

2.2 *Dataset Description*

The image dataset called FERET database is used in the paper. This database was created to develop, test, and evaluate face recognition algorithms. A set of around 994 images are used in the work for representing male and female genders. The images from the database are stored in a set of folders. A prediction model is computed and trained on the set of the images stored to perform the gender classification. The images are read from a given path and are resized and processed to obtain a model which is then used for predicting the gender. The dataset sample and the training images from the dataset are shown in Figs. 5 and 6.

Fig. 5 Dataset sample

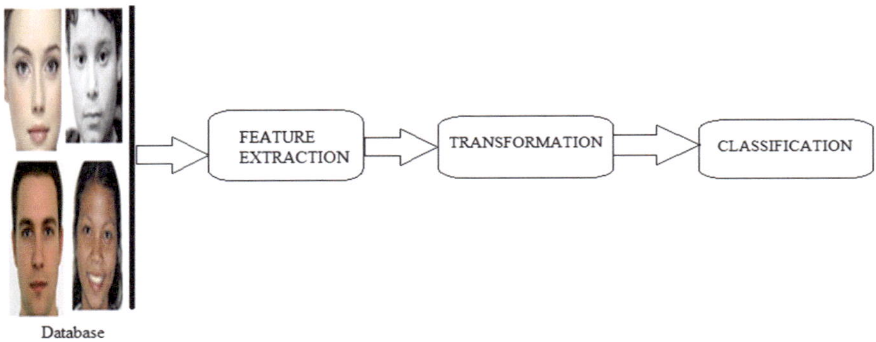

Fig. 6 Training images from the dataset

3 Face Detection

For optimal determination of the features of the observer, many computer vision methods are combined together. One among them is the face detection. Real-time video of the observer is taken using the camera, and the detection of face is done using the Haar cascade classifier.

3.1 Haar Cascade Classifier

It is originally an effective object detection method by Viola and Jones. In this, training of a cascade function is done from a lot of positive and negative images. This function detects the objects in other images. In training the classifier, the positive images will be the ones with faces and negative images will be those

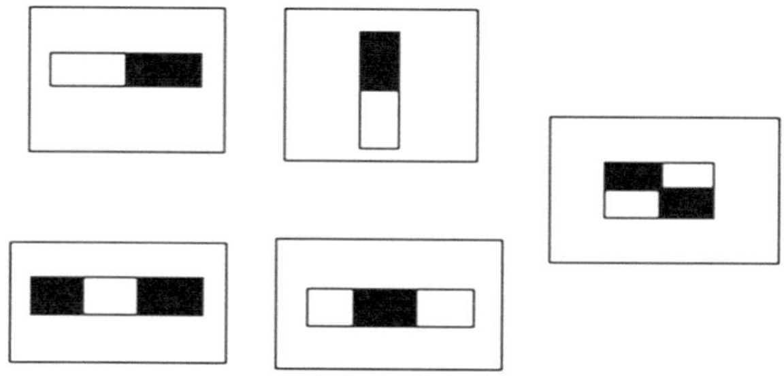

Fig. 7 Haar features

without faces. The features are then extracted from it. Each Haar feature is a single value obtained by considering the adjacent rectangular regions at a specific location in a window, and by summing up the pixel intensities in each region and calculating the difference between these sums. A window is moved over the input image during the face detection, and the Haar-like feature is calculated for each subsection. Comparing this difference to a learned threshold, which separates non-objects from objects, is done and is shown in Fig. 7.

The major limitation for this in real time is the varying illumination. Hence to increase the performance in detection, normalization is performed. The further processing is done once the face of the observer is detected.

4 Feature Extraction

The gender identification method is mainly based on the combination of feature extraction and classifier. The images in the dataset are learned and computed by extracting the features of the face. The images are read from a given path and resized. This returns images as a list of numpy arrays and their corresponding labels. The feature extraction method used in this paper is the Fisherfaces method. The dataset is loaded and a model is created using computer vision algorithms, which is then used for the prediction.

4.1 Fisherface Technique

The algorithm which was proposed by Belhumeur, Hespanha, and Kriegman uses both principal component analysis and linear discriminant analysis to produce a subspace. It is a combined algorithm of PCA and LDA. In PCA, the total variance in data is maximized. It does not consider any classes and hence does not contain any discriminative information in the components identified. Thus, LDA is adopted along with that to make the classification possible. Unlike Eigenface method, this method has the advantage in minimizing variation within each class while maximizing class separation [8]. In this method, the ratio of the between-class scatter to the within-class scatter is maximized. The projections are thus shaped by this result where the distances between samples of the same class are at a minimum while those between the classes are at maximum.

Algorithmic Description

- An image matrix is constructed with each image represented in each column where each image is assigned to a class.
- The image matrix is then projected into the subspace with the rotation matrix identified by PCA.
- The between-class scatter and the within-class scatter of the projection are then calculated.
- Their ratio is then maximized using LDA.

5 Classification

Before computing the data and labels for learning on the given image and their labels, the nearest neighbor classifier method is adopted for the classification into different classes. The feature and the classifier together form a predictable model, and it does the feature extraction and learns the classifier. The classifier is trained so as to output a list with predictable label and the classifier output on which the decision is based on.

5.1 K-Nearest Neighbor (K-NN) Classification

In this method of classification, Euclidean distances are measured in image space and the label of the closest point in the learning set is assigned for the data to be classified. The proximity between the data points is often determined by the Euclidean distance metric. It is the distance between all pixels in a dataset. It is based on the assumption that the objects which are close in distance are similar. The given testing data must be correlated with each image in the training set to reduce the computational time. Once the classifier is trained, the input image is analyzed and processed to get the classifier output which is the prediction value.

All pixels in a dataset are separated by a distance called Euclidean distance. The number of neighbors, i.e., k values is then found out from the distance matrix. These data points are then analyzed, and the most common class label is then determined among the set. This common class label is given to the data point being analyzed. Hence the classification is done.

6 Prediction

The images from the database are processed and converted into a model using the feature extraction method and the classifier. The trained classifier hence gives the output of the predicted label. Several restrictions are faced for the accurate determination of the gender due to the light illumination, variation in pose angles, etc. The FERET database is used to train the classifier of gender. The model can be reused easily, and hence learning it from the dataset all over again is not required. Thus, the processing and computation time period can be reduced.

7 Advertisement Broadcasting from Server

The camera captures the digital images in real time, extracting the demographic features of the observers. After predicting the gender of the viewers, the system automatically selects and broadcasts the content relevant to the specific detected observer from the server. The information targeted to both males and females is stored in the server and then displayed correspondingly to the digital signage display. Once the prediction is done, the advertisement images are retrieved from the mysql database in which they are stored. The insertion and retrieval of images from the mysql database are done using different table operations in Python language. Creation and deletion of tables, inserting images into database, and retrieving them from the server of specific IP address into the display system are performed using Python.

8 Results and Discussion

The primary aim of the system to provide observer-specific broadcast of the advertisement content is performed in a cost-effective way. This helps the advertisement industry to be more flexible with respect to the customer. The prediction of gender using the computer vision algorithms was tested using several testing images and found error rate of around 0.2%. This is shown in Figs. 8 and 9. The FERET database was used as the dataset for training and testing (Figs. 8 and 9).

The real-time classification of the gender is done with the camera connected to the Raspberry Pi. The image is read from the camera here instead of an image. Initially, no computed model would be available for classifying the gender. It is to be trained from the image set in the database. The image data from the dataset are read and processed, and the model is computed. The images are passed as numpy arrays. Once the prediction model is created, images are grabbed from the Webcam in real time and is shown in Fig. 10. The model created can be easily reused so that the computation process time can be saved.

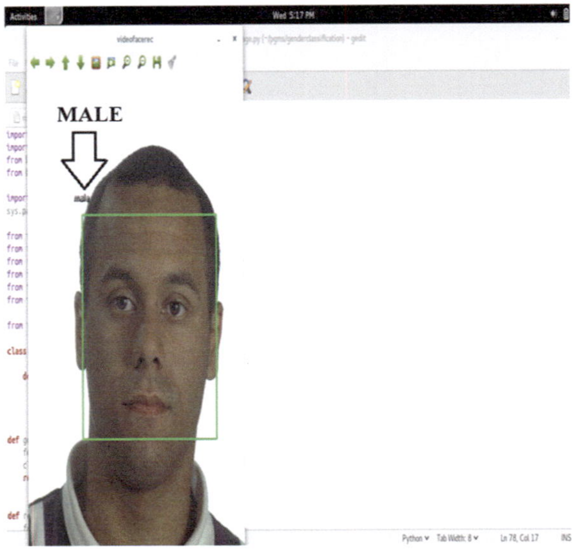

Fig. 8 Testing of male image

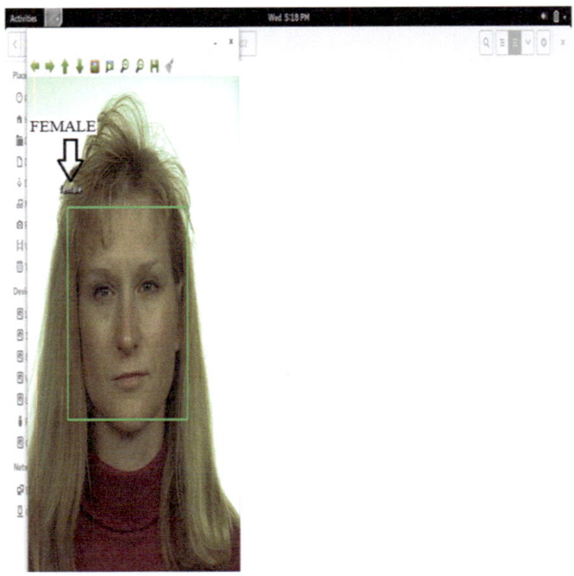

Fig. 9 Testing of female image

The identification of gender is followed by the corresponding advertisement display based on the gender. The contents to be displayed based on the demographic features are present in the server which is configured using an IP. The display enhanced by the camera analyzes the observers through their facial

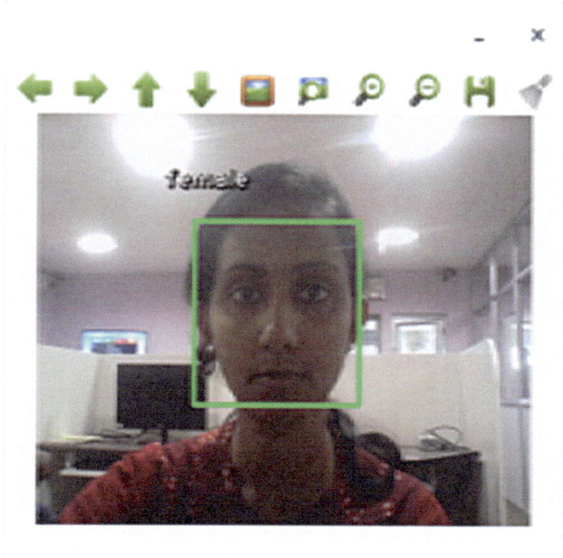

Fig. 10 Gender identification from real-time video

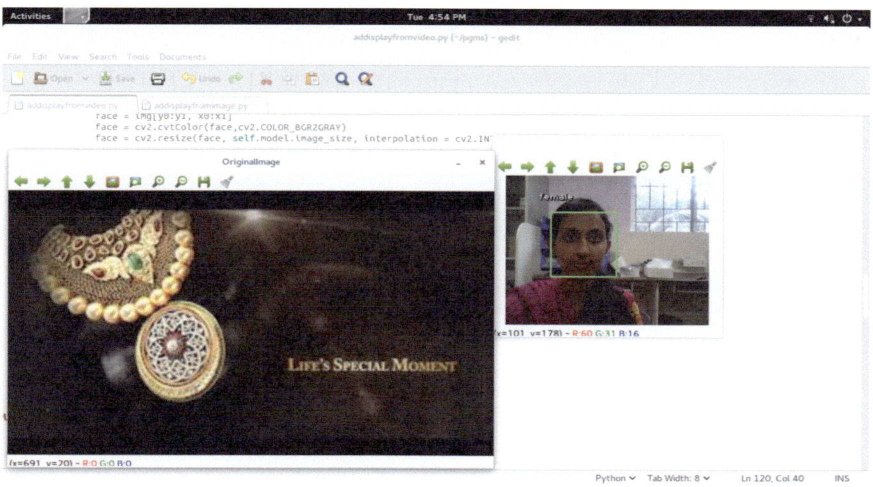

Fig. 11 Real-time ad display based on gender

characteristics and predicts the gender. Real-time adaptive content is then broadcasted in multiple locations as shown in Fig. 11.

More features can be included in the application to make the digital signage systems more flexible and adaptive to the observers. The gender classification is used to increase the user satisfaction and demand of the advertisement industry.

9 Conclusion

This paper presented an approach in the real-time classification of gender of the observer and advertisement broadcast on the digital signage display based on the gender. Identifying the demographic features of the observer was the major challenge. This was implemented using the various computer vision algorithms in Python language in Raspberry Pi. The application can be improved by analyzing more demographic features of the observer like age. Tracking the movement of the hands of the observer is another major research area which can make the system more flexible and efficient. The use of more complex algorithms can help in the observer tracking and proper classification. The cost-effective and adaptive nature of the system makes it more efficient along with the real-time capable computer vision algorithms.

References

1. Ravnik R, Solina F (2013) Interactive and audience adaptive digital signage using real-time computer vision. Int J Adv Rob Sys. 10(2):107
2. Alrashed HF, Berbar MA (2013) Facial gender recognition using eyes images. Int J Adv Res Comput Commun Eng. 2(6):2441–2445 June
3. Bera S (2014–2015) Gender recognition from facial images using support vector machine; Shirkey DM, Gupta SR (2013) An image mining system for gender classification & age prediction based on facial features. Int J Sci Mod Eng (IJISME) 1(6). ISSN:2319-6386
4. Basha AF, Jahangeer GSB (2014) Exploring a novel method for face image gender classification using random forest and comparing with other machine learning techniques. Int J Comput Sci Issues (IJCSI) 11(6):2
5. Zhang C-Y, Ruan Q-Q (2010) Face recognition using L-fisherfaces. J Inform Sci Eng 26: 1525–1537
6. Ravi S, Wilson S (2010) Face detection with facial features and gender classification based on support vector machine. Int J Imaging Sci Eng 23–28
7. Singh V, Shokeen V, Singh B (2013) Comparison of feature extraction algorithms for gender classification from face images. Int J Eng Res Technol
8. Belhumeur PN, Hespanha J, Kriegman D (1997) Eigenfaces vs. fisherfaces: recognition using class specific linear projection. IEEE Trans Pattern Anal Mach Intell. 19(7):711–720

Speech Recognition System Using Open-Source Speech Engine for Indian Names

Nitin Arun Kallole and R. Prakash

Abstract Speaker independence, continuous speech and huge vocabularies create most of the greatest challenges in automatic speech recognition. This paper describes Sphinx, a library that offers the feasibility of accurate, huge vocabulary, speaker-independent, continuous speech recognition. Using speech for device control is a proven hands-free solution. There are several products that use speech input for hands-free control. They usually cater the users with US/UK accent. In this paper, a speech recognition system is developed for the application of hands-free control system to be deployed in automotive environment for Indian users. This paper demonstrates the methodology and the challenges of customizing an open-source speech recognition engine for Indian users. It is demonstrated for the application of speech-based control of smartphone, and rear-view mirror rotation. Open-source package used is Pocketsphinx for speech recognition and festival for text-to-speech and pronunciation generation. All the implementations are done on a single-board computer, i.e. raspberry pi.

Keywords Voice signal · Automatic speech recognition · Sphinx Vocabulary

1 Introduction

Voice is one of the most important and spontaneous parameters of the human being. It is known as the individual property of everyone in the world. Each person will have their own voice characteristics. The information contained in the voice is linguistic, vocal information and emotion. Its frequency range is from 30 to 3000 Hz.

N. A. Kallole · R. Prakash (✉)
School of Electronics Engineering, VIT University, Chennai, India
e-mail: r.prakash@vit.ac.in

N. A. Kallole
e-mail: neeteen09@gmail.com

D. Thalmann et al. (eds.), *Intelligent Embedded Systems*, Lecture Notes in Electrical Engineering 492, https://doi.org/10.1007/978-981-10-8575-8_26

Recently, many researches also happened on the speech processing so that it is being truly usable. These progresses in the speech recognition domain developed so many techniques such that many works in our daily routine are done based on user voice commands. Recently available advancement includes capabilities like voice-enabled SMS, email and even navigation with voice.

Voice recognition has nowadays intensively popular and wide range of applications [1–5]. It reduces efforts as well as gives comfort to the user. For example, the person, who wants to the call will be made somebody else from his mobile phone, can just speak his name and call will be forwarded. It also plays an important role like controlling, serving as a median for physically handicapped, low mobility peoples [6]. Though it is popular, several technical challenges have slowed down the exploitation of such applications on embedded devices. The most difficult of these is the computational requirements of speech recognition for a medium-to-large vocabulary scenario. Finally, it requires quick and lightweight speech recognition.

In this paper, an automatic speech recognition system is introduced, i.e. based on speech recognition system, which will be helpful to capture and recognize the words spoken by speaker who is speaking to the microphone. It takes user commands in the form of voice signal and controls respective accessories automatically. The basic goal is to create hands-free control of some of the functions in the automotive environment, taking direction from the user's voice. The open-source software called Pocketsphinx is used, which depends on another library named sphinxbase. It is a version of Sphinx that can be ported and used in various embedded platforms.

It was developed at Carnegie Mellon University (CMU), with the objective of creating a speech recognition system for resource-constrained devices, such as hand-held computers [7]. Sphinx is a continuous speech and speaker-independent recognition system developed in 1988 at the Carnegie Mellon University (CMU) in order to try to overcome some of the greatest challenges within speech recognition, i.e. speaker independence, continuous speech and large vocabularies [8]. CMU Sphinx is a common methodology used to describe the group of the speech recognition systems. It makes use of hidden Markov model (HMM) and Mel-frequency cepstrum coefficients.

CMU Sphinx has a number of software systems as,

1. Sphinx
2. Sphinx 2
3. Sphinx 3
4. Sphinx 4
5. Pocketsphinx

Pocketsphinx, it is an embedded application tool. The entire system is written in C with the aim of having fast response and lightweight applications. For installing Pocketsphinx on single-board computer, there are some prerequisites that are to be installed in the system, e.g. gcc compiler, python, libasound dev, alsa utils, bison.

One of the main advantages of Pocketsphinx over other ASR systems is that it has been ported and executed successfully on different types of processors and several ARM processors [7]. After downloading and compiling the source code for Pocketsphinx, a test program using the default language model, acoustic model and dictionary was tested.

2 System Overview

The main aim of this paper is to develop a speech recognition system for controlling rear-view mirror and a voice call application. Latest versions of sphinxbase and Pocketsphinx, i.e. sphinxbase-0.8 and Pocketsphinx-0.8, are used. All processing are done on Linux-installed single-board computer systems.

To make a call, Android debug bridge (adb), i.e. a flexible command line tool, is used to communicate with a connected Android-powered device. A user will speak the phone number or contact name to the system as an input voice command, and that numbers/name will be recognized. To detect this mobile number/contact name, Pocketsphinx must be updated with the dictionary file, which consists of contact names and its phoneme representation. In other words, it must know the pronunciation. Contact name's file is retrieved from the user's phone. The phonetic dictionary will be generated consisting of pronunciations of these contacts. The recognized words/numbers will be shown on the mobile screen, and after that the phone call will be made automatically.

Each time dictionary file contents may vary for the different user. So dictionary file must be updated after receiving contact list file from the user phone. Festival software is used to generate phonemes of these Indian names.

Another application is to have control over rear-view mirror control and indicator. As described above, user has to utter command, e.g. up/down/right/left to choose and control one of the two mirrors.

3 Proposed Methodology

The methodology used in this paper is as shown in Fig. 1. The single-board computer, i.e. raspberry pi, is loaded with Pocketsphinx and festival software.

The functional behaviour is as shown in Fig. 2. Figure 2 shows the detailed block diagram of the complete system. The PC will have its own input to use headphone for giving input/output, while raspberry pi board have 4 USB ports. To connect headphone to the raspberry pi, it is essential to connect extra audio to USB sound card. It is connected to raspberry pi board through USB. The sound card will take input as the user's voice through two input headphone and convert it into USB compatible. The sound will be recognized by the board and shown on the terminal window.

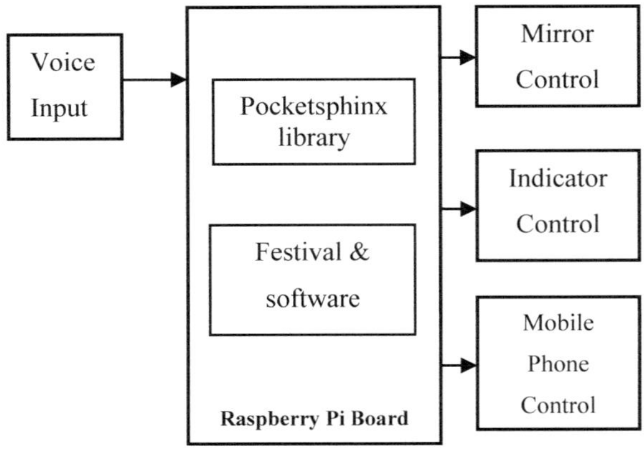

Fig. 1 Generalised block diagram

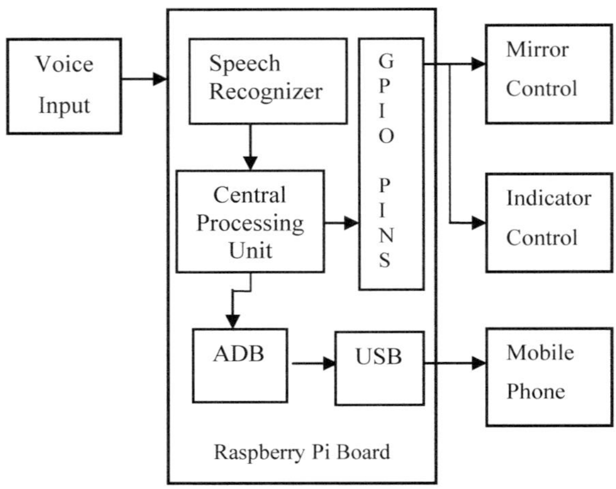

Fig. 2 Detailed block diagram

This detected keyword will be used as a command for the further controlling action. The result is compared with set of commands defined previously.

The development of application was done considering that the system can be used by a hands-free system that is located inside the car and can interact with the user. To interact with user, we use festival software which is basically text-to-speech generator. The system shall be able to listen and talk based on identified keywords that were pronounced by the speaker. Therefore, the system needs to be able to identify previously defined keywords from a spoken sentence, and it should be speaker-independent. Moreover, the system needs to be an interactive speech recognition system, in the sense that it shall be able to react based on an identified keyword.

Custom Language Model and Dictionary

Although Pocketsphinx can be used in applications supporting vocabularies composed of several thousands of words, the best performance in terms of accuracy and execution time can be obtained using small vocabularies [9]. While running the test program with default dictionary, the word accuracy for the application tended to be very low. In other words, most of the speech sentences used as inputs was not recognized correctly for Indian user scenario.

It is recommended to use a reduced language model and a custom dictionary with support for a small vocabulary. The smaller the vocabulary, the fastest Pocketsphinx will decode the input sentences as the search space of the algorithms used by Pocketsphinx gets smaller. Similarly the accuracy of the speech recognition becomes higher when the vocabulary is small. For example, there is an example application included with Pocketsphinx that uses a vocabulary containing only the numbers from 0 to 9. In this case, the overall accuracy is in the range of 90–98%.

On the other hand, it is recommended to use one of the default acoustic models included with Pocketsphinx. The main reason is that the default acoustic models have been created using huge amounts of acoustic data containing speech from several persons.

Custom Dictionary Generation

Although the default dictionary file (cmu07a.dic) integrated with Pocketsphinx includes more than 120,000 words, it is not feasible to use it in a small vocabulary application.

For this reason, we generated a custom dictionary with only words that are nothing but the voice commands which user will be uttering. The size of the custom dictionary is around 1/40th the size of the original dictionary. This allows the ASR application to handle the dictionary file much easier and perform the speech decoding faster.

As mentioned above, it is deployed for the Indian language; the phone contact list consists of Indian names. Obviously, the pronunciations of these will not be there in the default dictionary used by the Pocketsphinx.

The generation of a new phonetic dictionary might be much more complex as every word supported by the language model needs to be included in the dictionary. It is important to mention that every word needs to be translated into its corresponding representation in order to be used by Pocketsphinx. For example, Table 1 illustrates sample words in a phonetic dictionary for Indian names.

Table 1 Phonetic dictionary

Word	Pronunciation
Amit	AH MY IT
Abhishek	AH B HH IH SH IH K
Suhas	S UW HH AH Z
Akshay	AE K SH EY
Sharvari	SH AA R V AA R IY

As system is to be used in the automotive environment, generating new phonetic dictionary online was quite difficult as there will be no online sources. So, it has to be done offline. Festival software, a well-known text-to-speech tool, is used to generate the phonemes of each word. It is used for the generation of these phonetic dictionaries. These all phonemes will be accumulated in a single dictionary file.

A festival tool has different functions to play with words and its phonemes. Word and its corresponding transcriptions are generated with the help of these functions. After creating custom phonetic dictionary, the rest is to use it in real-time simulation. As mentioned earlier, the output is quite accurate because of less number of word and its phonetics.

As shown in Fig. 2, mobile phone control will be done by using Android debug bridge tool. To do so, Android mobile phone will be connected to the system via USB. The phone should be kept in USB debugging mode.

4 Flow Chart

The flow chart for the system described above is as shown in Fig. 3. Figures 4, 5 and 6 show the use case for general use case, mirror use case and mobile phone control use case, respectively.

1. **General Use Case**:

See Fig. 4.

2. **Mirror Control Use Case**:

See Fig. 5.

3. **Mobile Control Use Case**:

See Fig. 6.

5 Result and Discussion

Figure 7 shows the snapshot of dynamically generated dictionary with the help of festival tool as described earlier. It consists of Indian names and its phoneme arrangement. The same dictionary was used while simulation.

To evaluate system response for Indian names, thirty Indian users were selected consisting of 15 males and 15 females. All of them spoke some Indian names to the system through headphone. The recognized accuracy was different for different users.

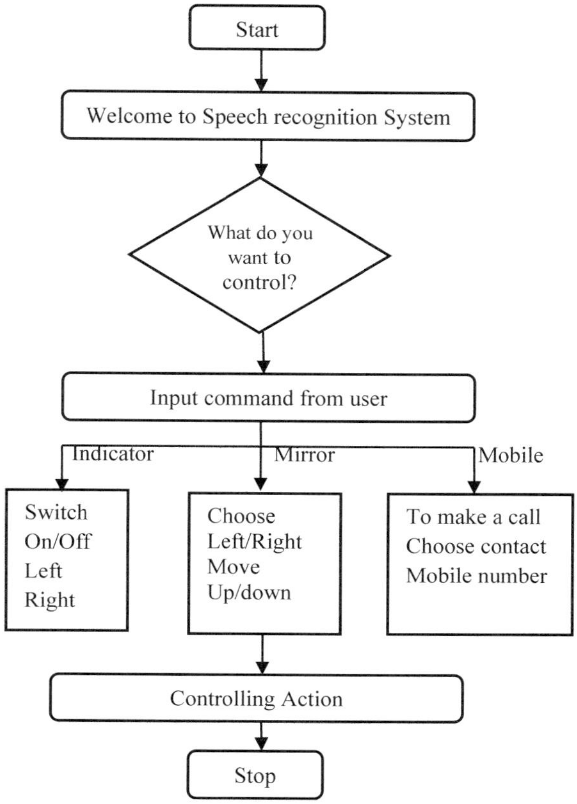

Fig. 3 Flow chart

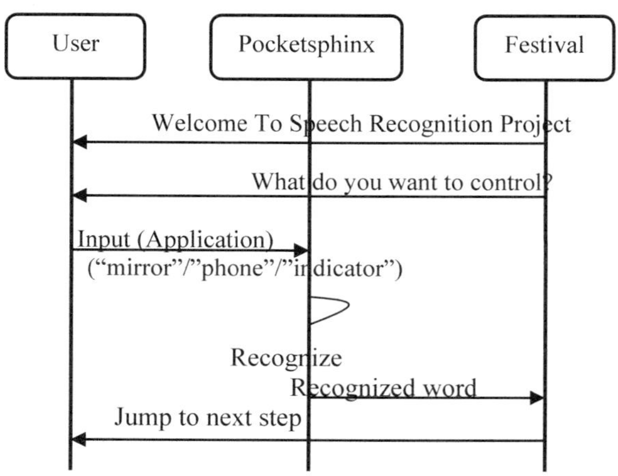

Fig. 4 General use case

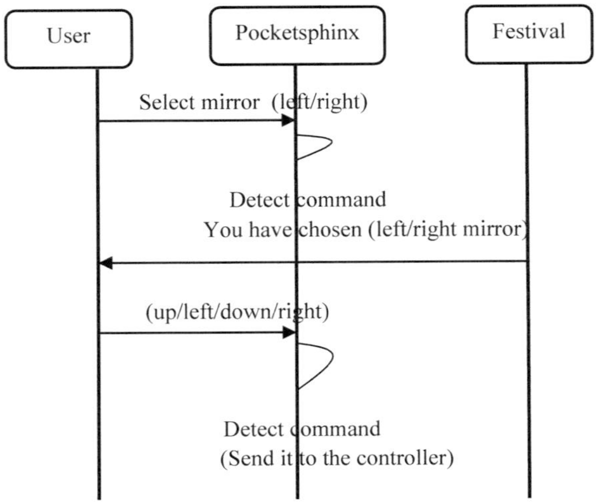

Fig. 5 Mirror control use case

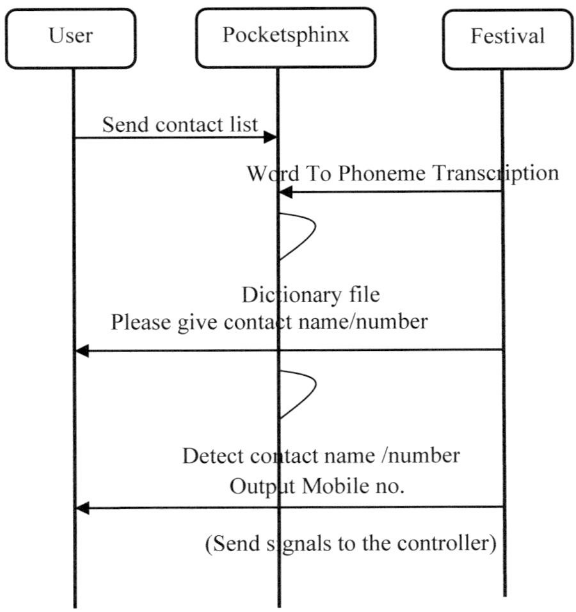

Fig. 6 Mobile control use case

Fig. 7 Dynamically generated dictionary

```
idea     AY D  IY  AH
recharge    R IY CH  AA R JH
snr    S N R
sabbir    S AE B  IH R
halkude    HH AH L K  Y UW D
rope    R OW P
somnath    S AA M  N AH TH
subhash    S AH B HH  AE SH
vashi    V AE SH  IY
calcutta    K AE L K  AH T  AH
date    D EY T
naval    N EY V  AH L
suhas    S UW HH  AH Z
nitin    N AY T  IH N
shrikant    SH R IH K  AH N T
dada    D AA D  AA
gaurav    G AO R  AA V
mama    M AA M  AH
devraj    D EH V  R AE
swapneel    S W AA P  N IY L
akshay    AE K  SH EY
devesh    D IH V  IH SH
didi    D IY D  IY
gharge    G AA R JH
nagendra    N AE G  EH N D  R AH
karan    K EH R  AH N
sukanya    S UW K  AE N Y  AH
sumit    S UW M  IH T
pranit    P R AH N  IH T
abhishek    AH B HH  IH SH  IH K
karnale    K AA R  N EY L  IY
raghu    R AE G HH  UW
sourabh    S R AH B
bidgar    B IH JH  ER
amar    AH M  AA R
omkar    AA M K  AA  ER
kumbhar    K AH M HH R
ajay    AH JH  EY
```

Some of the names (Gaurav, Akshay, Somnath, Subhash, Shrikant and Nitin) were recognized accurately at the very first attempt. Some of the other names (Neha, Sukanya, Amar, Devraj, Sumit and Prashant) were correctly recognized in the second and third attempt.

Table 2 Accuracy in percentage

Commands	Accuracy in %
Make a call	98
Mirror	97
Up	91
Down	97
Right	94
Left	90
Left indicator	98
Right indicator	96
Stop	99

Table 3 Indian name recognition analysis

Names	First trial	Second trial	Third Trial
sabbir	sumit		sumit
somnath	somnath	somnath	somnath
subhash	subhash	subhash	subhash
naval	nilesh	nilesh	
suhas	suhas	suhas	suhas
nitin	nitin	nitin	nitin
shrikant	shrikant	shrikant	shrikant
gaurav	gaurav	gaurav	gaurav
swapneel	swapneel	swapneel	swapneel
akshay	akshay	akshay	akshay
devesh	devesh	devesh	devesh
anant	anand	anand	anand
prathmesh	prathmesh	prathmesh	prathmesh
rahul	rahul	rahul	rahul
ankit	Ankit	ankit	ankit
omkar	Omkar	omkar	omkar
devraj	Devraj	devraj	devraj
amar	anant	amar	amar

On the other hand, same users were instructed to speak some of the controlling commands like, e.g. make a call, move, forward, backward. Table 2 shows the accuracy achieved.

All users were allowed to tell any 10-digit mobile number randomly; the system gave almost 95% accurate results (Table 3).

6 Future Work

The same voice recognition system introduced here can also be modified to the applications of controlling the appliances in the various fields. This extended system will be very helpful for physically challenged and people with low mobility.

Even though the overall accuracy of the speech recognizer application is fairly high, it still can be improved. For this reason, more testing is needed in order to identify areas of improvement. For example, it would be beneficial to quantify the effect of speech variability, such as the speaker's accent, on the overall accuracy of the system.

7 Conclusion

The speech recognition technology using open-source tool is demonstrated. The system will differentiate between two signals. The system will produce output in real time and also for stored .wav file. It is recognizing the words properly, whatever user has given to it in as a voice. In certain places, wherever the high security is involved, the system can be used.

Using available tools such as Pocketsphinx proved to be a good strategy as it simplified the process of creating an interactive ASR application. In this regard, even though Pocketsphinx is a free open-source library, its documentation is still somehow incomplete compared to other Sphinx decoders, most notably Sphinx 4. Nonetheless, one of the main advantages of Pocketsphinx is that it can be compiled for different platforms, such as Windows, Linux and Android.

References

1. Wilpon JG, Rabiner LR, Bergh A (1982) Speaker-independent isolated word recognition using a 129-word airline vocabulary. J Acoust Soc Atnrr 72(2):390–396
2. Jelinek F et al (1985) A real-time, isolated-word, speech recognition system for dictation transcription. In: Proceeding of IEEE international conference acoustics speech, signal processing, Mar 1985
3. Rabiner LR, Wilpon JG, Soong FK (1988) High performance connected digit recognition using hidden Markov models. In: Presented at the IEEE international conference acoustics speech, signal processing, Apr 1988
4. Cole RA, Stem RM, Phillips MS, Brill SM, Specker P, Pilant AP (1983) Feature-based speaker independent recognition of English letters. In: Presented at the IEEE international conference acoustics speech, signal processing, Oct 1983
5. Ravishankar MK (2005) Efficient algorithms for speech recognition. Ph.D. thesis, Citeseer
6. Shim B-K, Kang K-W, Lee W-S (2010) An intelligent control of mobile robot based on voice commands. Proc lEEE 98(8):1107–IIIO
7. Huggins-Daines D, Kumar M, Chan A, Black AW, Ravishankar M, Rudnicky AI (2006) IEEE international conference on PocketSphinx: a free, real-time, vol 1. I. IEEE, p I

8. Lee K-F, Hon H-W, Reddy R (1990) IEEE "an overview of the SPHINX speech recognition system", vol 38, no I, January 1990
9. Kumar A, Tewari A, Horrigan S, Kam M, Metze F, Canny J (2011) Rethinking speech recognition on mobile devices

Energy Estimation of Embedded Systems

Anagha Ram and M S Bala Murugan

Abstract Embedded systems are identified by the processors, on which application-oriented programs have to run. Most of the actions on the processor are controlled and coordinated by the software, and also, it has a fundamental role in the system design. In this paper, we are estimating and optimizing the energy consumed by the processor with the Lagrange principle, by considering the task and energy consumed by the particular task for a particular time instant by the core in Ubuntu as well as in arm development studio-5. This helps to improve the performance of the processors dynamically in the future generations. Now they are widely used in hand-held devices and many other portable consumer gadgets. The ARM processors provide tremendous achievement with less power consumption and compact size performance analysis.

Keywords ARM processor arm development studio-5 · Embedded Lagrange · Optimising · Portable gadgets · Ubuntu

1 Introduction

Nowadays, power consumption [1] has become a huge constraint in the case of microprocessor design. In the hardware and software implementation, the energy efficiency has major importance. Blooming field of microelectronics gives rise to the development of complex features in the sector of embedded systems. Majority of the computer systems including the embedded system [2, 3] has adapted different methods for the energy efficient. Reducing the energy consumption of real-time embedded system has become a great challenge in the designing field using the

A. Ram · M. S. B. Murugan (✉)
School of Electronics Engineering, SENSE, VIT University, Chennai, India
e-mail: balamurugan.ms@vit.ac.in

A. Ram
e-mail: anagha.ram@2014vit.ac.in

D. Thalmann et al. (eds.), *Intelligent Embedded Systems*, Lecture Notes in Electrical Engineering 492, https://doi.org/10.1007/978-981-10-8575-8_27

multiprocessors [4, 5]. There is an increase interests in the design of multicore processor in order to improve system throughput and reducing processor power [6].

The processors are the major reason for the power consumption of the embedded systems. Nowadays, multiprocessor platform plays an important role in computing system. Multicore processors have already been popular to improve the total performance. Multicore processors are providing more thresh out with low power consumption while comparing with the unicore processors. Dynamic voltage and frequency scaling (DVFS) and algorithm are developed for real-time task for operating frequency, and this method helps the processor to perform at different energy level with its capability of changing its voltage as well as frequency by which the energy is saved, and it is believed that the heterogeneous multicore is the best in power and performance trade-off. Minimizing vitality utilization is a standout amongst the most difficult themes for the configuration of installed and ongoing frameworks utilizing multicores. Numerous present-day processors could now work at different supply voltages, where diverse supply voltages lead to various preparing speeds.

2 Literature Survey

2.1 Power Model

The power consumption of instruction can be divided into the base power cost and the inter-instruction cost [7]. The power P consumed by the processor while executing is given by

$$P = \text{Average current} * \text{Average voltage}\ (I_{\text{DD}} * V_{\text{DD}})$$
$$E = P * N * \daleth$$

where E is the energy consumed by the program, P is average power, I_{DD} is the average current, V_{DD} is the average voltage, N is the number of cycles in the program and $\daleth$ is the period of the cycle. The ability to draw the current by CPU is very essential to determine the cost.

2.1.1 Power Cost

The parameters that provide information on power consumption of the instruction must be considered, and the effect of each of this instruction should be determined. The base power cost is determined by executing a loop having same instruction repeating in it and then, average current is measured. Thus, the average voltage V, the number of cycles of instruction N, the time period $\daleth$ is obtained. Base power costs of instruction with different operands are different.

2.1.2 Effects of Instruction

The inter-instruction effects [8] are considered when there are different sequences of instruction that are occurred in manner as follows: effect of circuit state, the power cost of an instruction in a program differs, effect of resource constraint which leads to stalls such as pipeline stalls and write buffer stalls, effect of cache misses. The energy consumption of the source and destination registers, conditionally executed instructions and different addressing modes has the major role in the power consumption.

2.2 *Power Model*

The energy consumed by a given program [9] can be given as

$$\sum(B_i * N_i) + \sum(O_i * N_i) + \sum(E_k)$$

where B_i is the ith instruction's base cost, O_{ij} is the overhead between instruction, E_k is considered as cache misses and stalls.

2.3 *Microcontroller Energy Consumption*

In this, the energy consumption does not depend on instruction type, else it depends on number of cycles executing. The measured results are given in Table 1 [10]. This is generated by considering different instructions executed with different addressing modes and arguments.

The evaluated result will be the product of the value with number of cycles of the guidelines.

=cp1.instructions1-cycle + cp2.instructions2-cycles + cp3.instructions3-cycles + …

Table 1 Energy table

Instruction groups	Coefficient	Average energy per cycle (μJ)
1-cycle	cp1	0.0535
2-cycles	cp2	0.0566
3-cycles	cp3	0.0585
4-cycles	cp4	0.0580
5-cycles	cp5	0.0576

2.4 Memory Energy Consumption

The energy dissipation of memory considers energy consumption for read and write access. A constant value for read and write access is given in Table 2 [11], and the product of this constant value with total number of read and write access will give the energy consumption.

While considering the consistent state current, its general vitality appropriation ought to be ascertained amid the execution time of the program where read and write access is performed. The table is created for the program by determining and measuring different memory access with different addressing modes.

This work primarily emphasizes on the achievement of the execution of the product code. The loop change technique has been considered and analysed to have a perfect software code since the loops have major role in the time-consuming. The loop transformation should sustain the value of the original code. Various loop transformation techniques include loop reversal, loop rolling and loop fusion. Here c program with loop reversal is considered for the software, and loop fusion is done for the hardware implementation.

The arm ds-5 platform is selected for doing this work since it supports the development of the software by including the tool chain and all the stages of the development.

Figure 1 depicts the fully featured debugging environment of the original program of incrementing code, and in this work instead of incrementing function the decrementing code is used without changing any other operations since in this method the loop counts the count number from n to zero by using the decrement operator which performs the operation faster than by using the increment operator in the loop.

Table 2 Energy consumption during memory operation

Memory operation	Coefficient	Energy (μJ)
Read access	cm1	16.2
Write access	cm2	16.6

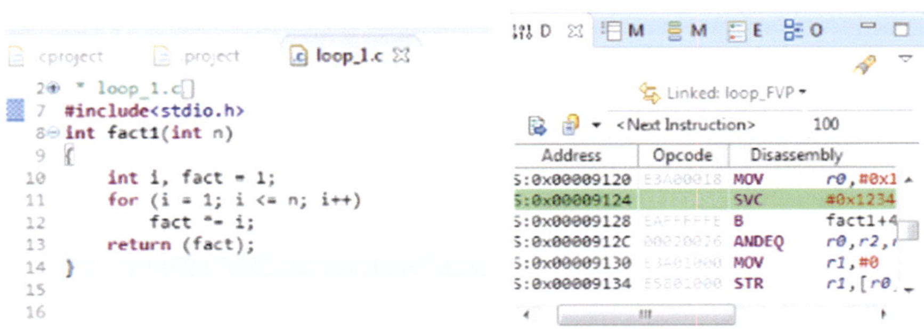

Fig. 1 Graphical debug environment of original code

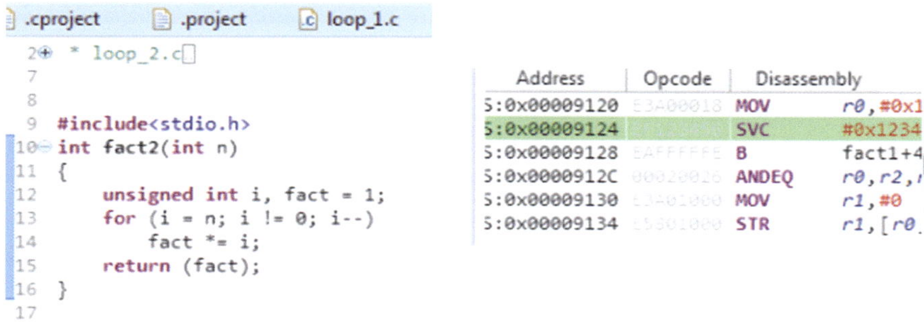

Fig. 2 Graphical debug environment of loop reversal code

Fig. 3 Analysis of the instruction usage

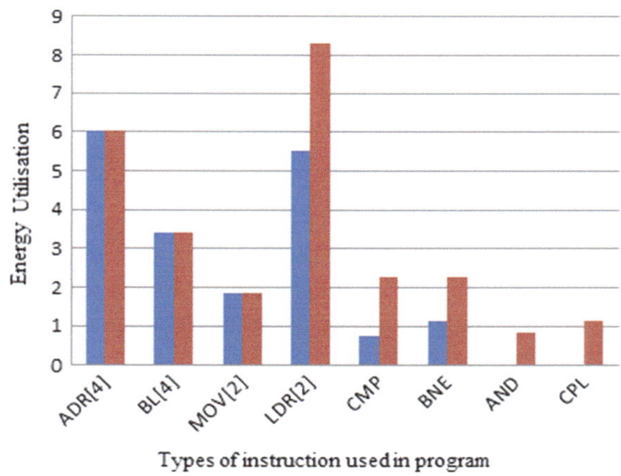

Figure 2 shows the fully graphical debugging environment with loop reversal mode. The arm development studio provides the assembly instruction associated with the written c program and shows the assembly instructions came in associated with the program. From the generated assembly code, it is very easier to determine the instruction that comes into act and identifies the energy consumed by each instruction with the help of the given data table; thus, energy estimation of the instruction in the program is determined.

Figure 3 depicts the number of repetitive instructions that are occurring in the programs that analysed above using the loop reversal technique.

The hardware set-up is being done with the help of BeagleBone Black. In this, the board is powered by the PC by using the USB cable. Initially, powering the board with the PC, then the drier is installed that is required for the operating system. The debian 7.5 is installed after this the online browsing process is performed where the BeagleBone itself acts like a mini computer system. With the help of a remote system explorer, the communication is set up between the arm ds-5 with the board. Figure 4 shows the window of remote system explorer, and it will also provide terminal platform for the execution of the program. The programs with the loop fusion technique is being executed and then the generated output file with the extension of .awk is obtained, and this generated output file which is then being copied into the root folder under home in the sftp file and thus generated the xaos pattern to determine the energy consumed by the original code as well as the fused code technique.

3 Work Done

As the quantity of transistors on a chip builds, the adaptability to decide a processor design too increments, the present pattern is to utilize them to coordinate various centres on a chip. In the case of the single core processor, its area performance reduces by the Pollack's rule, by which processor performance depends on square root of area which is depicted by $I_{\text{pollack}} = \sqrt{N}$, where N gives N times larger in area compared with the base. Homogeneous multiprocessor platform energy-aware scheduling is receiving wide research interest. A dynamic voltage and frequency scaling (DVFS) scheme is often applied to reduce the processor's power consumption. DVFS is now commonly implemented in various computing systems, both at hardware and software levels. A homogeneous platform achieves scheduling with minimum energy consumption. In the scheduling process, more number of tasks are taken into account than the processors and each processor is equipped with particular task, and all processors have given the same frequency, whereas varying frequency will vary with the time, and the power consumption can be obtained as

$$\sum_{i=1}^{k}(k-i+1)P^{\text{atu}}((V_i - V_{i-1})Dfq^{\max}/\tau_i)\tau_i$$

where P^{atu} is the actual power consumption of the processor, and its value will be null if it is in the idle state. The τ_j indicates the time interval and $(V_i - V_{i-1})Dfq^{\max}/\tau_j$ is the frequency utilized by the processor during the interval.

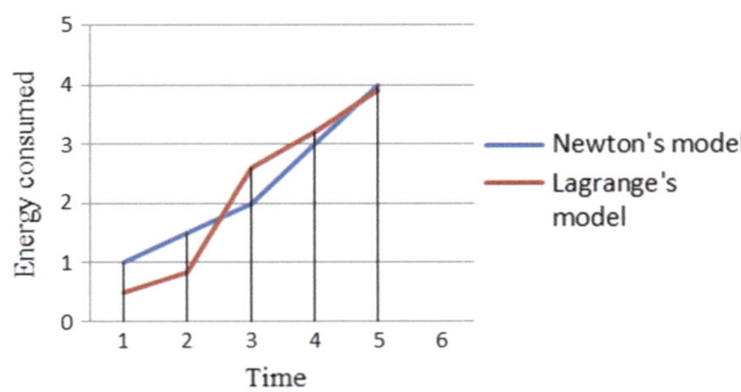

Fig. 4 Analysis of the Newton's and Lagrange's model

In homogeneous independent multiprocessor platform, the largest task first strategy is used. The optimization problem obtained in the method can be reduced with the Lagrange interpolation system.

In case of equally spaced x value, a number of interpolation methods are available such as Newton's forward and backward interpolation, Gauss's interpolation, Bessel's formula, Laplace-Everett's formula. But all these methods fail when spacing of x is unequal. In Newton's method, the coefficients of the polynomial are calculated using divided differences where the amount of computation is very large; when a number of interpolation points are changed the previous results cannot be changed. Laplace-Everett's estimate the value of function of independent variable which is equally spaced points of the central differences of the function of even order only. In such case, Lagrange interpolation is one of the best options.

Interpolation is the process of estimation of an unknown data by analysing the given reference data. The word interpolation refers a method to determine a function from a given set of the data. The polynomial obtained from the interpolation is recognized as the best from the data points which is proven mathematically. The interpolation produces the results between the known data points. Lagrange interpolation method uses a set of $\boldsymbol{x}$ and $\boldsymbol{y}$ values. It creates a polynomial such that for each $\boldsymbol{x}$ value assumes the corresponding $\boldsymbol{y}$ values.

Algorithm

Step 1.
Read $\boldsymbol{x}$,n
Step 2.
for $i = 1$ to $n + 1$ read the x_i and y_i values
Step 3.
sum = 0
Step 4.
temp = 1, $k = i$
Step 5.
for $j = 1$ to $n + 1$
Step 6.
If $(j \neq i)$ then temp = temp$(\boldsymbol{x} - x_j)/(x_k - x_j)$
end
Step 7.
Sum = Sum + y_i (sum is the value of y at x)
end
Step 8.
Write x, sum
Step 9.
Stop

The drawback in Newton's interpolation method is that it requires uniform interval between the x values of given set, while the Lagrange method does not require the uniform interval between the $\boldsymbol{x}$ values of given set.

For large set of a given n data points, the interpolating polynomial obtained will be in the order of n. The obtained high-degree polynomial passes over every given data points, and this provides the better energy estimation [12, 13]of the processors.

$$\sum_{i=0}^{n} y_i(x) = \sum_{i=0}^{n} y_i \left[\prod_{i=j,j\neq 0}^{n} (x - x_j) \div (x_i - x_j) \right]$$

The difference between the points of y_i which is given as $y0$, $y1$, $y2$ ..., etc., is the same as value of the functions of x_j. The parameters of the equation are related such as the value of the x_i refers to the task and y_i is the amount of energy used by the core for that particular task. The product of x_j subtracted from x_i and x_j get subtracted from x, which is the weighting function of x. In the expanded version, the product of lower limit and upper limit on j where j equal to 0–n, also j is not equal to i. The term $x0$ subtracted from x divided by xi is the first value of the j to have, and the condition stays true for every instance where it is get subtracted with $x0$ (Figs. 5 and 6).

In the next term, the x is subtracted from $x1$ which is get divided by term $x1$ subtracted from x_i, and this continuous till the n data points. The Lagrange type of

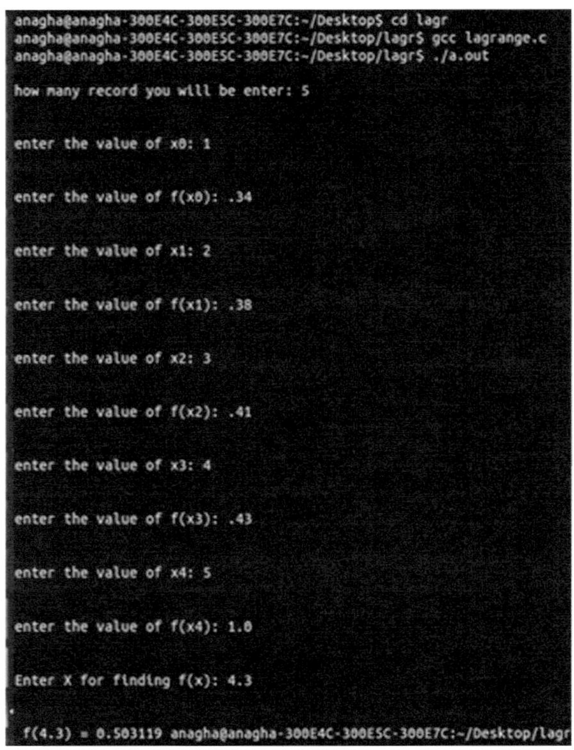

Fig. 5 Executed result for Lagrange code

Fig. 6 Threads while executing the code

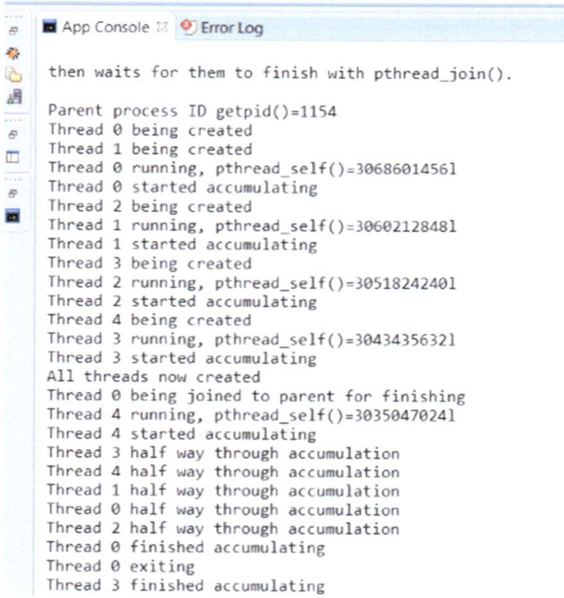

the insertion technique is normally favoured in the confirmation and hypothetical contentions, in view of the ceaseless way of the polynomial subsidiary. The estimations of the separated contrasts shape a progressive, repetitive relationship between the past two coefficients. Computationally, the partitioned contrasts can be composed as a table that essentially disentangles the algorithmic usage.

4 Conclusion

The studies on homogeneous processors are still very important without by which the achievement of the multicore or heterogeneous processor to improve their upcoming performances in future generations are not possible. In this, it aims to improve the performances by dynamically introducing the executing the resources of processors and by increasing the frequency.

Acknowledgements Authors would like to thank the anonymous reviewers for their comments in improving the paper and also, we extend our gratitude to VIT University, Chennai, for their support.

References

1. Burd TD et al (2000) A dynamic voltage scaled microprocessor system. IEEE J Solid-State Circ 35:1571–1580
2. Konstantakos V, Chatzigeorgiou A, Nikolaidis S, Laopoulos T (2008) Energy consumption estimation in embedded systems. IEEE Trans Instrum Meas 57:747–794
3. Lee S, Ermedahl A, Min SL (2001) An accurate level energy consumption model for embedded processor. In: Proceedings of 5th ACM SIGPLAN workshop LCTES, Aug 2001
4. Devi UC, Anderson J (2005) Tardiness bounds for global edf scheduling on a multiprocessor. In: IEEERTSS
5. Hsu H-R, Chen J-J, Kuo T-W (2006) Multiprocessor synthesis for periodic hard real-time tasks under a given energy constraint. In: ACM/IEEE conference of design, automation, and test in Europe (DATE), pp 1061–1066
6. Kavvadias N, Neofotistos P, Nikolaidis S, Kosmatopoulos K (2004) Measurement analysis of the software related power consumption, Aug 2004
7. Tiwari V, Malik S, Wolfie A (1994) Power analysis of embedded software: a first step forward software minimization. In: IEEE transaction on VLSI systems, vol 2, no 4, pp 437–445, Dec 1994
8. AT91SAM7X256 Data Sheet, Atmel, 2009
9. Lee M, Tiwari V, Malik S, Fujita M (1997) Power analysis and minimization techniques for embedded DSP software. IEEE Trans VLSI Syst 5(1):123–135
10. Russell JT, Jacome MF (1998) Software power estimation and optimization for high performance, 32-bit embedded processors. In: Proceedings of ICCD: VLSI computers and processors, Oct 1998, pp 328–333
11. Vazirani VV (2001) Approximation algorithms. Springer
12. Konstantakos V, Chatzigeorgiou A, Nikolaidis S, Laopoulos T (2006) Energy consumption estimation in embedded systems. In: Proceeding of IEEE IMTC, Apr 2006, pp 235–238
13. Konstantotos V, Laopoulos T (2008) Energy consumption estimation of the embedded systems. IEEE Transac Instrum Electr Meas Syst 57(3)

Design of Communicating Power Supplies and Controlling the Electronic Devices Using Internet and Mobile Application

Gunta Krishna Kishore and M S Bala Murugan

Abstract Conserving power inside houses is normally hampered by the lack of in-depth details about what exactly is using the power, how much it can be employing, along with how to remotely and automatically control electronic devices. This issue is especially for many smaller, power-consuming devices which can be used for both commercial buildings and residential. A large number of electronic devices use switching alternate current to direct current supply to operate electronic devices. The "communicating power supplies" (CPS) to enable the respective connection of one's along with control data between device along with a making administration system or maybe additional central organizations. Here developed the proof-of-system concept involving Internet-connected communicating power supplies along with proven both equally power confirming along with control by using a customized, cloud-based data clarifying property. In the event that CPS technology started to be widespread inside devices, a mix of computerized along with human being interactive remedies might allow substantial number of power financial savings and the security system.

Keywords Automation and security · Communicating power supply
E-mail alert · Power reporting · Power management

G. K. Kishore · M. S. B. Murugan (✉)
School of Electronics Engineering, SENSE, VIT University, Chennai, India
e-mail: balamurugan.ms@vit.ac.in

G. K. Kishore
e-mail: kishore0432@gmail.com

D. Thalmann et al. (eds.), *Intelligent Embedded Systems*, Lecture Notes
in Electrical Engineering 492, https://doi.org/10.1007/978-981-10-8575-8_28

1 Introduction

Electricity is the simplest need of every person in this modern world. Power intake graph is increasing daily wherein because the sources of strength are diminishing parallel. Utilization of strength is developing appreciably paving the manner for strength green technologies and digging for renewable electricity assets. Considering the fact that prevention is higher than treatment recognition of electricity intake needs to be added into each place earlier than resources get extinguished. Industrial users consume about 50% of the entire energy, personal and commercial transportation consumes 32%, whereas residential appliances devour 12% [1] and industrial uses amount to 5% of the total power and last 27% of the world's electricity is lost in power transmission and era. During the year 2014–15, electricity consumption was 747 in kWh per capita, and electricity generation in India was 1,010 kWh per capita. Electricity consumption in India is expected to rise to around 2,280 BkWh by 2021–22 [2]. Figure 1 shows that electricity consumption in kilowatt-hours per capita was plotted against year [3]. The designed device will help in decreasing the power wastage [4, 5] through constantly tracking and controlling the electrical home equipment and also provide the security alert and fire alarm systems.

Among all the microcontrollers, Raspberry Pi is usually selected due to the features it is got like convenience, Linux compiler, secure start-up along with peripheral your local library. Since Raspberry Pi provides 10/100 MBit Ethernet compatibility, it is usually interfaced to be able to Ethernet modem to be able to implement IoT or interface Wi-Fi USB modem to Raspberry Pi. The checked values from sensors might be continuously stashed and updated in a MySQL and Sparkfun database. There are many available source impair platforms for instance Ubidots,

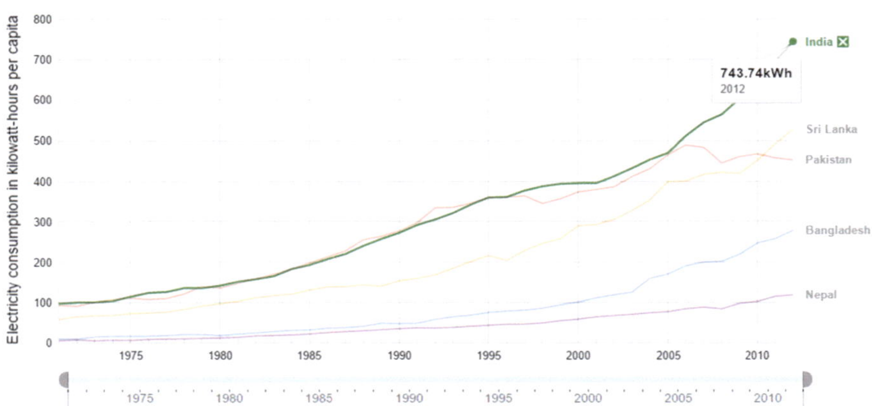

Fig. 1 Year wise electricity consumption in India kWh per capita. *Source* World Development Indicators

xively, along with ThingSpeak and many others. For various dashboard equipment, Sparkfun gives libraries along with BSP files to Raspberry Pi. It is for picking Sparkfun as storage software for checked data from current calibrating sensors. Controlling with the devices may be the other task that might be done to save energy. Relays can be employed as actuators to be able to turn upon and shut off the appliances according to the needs.

Automation technique [6] online helps make user to function the system regardless if user is not in vicinity in the automation technique. In this particular context, IoT concept has become initiated. IoT symbolizes integration associated with devices by means of Internet which means that the products utilize Internet protocol (IP) target as exclusive identifier. When interfaced to Ethernet, just about every Raspberry Pi builds unique IP target. Depending upon having a amount of rooms within house, user can offer controllers to each place.

2 System Concept

The reduction in the consumption of electricity and the use of plug loads is very important. This is because the requirement of the number of electronic devices is growing up for more efficient buildings. Nowadays—the solution for saving energy is based on communicating power resources [1]. The communicating power resources system concept measures the power usage of the electronic devices and then reports the information of devices identity and consuming power over a network to a central entity and the receiver receives from same control unit. On this processes, still the user also be able to control electric devices directly by using Web server and mobile application, so the control operations are continued. The concept of communicating plug loads is shown in Fig. 2. Here, four electronic devices are powered by Hall Effect current sensors and the main supply converts the alternate current to the direct current by using bridge rectifier circuit. The Hall Effect current sensors measure energy usage and store the information in cloud-based Web service. The cloud calculates the data and passes control information back to the sensor devices. Based on this information, the Hall Effect current sensor forwards controls to the electric devices and changes the power state accordingly. Power data information is reported at every regular intervals.

In this concept, controlled the electronic devices based on the Web server. At the same time when the device is switch ON/OFF, that information will be sent to the respected owner through SMS and e-mail. The security and fire alert system are also major requirements for households and industries. The solution for reducing loss of lives and property is to quick respond in emergency situation. So, there comes the necessity of standalone smart fire detection systems. This concept renders the works

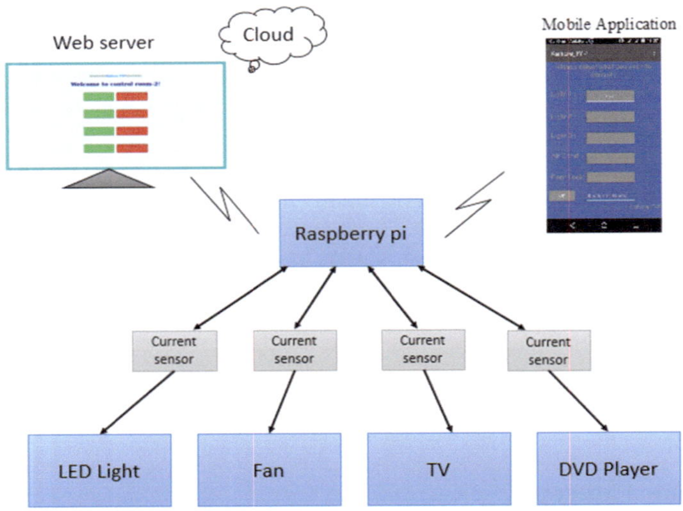

Fig. 2 System concept for communicating electronic devices, report data to cloud and controlling through Web server

of as quick as possible, send SMS and e-mail (to owner, fire, police, and ambulance), and alarm notification [3]. The concept equipped with MQ-2 sensor and DS18B20 sensor can detect unfavorable emergency situations, as it happens, and with the help of Web server can alert instantly for undertaking cautious measures. In these fatal situations, early detection and faster alert information will yield lesser losses of life and property.

Security system, one of the major issues for houses and industries, the system equipped with camera, PIR motion sensor, and ultrasonic sensor to detect the person and distance between the person and restricted area, if anyone found on that area, sensors will detect and capture the image and send information through e-mail. The system concept is showing in Fig. 3.

3 Related Work

3.1 Communicating Power Supplies

Lanzisera [1] proposes a power-efficient solution using new idea of communicating power resources to facilitate the data switch approximately power and control the data among the device and constructing power management system. The

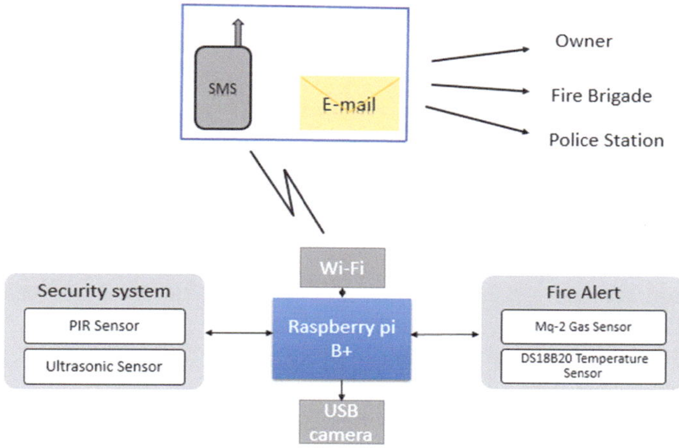

Fig. 3 Security system concept

components of CPS are an MBed controller to control all the data and a RF transceiver to speak to person. All the records acquired can be stored within the cloud records base using IoT platform. The gadget is tested on three gadgets, i.e., television, video participant, and LED light. CPS gadgets are incorporated into the product to offer local controls and mechanically include product identification data.

3.2 Renewable Energy Gateway

Han [7] planned a photovoltaic system management to enhance the home electricity management based on the PLC that consists of PLC modem, renewable energy gateway, and also smart product source. The PLC modems contact the REG from the power series which transports the dc strength generated simply by PV modules to the grid-connected inverter. The REG outlets the status, the clever device offers the status to the entire solar farm. This method permits the client to limit the failure and quickly resolve it.

3.3 Smart Home Energy Management System

Hu [8] initiated a hardware system that includes clever domestic strength management device such as applications including communique, sensing technology, and a gadget-learning algorithm. SHEMS consists of sensors which can be used stumble on human sports, and with the assist of this information, machine-learning algorithm is carried out therefore. This execution reduces the entire energy payments for clients with none need of people presence.

Han [9] started smart household interfaces and also device definitions to permit interoperability involving ZigBee devices manufactured by various producers of power equipment, yards, and clever energy empowering products. ZigBee can be employed for transferring the info about the energy and electricity of devices. For keeping track of the solar power panels, power-line verbal exchanges is employed. This standard protocol establishes this wireless circle, based about the Kruskal's protocol value measured from your RF radio.

Assaf [10] brought domestic control and protection device based on the sector programmable array. The FPGA used is Nios development board cyclone-II edition which offers hardware platform for developing eRaspberry-Pi systems based totally on Altera cyclone-II devices.

The model of the suggested system was created and your correlation connected with software and hardware is performed. The logics with regard to controlling are made in FPGA and also communicated to the user by means of Web server. The net server is established by utilizing HTML or perhaps Java-dependent script. User alerts pick up through net server made in php and also thereby setting switch adventures and controlling them by means of controller offers the entire safety system.

4 System Architecture

4.1 Raspberry Pi B+ Controller

The Raspberry Pi is a card-sized single board which was developed in UK and has Broadcom BCM2835 system on chip (SoC) computer. This chip includes an ARM1176JZF-S of 700 MHz. It also has 512 MB of RAM, and the CPU speed ranges start from 700 MHz. There is a Video Core IV GPU and 10/100 MBit Ethernet compatibility [11]. The disadvantage is that it does not have a built-in hard disk, whereas it uses an SD card for booting and to store the data. For the Ethernet connection, we have to connect a cable to the router and then it works automatically. For USB Wi-Fi dongle, use the GUI that comes with Raspbian operating system to find your wireless network and enter your WEP/WPA password.

4.2 Hall Effect Current Sensor

Hall Effect current sensor (ACS712) is a sensor which is used to sense the current flowing through the electrical devices. It consists of a magnet in which the main wire is used to connect the device. Then, this device is passed through the magnet. Hence when the device is powered due to the current flowing in the wire, then there is a magnetic induction due to which a small voltage is generated at the other end of the sensor.

4.3 PIR Sensor

PIR sensors are the one which allows you to sense human motion. Hence, it is used to detect whether a human is present in or out of the sensors range. The main advantages of a PIR sensor are that they are small, consume low power, are inexpensive, and do not wear out easily. Thus, PIR sensor has been used in the system. The PIR sensor is referred as IR motion, passive infrared, pyroelectric sensors. It can be used to detect levels of infrared radiation. The PIR has two slots in it, and each slot is made of a special material that is sensitive to infrared radiation [12]. When the human is not detected, both slots detect the same amount of infrared radiations. When a human passes by, it first intercepts one half of the sensor, which causes a positive differential changes between the two halves. When the human body leaves the sensing area, the reverse happens.

4.4 Ultrasonic Sensor

Ultrasonic HC—SR04 is used to calculate the distance of human from a restricted area. It is also able to sense 2–400 cm and a ranging accuracy can reach up to 3 mm. The modules in this sensor include the ultrasonic transmitters, control circuit, and receiver. The basic principle of its working is that it uses input–output trigger for at least 10us high-level signal. The sensor module can automatically send eight 40 kHz and detect whether the pulse signal is back or not.

Test distance = (high-level time $\times$ velocity of sound (340 M/S))/2.

4.5 Gas Sensor (MQ2)

This gas sensor (MQ2) module is used to detect leakage of gas in household and industries. It can able to detect the LPG, smoke, I-butane, methane, propane, alcohol, hydrogen. Based on its immediate response, time measurements can be taken as fast as possible. And also the sensitivity can be adjusted by the potentiometer.

4.6 DS18B20 Temperature Sensor

The DS18B20 temperature sensor has the ability to provide 9- to 12-bit temperature. Here, the information is sent to the DS18B20 sensor over a 1-Wire interface. This wire is connected from a Raspberry Pi to a DS18B20 temperature sensor. The power to write, read can be derived from the data line itself. Multiple DS18B20s sensors can exist on the same 1-Wire bus as each DS18B20 sensor contains a unique silicon serial number. This allows to place the temperature sensors in many different places for our need. Applications where this system is useful include the HVAC environment, temperature sensing inside buildings, machinery, or equipment.

4.7 Relays

Relay can easily control just about any appliance while using magnetic circuit within it. A eight-channel relay can generally control eight appliances. It needs 12 V power. When relay gets unwraps, the permanent magnet ciruit inside the relay willl turn off the device. Eight-channel relay can easily control a series of devices at a time.

5 Proof-of-System Concept Demonstration

The proposed system allows to monitor the power and control the electronic devices using Web server. To illustrate this concept, I rapidly prototyped a Raspberry Pi-connected sensors and electronic devices controlled with Web servers, which consists of two control units. In this section, I explain the hardware and software implementation and its working principle.

5.1 Hardware Implementation

The complete system design and controlled with Raspberry Pi B+. It is a credit-card-sized single board having Broadcom system on chip (SoC). The Raspberry Pi can be used to automate industrial and home appliance at a relatively

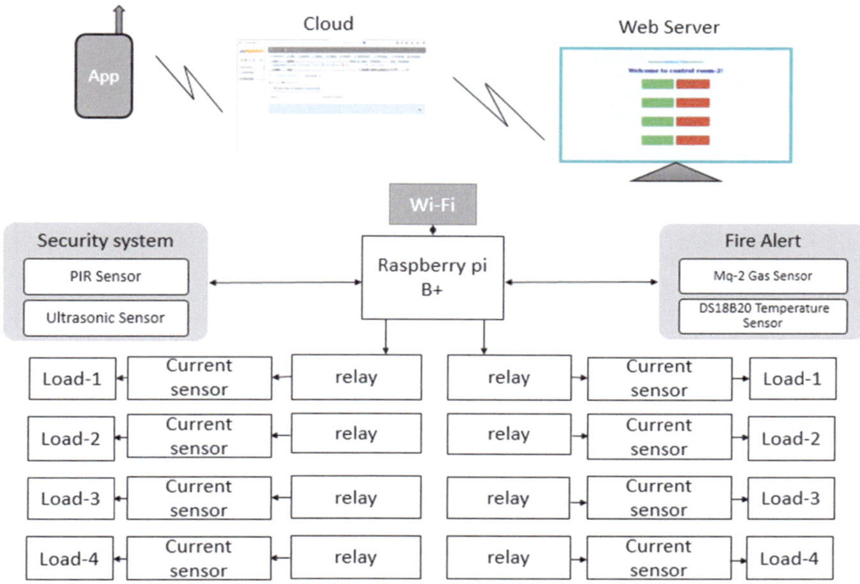

Fig. 4 Proposed block diagram for complete system

low cost. It is run on the concept as the Internet of Things (IoT). It supports variety of Debian-based OSs, primarily Raspbian OS. Adding wireless display features to it supports as miracast. It should be configured to Wi-Fi dongle. Latest wheezy image will support streaming and this image is made specially for streaming of data. The complete system proposed block diagram as shown in Fig. 4. The system consists of two control unit. Each unit contains of four loads powered by current sensors to calculate the power and upload the data in cloud. Based on the power data, the devices are controlled; at the same time, the devices will be controlled using Web server and mobile application.

The security system contains sensors which are controlled by Raspberry Pi, and the sensor data will be upload in phpMyAdmin and Sparkfun; if any disturbance occurs, then the Raspberry Pi immediately responds and sends information through e-mail via ssmtp and SMS via way2sms using Web server. Figure 5 shows the security system hardware setup. The complete system photograph is shown in Fig. 5.

The complete hardware setup for proposed concept is shown in Fig. 6.

Fig. 5 Hardware setup for security system

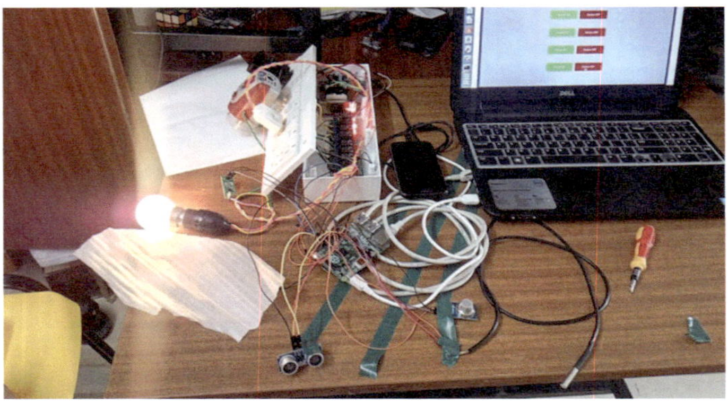

Fig. 6 Photograph for complete hardware setup of proposed system

5.2 Software Implementation

The software environment of the complete system is frameworked with python libraries. The hardware of the system runs in the multi-tasking environment. One of the tasks hosts the Web server templates into the server while the other tasks take the responsibility of pushing the calibrated sensed value in the existing IoT Platform Sparkfun. Being running parallel, Sparkfun and Web server template, both the tasks take the input from the same sensors and display the status of security system, electricity consumed, and electricity bill; at same time, the devices are

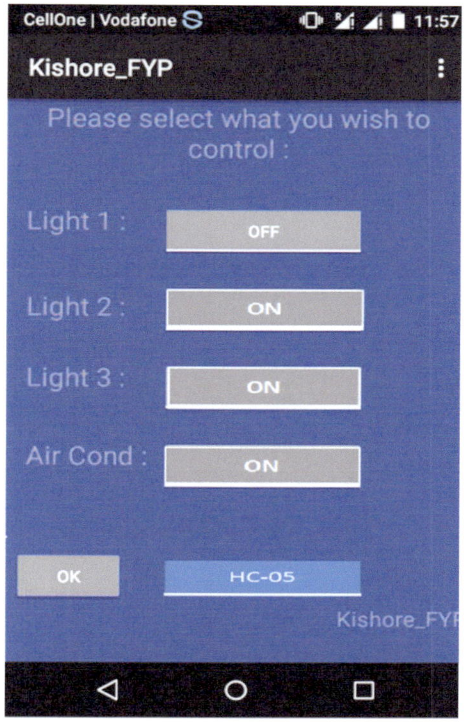

Fig. 7 Mobile application of this system

controlling through the mobile application and Web server and send the system status via ssmtp and way2sms. Figure 7 shows the mobile application of this system.

The Web server was developed by using php, HTML, JavaScript, CSS pages, and python running in the background hosts the Web pages into the Web server. Figures 8 and 9 show the control units 1 and 2, respectively, of the system.

These two Web servers controlled the electronic devices, and when device is ON/OFF, that status will be sent to the owner via SMS and e-mail. Figure 10 shows the screenshots of received information.

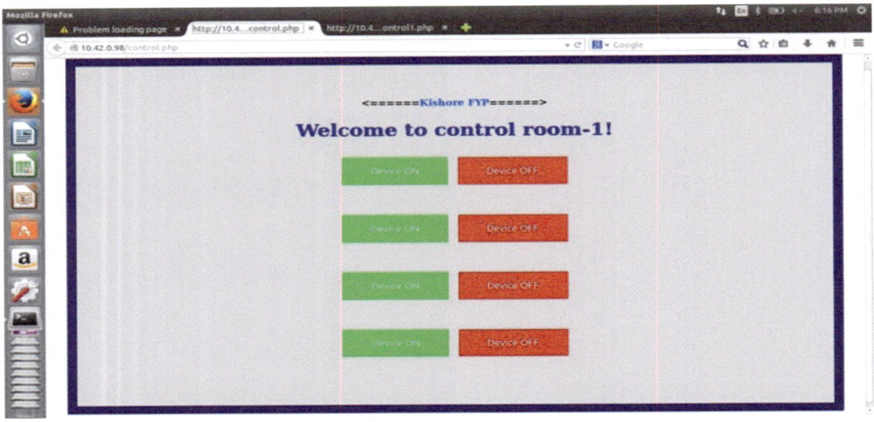

Fig. 8 Web server for control unit-1

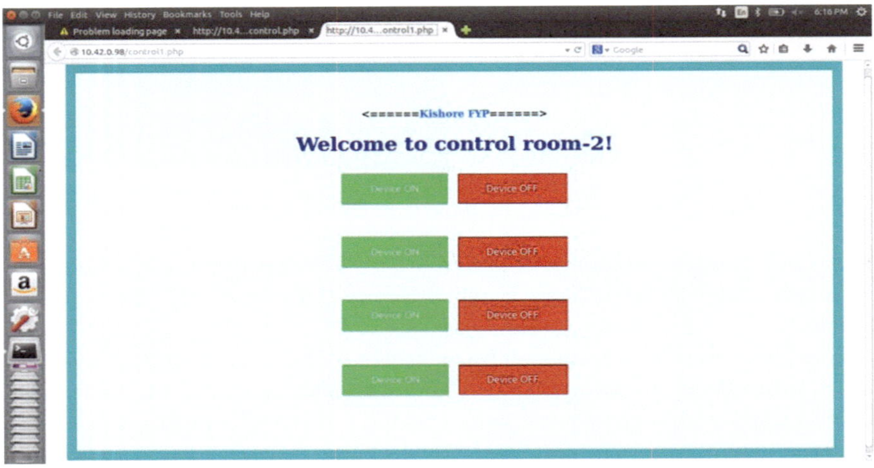

Fig. 9 Web server for control unit-2

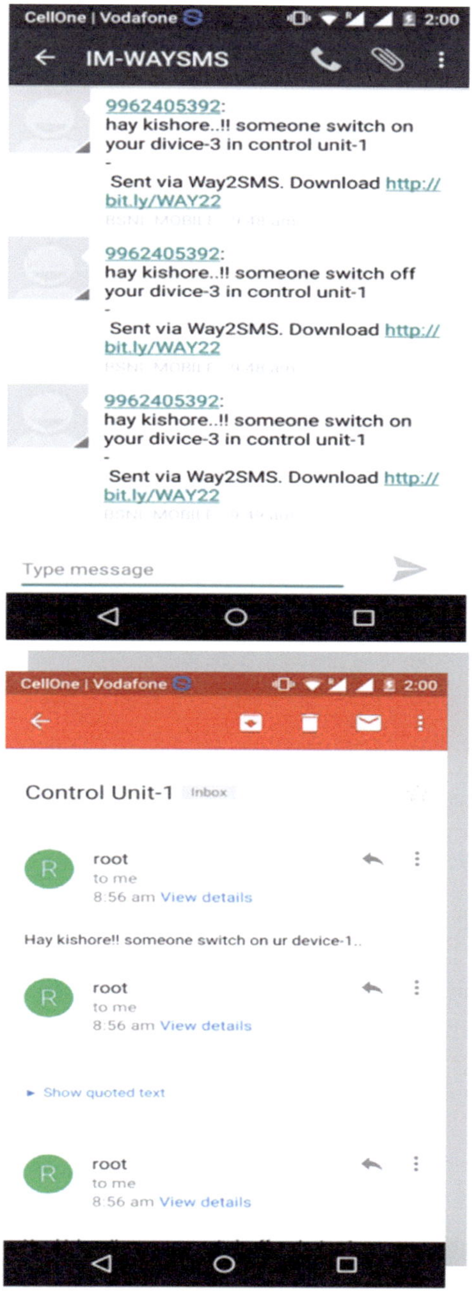

Fig. 10 Screenshots 1 for received messages from Web server

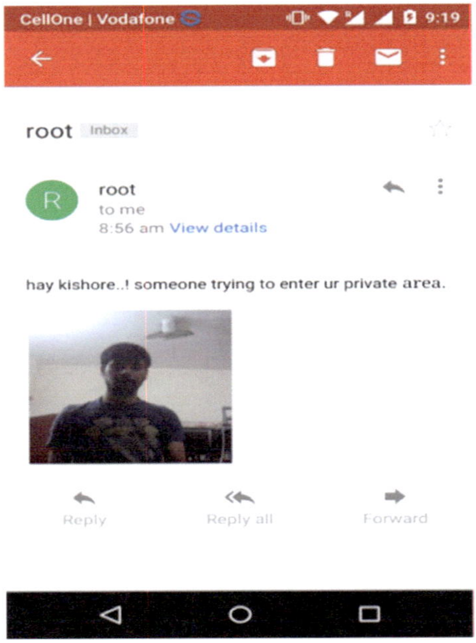

Fig. 11 Screenshots 2 for received messages from Web server

When someone enters the private area, the security system immediately capture the picture and send the information to owner. Figure 11 shows the reserved mail with attachment from the security system

5.3 *Monitoring the Sensors Data in Sparkfun*

The measured data from the sensors can be stored in local cloud and Sparkfun database. For each channel to be measured, Sparkfun gives a feed ID and API key. Feed ID and API key are provided to Raspberry Pi and then code is executed in order to update the values in Sparkfun. The outputs in Sparkfun can be seen in Fig. 10.

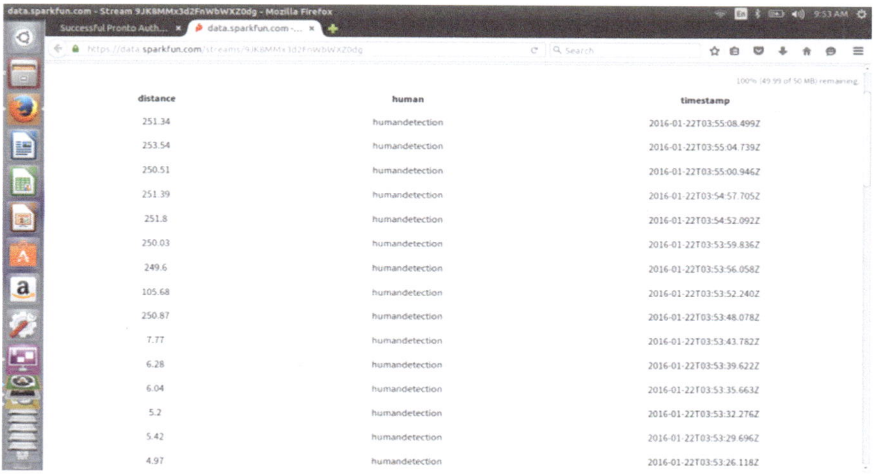

6 Conclusion

The paper presented Web server-based appliance for home and industries that permits improved awareness of power consumption in electronic devices and users. This technology is the future of industries and home power monitoring for switching loads and the security services. The complete system has shown here that less complex, safe, and valid at low price. By using this system to quick respond for the security services like alarm, police, ambulance and fire brigade of a nearby region and informed about emergency instantly. The low price, easy to configure, and tight coupling with the electronic product make communicating plug loads a good application of Internet of Things concepts.

Acknowledgements This paper was put forward and supported by VIT University under the final year project. I would like to thank Prof. M. S. Balamurugan who has given his full cooperation and technical support for this project and in writing this paper.

References

1. Lanzisera S, Weber AR, Liao A, Pajak D, Meier AK (2014) Communicating power supplies: bringing the internet to the ubiquitous energy gateways of electronic devices. IEEE IoT J 1(2)
2. Garg P (2012) A treatise on "electricity and magnetism". J Sustain Energy Environ 3:7–17
3. Asif O, Rahman MT, Muhammad EH, Belayat Hossain Md, Hasan M, Chowdhury ME (2014) Fire-detectors review and design of an automated, quick responsive fire-alarm system

based on SMS. Int J Commun Netw Syst Sci 7:386–395 (Published Online September 2014 in SciRes)
4. Cho HS (2009) Determining location of appliances from multi-hop tree structures of smart power meters. IEEE Trans 55(4)
5. Electric power consumption (Wh per capita) IEA Statistics IEA/OECD 2014
6. Bai Y-W, Lien C-H, Lin M-B (2007) Remote-controllable power outlet system for home power management. IEEE Trans Consum Electron 53
7. Han J, Lee I, Choi C-S, Kim S-H, Park W-K (2014) PLC-based photovoltaic system management for smart home energy management system. In: IEEE conference on consumer electronics (ICCE), 2014
8. Hu Q, Le F (2013) Hardware design of smart home energy management system with dynamic price response. IEEE Smart Grid 4
9. J-H Lim, D-M Han (2010) Smart home energy management system using ZigBee and IEEE 802.15.4. IEEE Energy Manag
10. Assaf MH, Mootoo R, Das SR, Petriu EM, Groza V (2012) Sensor based home automation and security system. IEEE 2012
11. Raspberry Pi B+ model from Wikipedia, the free encyclopedia. https://en.wikipedia.org/wiki/RaspberryPi_B+model
12. PIR sensor date sheet theory. http://adafru.it/aKh

Real-Time Human Detection and Tracking Using Quadcopter

Rana Praful George and V. Prakash

Abstract **Background**: Detection and tracking of humans in real time is a challenging and important field of research. Human detection has many applications in detection of human and video surveillance. Nowadays, quadcopters are used in different fields like research, military and law enforcement, commercial fields for navigation, searching, and areal imagery. **Methods**: In this paper, the primary objective of the system is to detect and locate people who got isolated in some areas due to natural disasters like floods and earthquake. It is carried out using a single-board computer Raspberry Pi (RPI), and the image processing part is carried out with the help of OpenCV. The detection and tracking of humans is done using quadcopter. Quadcopter is equipped with camera, the video capture is processed using RPI, and live streaming along with human detection is sent to the base station over the wireless network. **Applications**: It is used in effective human detection in real time video streaming. It helps the rescue team to locate people and to know their current situation and thus to take necessary actions including medication and evacuation. **Development/future work**: A prototype model for human detection and tracking using quadcopter is designed and developed. Better battery pack for longer flying time and use of 3G USB dongle for large area surveillance.

Keywords Human detection · Raspberry Pi · Image processing
OpenCV · Quadcopter

R. P. George · V. Prakash (✉)
School of Electronics Engineering (SENSE), VIT University, Chennai, Tamil Nadu, India
e-mail: praksh.v@vit.ac.in

R. P. George
e-mail: rana.prafulgeorge2014@vit.ac.in

D. Thalmann et al. (eds.), *Intelligent Embedded Systems*, Lecture Notes
in Electrical Engineering 492, https://doi.org/10.1007/978-981-10-8575-8_29

1 Introduction

Detection and tracking of dynamic objects and humans has become an important part in many computer vision applications. Many techniques are being developed for performing computer vision application without the direct human intervention. This allows humans to interact with dangerous environments without risking any loss to human life. Computer vision algorithm provides valuable information from a single image or series of images. The information extracted from the images can be more helpful than that extracted from the same image by a human.

Quadcopter [1], also called quadrotor, is a helicopter with four rotors with which it is lifted and propelled. Rotors are placed in a square configuration which is equidistant from center of mass. They are called small unmanned aerial vehicles (UAVs) due to their simple structure. It is navigated by controlling the angular velocities of the rotors mounted on it. It used for rescue, supervision, surveillance, and in military operations, as it can hover around and track the people on the ground.

Though there are many applications of personal robots that are developed to do useful tasks autonomously and without any human interaction, taking instructions from a human as the tasks get more and more advanced will become crucial. Therefore, developing systems that will allow a robot to navigate toward and track the humans in the environment allows in implementing advanced robotics tasks in the future. An automated system for human detection, recognition, and tracking can be used in a wide range of applications especially in surveillance systems. Surveillance applications also include detection of humans who got trapped in remote areas due to some disasters like floods. This system helps to understand the real-time conditions of the humans. With expanding terrorist exercises and enlarging demand for video surveillance, it turns into the need of an hour to formulate an efficient and quick human detection and tracking algorithm.

This paper deals with a better human detection and tracking technique using histogram of oriented gradients (HOG) [2] algorithm along with OpenCV. Main purpose of this project is to detect and track the humans who got isolated in remote areas due to natural calamities like flood. The image processing part includes human recognition, detection, and tracking which is being carried out using Raspberry Pi board. The USB camera captures the video, and images are processed under HOG algorithm according to the frame rate given by the controller. Each image is processed, humans are detected, and the detected numbers are send to the base station along with the live video streaming. This is carried with the help of quadcopter. The data is sent to base station over the wireless network.

There are many techniques for detection and tracking of human and objects in a live video stream. It includes many classical methods such as color matching, template matching, feature detection, and matching and motion estimation [3]. Color-based detection methods [4] are also used in object detection process. Another object tracking method called Tracking-Learning-Detection (TLD) [5] is also used as the object detection and tracking becomes difficult from drone as the area of interest changes along with the drone [6]. Monocular vision system for object tracking in closed-loop controls [7] is also used for object detection and tracking.

The paper is arranged as follows: Sect. 2 discusses the system design using Raspberry Pi board, USB camera and quadcopter, and the basic methodologies; Sect. 3 outlines the hardware components. Section 4 discusses the software design which we use in this paper. Section 5 explains the experimental setup for realization of the project. Section 6 presents the results obtained. Section 7 reports the conclusion and future work which can be included for high-level application.

2 Block Diagram

The critical part in human detection and tracking process is to detect the human from the image captured by the quadcopter camera. Human detection using quadcopter is shown in the block diagram provided in Fig. 1. The USB camera takes live video using Raspberry Pi. Human detection algorithm using HOG descriptor is performed for detecting the humans in the video stream captured by the camera. It captures the images and gives the number of humans detected in each image frame. The image processing is done using OpenCV. Quadcopter is used to perform the surveillance over an area to detect the people. It is controlled using a RC controller designed at 2.4 GHz frequency. It is navigated by controlling the speed of servomotors to which propeller blades are attached. A GPS module is also mounted on to the quadcopter to track its position as it moves.

The data is transmitted from quadcopter to the controlling base station wirelessly. The live video taken is also sent to the base station along with the processed images and the GPS coordinates over an IP network [8] using Wi-Fi connectivity

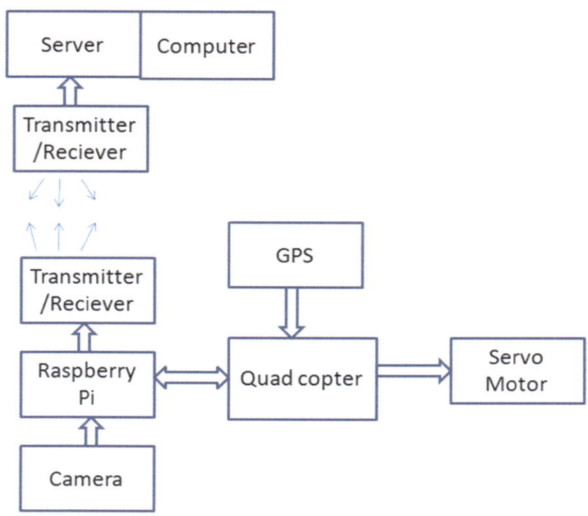

Fig. 1 Human detection and tracking using quadcopter system

which helps the rescue workers to locate the isolated people in remote places and to take rescue measures accordingly.

2.1 Methodology

The proposed method is more efficient method using histogram of oriented descriptors which is an excellent performance comparative to feature sets which are used in other applications including wavelets [9]. The proposed descriptors have resemblance to edge orientation histograms, SIFT descriptors [10], and shape contexts [11], but they are computed on a dense grid of uniformly spaced cells and they use overlapping local contrast normalizations for improved performance. The histogram of oriented gradient descriptor method measures or checks the occurrences of gradient orientation in localized portions of an image like detection windows or region of interest (ROI). In human detection process, the input image is analyzed and the features like location, position, number, size, and orientation are calculated. The human is distinguished from all other patterns present in the scene by displaying a rectangular window around the human. This determines the location and presence of a human. Figure 2 shows the methodology for implementing the human detection algorithm.

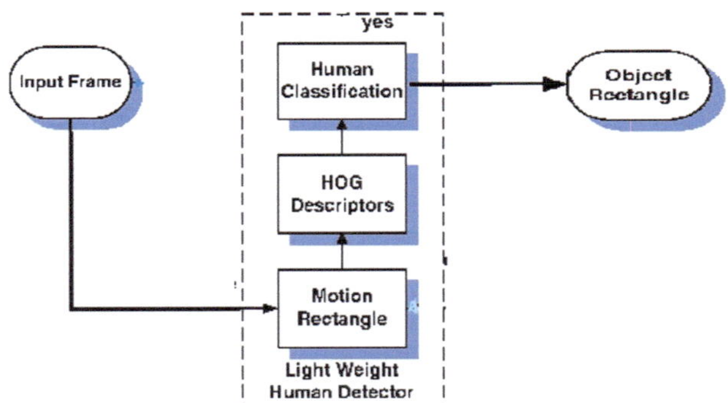

Fig. 2 Methodology using HOG descriptors

2.2 HOG Detection Algorithm

The HOG is used in computer vision and image processing applications as a feature descriptor for the purpose of detecting the objects. This technique counts the occurrences of gradient orientation in localized portions of an image. This method of HOG is similar to that of edge orientation histograms, shape contexts, and scale-invariant feature transform descriptors, but differs in that as it computes over a dense grid of uniformly spaced cells. For improving the accuracy, it uses overlapping local contrast normalization.

HOG descriptor algorithm implementation [12] is as follows:

- Gradient computation: Divide the image into cells which are small connected regions, and compute histogram of gradients or orientation of edges for each cell.
- Orientation binning: Each cell is discretized into angular bins according to orientation of cells. Pixel of each cell contributes a weighted gradient to its corresponding angular bin.
- Descriptor blocks: Adjacent cells are grouped into blocks. Further normalization process is carried out based on the blocks formed by arrangement of cells.
- Block normalization: The block histogram is represented with the help of group of normalized histograms. The descriptor is represented by the normalized group of histograms.

3 Hardware

This section discusses the Raspberry Pi board, quadcopter, and other peripheral modules needed for this prototype implementation.

3.1 Raspberry Pi Board

Raspberry Pi is a credit card-sized single-board computer developed by Raspberry Pi Foundation, based on Broadcom BCM2835 system on a chip (SoC) [13]. It uses standard computer keyboard and mouse and can be plugged to computer monitor or television. It can do everything a desktop computer could perform from browsing the Internet to playing high-definition videos and games to the ability to interact with outside world with the advantage of having ARMv7 Quad-Core Processor running at 900 MHz with 1 Gb RAM. It has been used in implementing many projects and real-time applications which require GPIOs and other peripheral features. It is capable of performing operations like human detection, human tracking on the image captured by the camera.

3.2 Quadcopter

A quadcopter is designed and assembled for surveillance purpose. Quadcopter parts include a customized flight control board, servomotors, propeller blades, and skeleton frame.

The control board used is LH X6 customized circuit and receiver board. It has transmitting and receiving IC's and antenna for its control through a RC remote controller. It is also provided with contact breakers for circuit protection. The power required for the board is provided using a rechargeable Li-Po battery. It has inbuilt gyroscope for balancing as it navigates. There are many other boards like KK board.

Servomotors are used to rotate the blades which help in lifting and its navigation. According to the speed variations of motor, the drone is moved in different directions. This paper uses motor and gear setup for motion.

A quadcopter uses two clockwise and two anticlockwise propellers. They are classified based on length and pitch. A high pitch propeller moves greater amount of air but could create turbulence during hovering. While considering length, a small increase in length can increase the efficiency, but small propellers are easy to stop and speed up than large propellers. Propellers and motor should be selected such a way that it can improve the efficiency with improved battery life and control experience.

The skeleton frame is the structure which holds the components together. It should be rigid, lightweight, and be able to reduce the vibrations. It contains a center plate where flight controller board is fixed and four arms on to which motors are connected at its end. Mostly available materials for frame are carbon fiber, aluminum, wood such as plywood or MDF.

3.3 USB Camera and Wi-Fi Dongle

Camera is used for capturing the images as the quadcopter flies. The proposed method uses Logitech USB camera for image capture. It is directly connected to the Raspberry Pi board, and images are processed. The data is transmitted to the base station over wireless network. The preferred network for transferring the data is wireless rather than wired [14]. For data transfer over the Wi-Fi [15], a Wi-Fi dongle (Ralink) is connected and programmed to the Raspberry Pi board.

4 Software

4.1 OpenCV

OpenCV [16] is a open-source computer vision and machine learning software library used for computer vision applications. This library uses more than 2500 algorithms for human detection, object tracking, 3D modeling, motion estimation, object identification, image comparison, image processing for augmented reality, etc. It is a comprehensive set of both computer vision and machine learning algorithms. OpenCV contains library functions aiming at real-time applications. It has python, C++, C, Java, and MATLAB interfaces which are used to support Linux, Windows, Mac, and Android OS.

4.2 Python

Python is an open-source programming language which helps to work quickly along with the efficient integration of systems [17, 18]. It is a high-level, general-purpose dynamic programming language which emphasizes on readability. The main advantage of python is its simplicity and its ability to support many operating systems. Python is used in scientific programming with the help of Matplotlib and NumPy libraries. It is also used in artificial intelligence and also used as scripting language for many Web applications.

5 Experimental Setup

The hardware setup for human detection using Raspberry Pi is mounted on to a quadcopter as shown in Fig. 3. Raspberry Pi is powered up by using an external power source of desired power requirements. USB camera and Wi-Fi dongle are connected to the USB ports of Raspberry Pi for image capturing and processing and also for the wireless transmission to the base station.

The frame of quadcopter is modified for mounting the Raspberry Pi and other modules. A separate holding plate is made using aluminum and lightweight board for mounting the above hardware setup. Modification is done in such a way that it should not affect the lifting capacity, orientation, and control of the quadcopter.

6 Results and Discussion

Human detection and tracking algorithm using HOG descriptors is sufficient enough to detect the humans and the number of people in each image. The live video taken by the camera is processed, and when humans are detected, bounding

Fig. 3 Hardware setup for human detection using quadcopter

boxes are drawn by the HOG descriptor algorithm. Then images are being saved to the base computer, and a window listing the number of bounding boxes detected is also displayed in base computer. The number of people detected with bounding box is shown in Fig. 4. The number of bounding boxes in the processed image will be equal to the number of humans detected. So a window is created which displays the detected number of humans in each image. The window listing the number of bounding boxes is shown in Fig. 5.

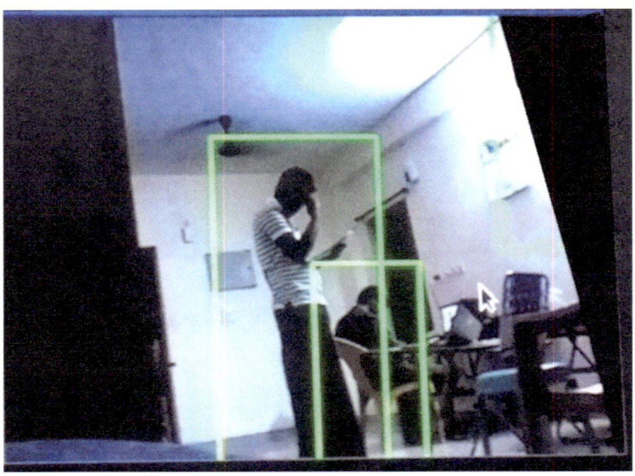

Fig. 4 Human detection with bounding boxes

Fig. 5 Window showing number of bounding boxes

The implementation of HOG algorithm at varying heights is given in Fig. 6. It shows that human detection can be done with varying heights by using the HOG algorithm which is implemented in quadcopter-based human detection system.

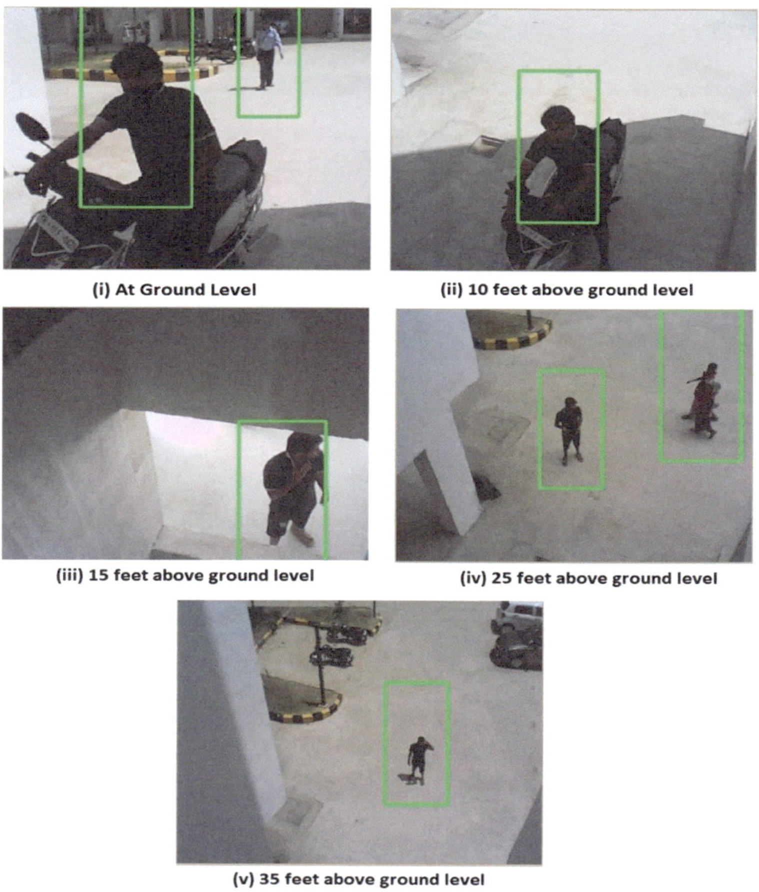

Fig. 6 Human detection with varying heights

Along with the human detection, the proposed method performs movement indicator function. There occurs a chance of non-detection of humans due to the small delay time during the processing of algorithm for human detection. Since the system has to be used for the rescue purposes, the system should have a higher efficiency. So in this system, HOG-based human detection is combined with movement indicator algorithm for increasing the efficiency. The movement indicator function along with human detection is shown in Fig. 7 in a red circle.

Fig. 7 Movement indication along with human detection (in red circle)

Fig. 8 Live video streaming over network

The video captured can be streamed over the local network with the help of Wi-Fi dongle within its range. It can be viewed from a remote computer or using a mobile phone in the local network. Live video streaming carried out using a mobile phone by a specific IP address is shown in Fig. 8.

Human detection and tracking along with number of humans is done using HOG algorithm in OpenCV, and the entire hardware set is mounted on a quadcopter which can be used for surveillance actions during disasters like floods and earthquake. It helps the rescue workers to track and to realize the current situation of isolated people. Thus, a prototype model for human detection and tracking using quadcopter is designed and developed.

7 Conclusion and Future Work

A new methodology for detection and tracking of surviving humans in a destructive environment is implemented. The system integrates the existing image processing techniques with motion tracking algorithm for detecting humans within the camera's range of view. A quadcopter prototype is designed and tested with separate boards for flight control and image processing. HOG descriptor algorithm technique is used in OpenCV for detecting humans and to process them further in order to send the data to the base station which includes human numbers, storing and displaying of images from quadcopter. A prototype model for human detection and tracking using quadcopter is designed and developed.

This quadcopter-based detection system is not able to perform for longer period of time due to the limitation of Li-Po battery. The proposed system uses lightweight cardboard stand for mounting RPI and other peripherals. Increase in the capacity of the power source and metal stand results in increased weight, which affects the payload capacity of the quadcopter. In future, this model can be modified with better hardware design including high-powered servomotors, a metal stand, and better battery pack for longer flying time for better surveillance and rescue operations. It can be powered with USB 3G dongle for wireless connection so that it can be deployed for large area surveillance and rescue operations.

Acknowledgements I would like to express my special gratitude and thanks to my internal guide Prof. Prakash V. whose guidance and immense support encouraged me to complete the project successfully.

References

1. Gheorghita IV, Mirea L, Braescu C (2015) Quadcopter control system. In: 19th international conference on system theory, control and computing (ICSTCC), 2015
2. Pang Y, Yuan Y, Pan J (2011) Efficient HOG human detection. Signal processing, vol 91. Elsevier, pp 773–781

3. Szeliski R (2010) Computer vision: algorithms and applications. Springer
4. Dang C-T et al (2013) Vision based ground object tracking using AR. Drone quadrotor. In: International conference on control, automation and information sciences (ICCAIS). IEEE, Nha Trang, pp 146–151
5. Kalal Z (2011) Tracking learning detection. Ph.D. thesis, University of Surrey, Faculty of Engineering and Physical Sciences, Centre for Vision, Speech and Signal Processing
6. Barták R, Vykovský A (2015) Any object tracking and following by a flying drone. In: Fourteenth Mexican international conference on artificial intelligence, 2015
7. Kendall AG et al (2014) On-board object tracking control of a quadcopter with monocular vision. In: International conference on unmanned aircraft systems (ICUAS), 2014
8. Al-Majeed S, Fleury M (2014) Wireless handover with application to quadcopter video streaming over an IP network. In: IEEE international black sea conference on communications and networking (Blackseacom), 2014
9. Mohan A, Papageorgeiou C (2001) Example-based object detection in images by components. IEEE Transac Anal Machine Intell 23(4)
10. Lowe G (2004) Distinctive image features from scale-invariant keypoints. Int J Comput Vis 60(2):91–110
11. Belongie S, Malik J, Puzicha J (2001) Matching shapes. In: IEEE international conference on computer vision (ICCV), 2001
12. "HOG Descriptor". https://software.intel.com/en-us/node/529070
13. Raspberry Pi. https://www.raspberrypi.org/help/what-is-a-raspberry-pi/
14. Debnath T, Cranley N, Davis M (2006) Experimental comparison of wired versus wireless videostreaming over IEEE 802.11b WLANs. In: Irish signals and systems conference (IET), 2006
15. Yadvendra P (2013) The video streaming over Wi-Fi network over android platform. Int J Comput Sci Inf Technol Secur (IJCSITS) 3(4)
16. "OpenCV". http://opencv.org/about.html
17. "Python". https://www.python.org/
18. "Python (Programming Language)". https://en.wikipedia.org/wiki/Python_(programming_language)

Sonar Data Processing Using Multicore Architecture Processor and Embedded Linux

Varun K. Jayan and A. K. Mohamed Husain

Abstract **Background**: Traditionally, sonar signal processing is performed on dedicated signal processors to meet the real-time processing requirements. Recently, the computational capability in embedded general purpose processors has also multiplied with the introduction of multiple cores and vector processing units. Embedded Linux is the usage of Linux kernel and various open source components in embedded systems. The key advantage of Linux and open source in the embedded system is the ability to reuse components which allows to design and develop complicated products based on existing components and to have full control of the software part of the system. This project is aimed at bringing up a bare multicore power architecture processor board with Embedded Linux, benchmarking it for performance and implementing a digital signal processing application exploiting multiple cores and SIMD units in each core. **Methods**: The T4240RDB processor is developed using Yocto Project. Yocto is an open embedded project used to build custom-based Linux images. Next, the processor is benchmarked for its processing capabilities and then, signal processing application is done on the processor. **Findings**: Yocto Project can be used to develop architecture-specific images as well as images can be modified and additional recipes and packages can be added. The T4240RDB can easily do signal processing application. **Application**: The T4240RDB can be integrated with the sonar subsystem and can be used for real-time data processing.

Keywords PowerPC · T4240RDB · Yocto · Embedded Linux

V. K. Jayan (✉)
School of Electronics Engineering, VIT University, Chennai, India
e-mail: varun.kjayan2014@vit.ac.in

A. K. Mohamed Husain
NPOL, DRDO, Kochi, India
e-mail: husainak@gmail.com

D. Thalmann et al. (eds.), *Intelligent Embedded Systems*, Lecture Notes in Electrical Engineering 492, https://doi.org/10.1007/978-981-10-8575-8_30

1 Introduction

Traditionally, sonar signal processing is performed on dedicated digital signal processors (DSP) to meet the real-time processing requirements. Recently, the compute capability in embedded general purpose processors has also multiplied with the introduction of multiple cores and vector processing units. Although DSPs remain a suitable choice for many systems, developers are finding that dedicated DSPs are no longer required to run real-time digital signal processing application. Instead, developers are moving on to multicore processors [1].

The Freescale T4240 is an advanced RISC architecture (power architecture) processor with twelve 64-bit cores (e6500) running at 1.8 GHz. Each core is dual threaded (24 virtual cores) and has SIMD capability with the advanced vector processing technology (Altivec). T4240RDB is a reference design board with the T4240 processor. The board is designed for evaluating the processor for applications in networking, telecom/datacom, wireless infrastructure and military/ aerospace [2].

Embedded Linux is the operating systems based on Linux kernel that is used in embedded systems. The main advantage of using Linux kernel and open source in the embedded system is the reusability of components which allows you to develop complex products based on existing components and to have full access of the software part of the system. Yocto is a set of tools and methods to build customized Linux embedded systems. It is an open source project founded by the Linux Foundation in 2010.

2 Block Diagram

Figure 1 shows the block diagram of an active sonar system. The transmission signal generator (TSG) is used to generate pulses to power amplifiers. The power amplifiers powers the transducers at the transmission side, the transducer sends pulses to hit the objects. The echoes are received by the transducers at the receiver side, and received signal is processed using the T4240RDB. The processed data is displayed on a GUI.

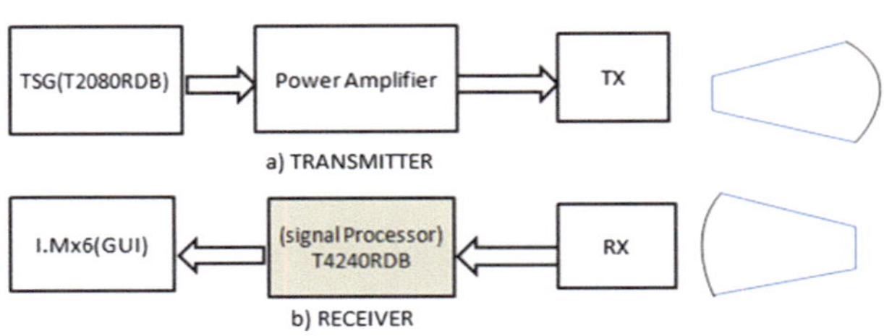

Fig. 1 Block diagram of: **a** transmitter and **b** receiver

3 Hardware

Figure 2 shows the T4240 block diagram. The T4240 consists of twelve e6500 cores. Each e6500 core has its 32 KB L1 instruction and data cache. A cache coherent interconnect manages the coherency among L1 cache, L2 cache, platform cache and memory. The processor also contains the 50 gbps networking and hardware acceleration. The hardware accelerators consist of the frame manager, queue manager, buffer manager, RapidIO Manager, SEC security, pattern matching and data compression saving CPU cycles for high value works. In addition, each e6500 core implements the Freescale Altivec technology SIMD engine, which achieves DSP-like performance for math-intensive applications [3, 4].

4 Software

4.1 *Yocto*

The Yocto Project is an open source project emphasizing on Embedded Linux-based system development. The project provides a build system, which pointed out to as an open embedded build system. The Yocto Project helps a software developer to develop their own file system and features for their corresponding architecture. Yocto provides only the Linux form of file system for development. Several board support packages are provided for various architectures by their own vendors.

Bitbake is a build engine that schedules tasks for the build. The reason for such a kind of build is that it reduces the task dependencies. The build environment provides configuration files where we can configure it according to user-specific applications. The build performs according to the configuration given and can build specific packages such as *compiler* and *make*. Open embedded core is a set of base

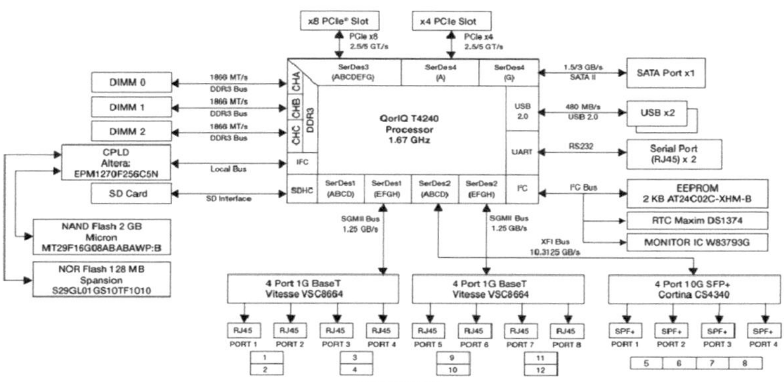

Fig. 2 T4240RDB block diagram

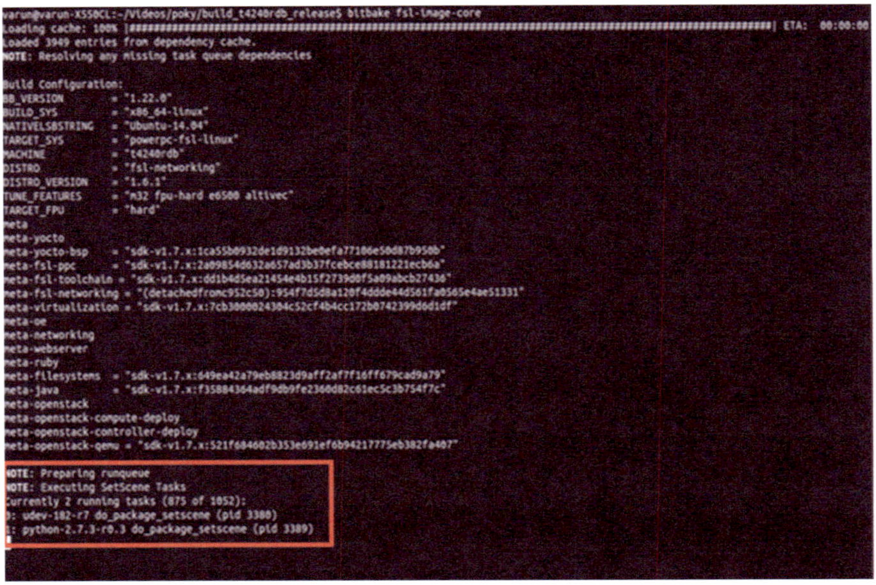

Fig. 3 File system image builing using Yocto

layers, set of recipes, layers and classes which are shared among all open embedded-based systems [5]. Poky is a reference system, and it is a collection of utilities, which is used to create a new distribution based on the Yocto Project. The build system of Yocto consists of source mirrors that are connected to the git repositories, and the source code is fetched and then necessary patches are applied to it, then it is compiled again; the files are filtered and analysed and then converted into the .rpm and .deb and .ipk formats and finally, quality analysis is done and then images are generated for the required architecture. Figure 3 shows the file system image building using Yocto.

4.2 *Software Deployment*

The Linux file system image that is built using Yocto is deployed to the target architecture (T4240RDB). Linux 3.2 is version built. It has features such as the GCC 4.8.1 and makes as configured during the build process. The size of the file system is around 322 MB. The increased size of the image file is due to the inclusion of core Linux features. The image built can be deployed using the TFTP or NFS deployment. TFTP is the method used here. TFTP server must be installed on the host PC; the server must be initialized and a serverip address must be given to the target board. (The board is assumed to have the uboot command shell up on a serial port.) The UImage file, the device tree and the file system are transferred

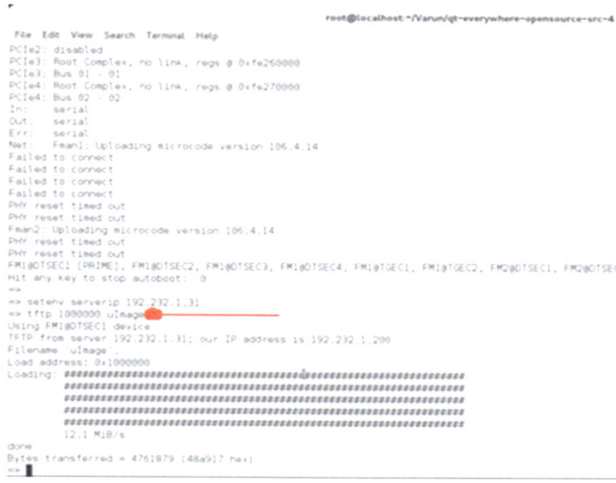

Fig. 4 Software deployment of the built image using TFTP

using the TFTP command. The target is booted using the bootm command with the images built using Yocto. The target board and host system are connected using an ethernet cable. The target board is connected to host system using serial port. Figure 4 shows the software deployment of the built image using TFTP.

5 Performance Analysis

5.1 Coremark

The coremark is a benchmark aimed to measure the performance of the central processing units used in embedded system. It is a set of parameters that is used to calculate the core performance, and it is measured in precise number of iterations.

In Fig. 5, the coremark performance parameters increase as the number of hardware threads increases in linear manner and hence can be concluded that the multithreading significantly improves the performance. The coremark performance test has been used by many processor manufacturing companies to test their processor performance.

5.2 Dhrystone

Dhrystone is a synthetic benchmark program designed to test the integer programming capability of the system. The Dhrystone consists of Dhry.h, Dhry_1.c, Dhry_2.c source files. The Dhry.h program is a set of header, global timer and structure definitions. The header describes the structure and mechanism of the Dhrystone program in detail. The Dhry1.c program is a set of wrapper and the main

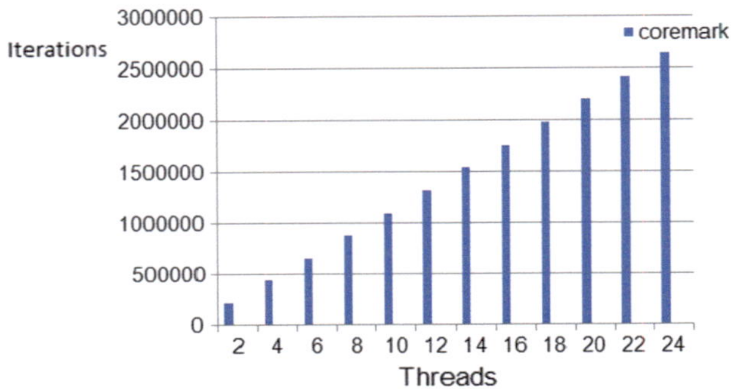

Fig. 5 Coremark scores for 24 threads

Table 1 Dhrystone result

Iterations	Dhrystones per second	DMIPS/Mhz
2,000,000	1666666.6	0.526
5,000,000	1818181.9	0.574

function. The wrapper contains timing definitions and variables that have been added for compilation.

Table 1 shows the result of the Dhrystone calculated for different number of iterations. The Dhrystone is calculated for 24 virtual cores without optimization the DMIPS/Mhz which is calculated by dividing the Dhrystones per second with 1757 and then by the frequency of the processor that is 1800 MHz resulting in an average of 0.55 DMIPS/Mhz.

5.3 *Floating Point Benchmarks*

Whetstone benchmark is a floating point benchmark that set industry standard for computer system performance. Whetstone benchmark performance is measured in Million Instruction Per Second (MIPS) and gemetric mean of floating point operations per second.

Fbench benchmark is a floating point operation completely depending on optical ray tracing algorithm and real-world applications. Figure 6 shows the floating point benchmark test consisting of fbench and whetstone benchmarks

5.4 *Altivec*

Altivec is a floating point and integer SIMD unit developed and owned by Apple, IBM and freescale semiconductor [5]. The vector unit has 128-bit registers that can

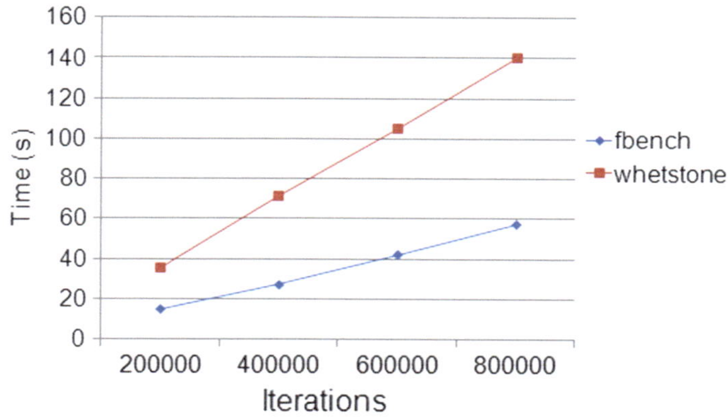

Fig. 6 Floating point benchmark performance test

Table 2 Altivec performance result

Multiply add matrices	Time taken in seconds			
$A = A + B * C$	Thread $N = 1$	Thread $N = 1$ Altivec yes	Thread $N = 12$ Altivec no	Thread $N = 12$ Altivec yes
1280	2.93891	1.2568	0.44421	0.129013
2560	23.3892	9.91572	2.548	0.887495
5120	186.651	79.2135	19.444	6.5750

be used as 16 8-bit signed or unsigned characters, 8 16-bit signed or unsigned shorts, 4 32-bit ints or 4 32-bit floats. The vector unit is specifically used for floating point operations that run DSP algorithms and other computationally intensive tasks [6].

From Table 2, it can be concluded that Altivec along with openmp parallelism provides significant performance improvement and can be effectively utilized for the high-end signal processing applications [7].

5.5 FFT Parallelization

Figure 7 shows the run-time performance of the FFT. Openmp directives are used in the FFT algorithms and are measured for the 2–24 processors [7]; the FFT algorithm gives a linear rise till 12 threads beyond that it is saturated. This is because the T4240 has 12 cores each dual threaded to give 24 virtual threads (threads in the same core share functional units such as floating point unit) [8].

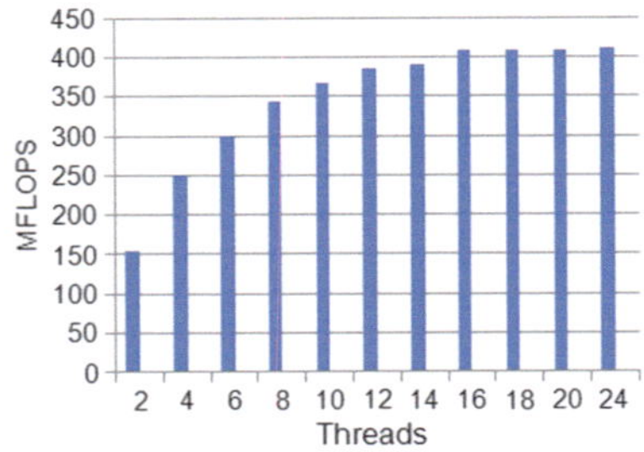

Fig. 7 FFT for different threads

6 Active Sonar Signal Processing

Active sonar signal consists of both transmitter and receiver; the transmitter transmits signal while receiver receives it and processes the signal. The active sonar signal processing system consists of transducers; when single transducers are used, it is called as monostatic and when an array of transducers are used, it is called as multi-static sonar. Compared to passive sonar which only has receiver section, the active sonar has both transmitting and receiving sections. The data processing has 3 units. Data reception consists of acoustic data reception; next, pre-processing consists of unpacking the data, data conversion, circular buffer filling, N-point-overlapped FFT computation, output to beamformer unit and beamformer unit; it consists of steering vector computation, pre-steering computation using steering and the FFT data from pre-processing unit, with the addition of data to form beam.

Functions required for the detection process

FFT computation, IFFT computation, complex kronecker product, complex QR decomposition, complex vector dot product, complex–complex matrix multiplication, complex hermitian computation; these functions are realized using the Eigen library; the Eigen is a high-level C/C++ header files for linear algebra, matrix and vector operations, geometrical transformations, numerical solvers and related algorithms [8].

The complex kronecker product, complex QR decomposition and the complex matrix product and the dot product for 640 matrices are measured in different number of cores using openmp. Figure 8 shows that matrix dot product, and matrix operations are parallelizable while QR decomposition and kronecker product are non-parallelizable.

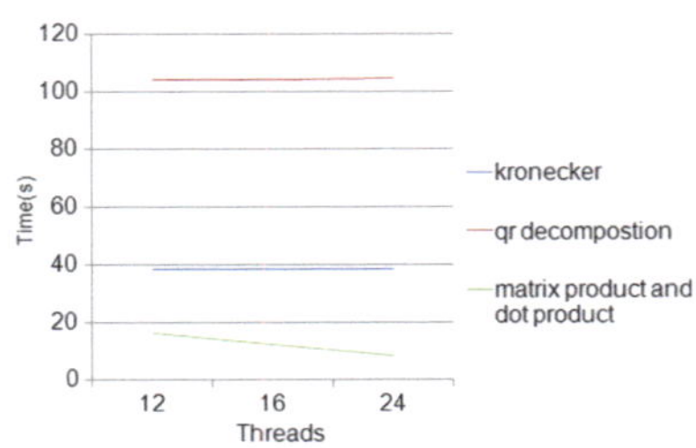

Fig. 8 Different functions on multicore

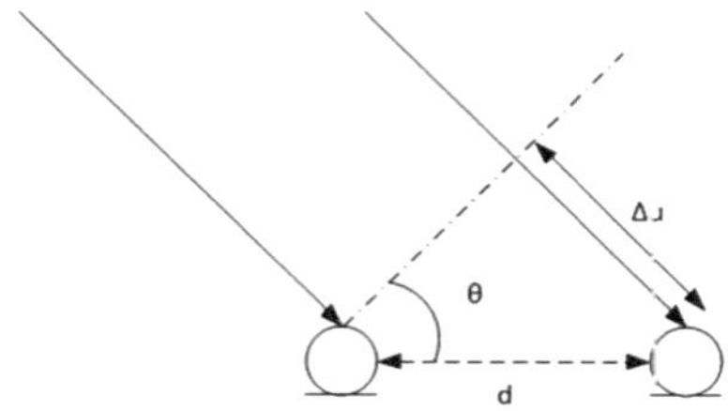

Fig. 9 Signal arriving at the transducers

Focusing signal on a particular direction is called as beamforming. The microphone/hydrophone should differentiate between various components coming from different directions, and it needs to suppress the signals that are not part of the target to analyse the signal. Beamformer uses an array of microphones, and the array of microphones is placed in a linear way. Suppose a signal is arriving at an angle of 60° to the linear array, the first microphone receives the signal first and the other microphone is delayed by a distance d between the array [9, 10].

$$\tau = \frac{d}{c}\sin(\theta).$$

Figure 9 shows the two sensor array arrangement and the delayed arrival of the signal to the sensors. Where 'd' is the distance between the two microphone. The speed of the sound is calculated using delay τ.

The steering vector is responsible for steering the array of elements in a particular direction. The direction to look upon is determined by steering vector and also finding the output power in the direction it is steered. The main process in the steering vector determination is the multiplication of matrix, which results in $O(n^2)$ operations. For every steering direction, this operation should be performed, which increases the execution time linearly as the number of steering directions increases. With a fixed number of steering angles, the computational complexity of the steering vector is less than that of the matrix inversion, *and* it will consume most of the execution time of the algorithm for sonar arrays where the number of array elements is less than the number of steering angles [11, 12].

7 Results and Discussion

Figure 10 shows the polar plot of the 4 element array, it has 2 grating lobes and one main lobe. The grating lobes are the unwanted frequencies and the main lobe indicates the direction of the maximum radiation. While Fig. 11 shows the polar plot of an 144 element array, it shows sharp spikes which is the direction it is looking. Fig. 12. Shows the plot for 12.8 kHz frequency for 4 elements. It shows the gain for

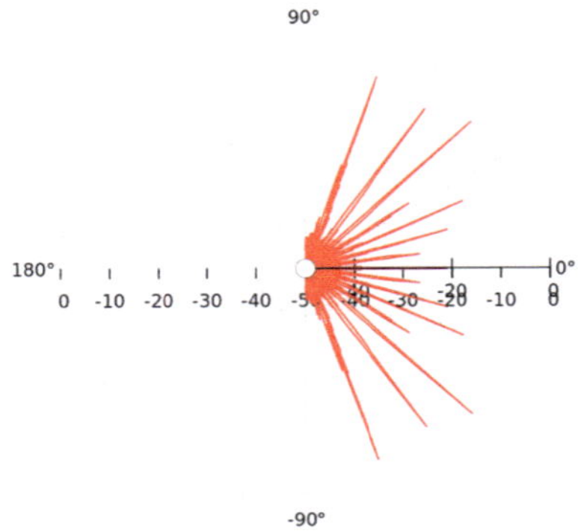

Fig. 10 Polar plot of 4 element array

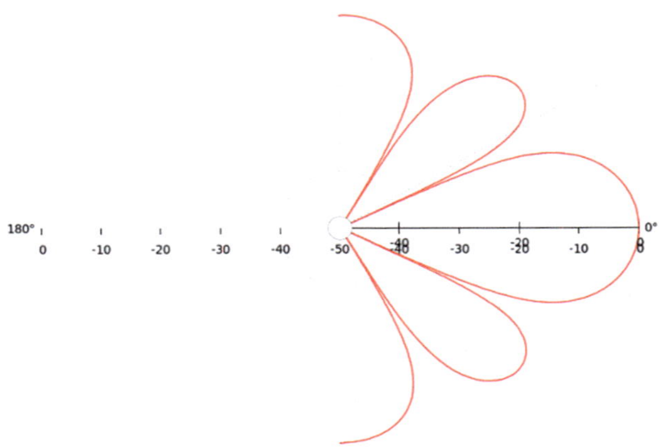

Fig. 11 Polar plot of 144 element array

arrival angles. The peaks in the figure shows increased gain in a particular direction. Fig. 13 shows the Beamformed output of an 144 element array at 12.8 kHz frequency. Higher gain in a particular arriving angle indicates the maximum radiation in that direction.

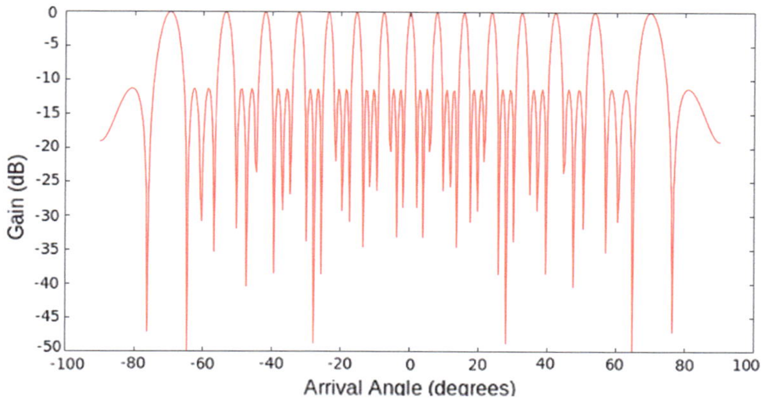

Fig. 12 Plot showing for 12.8 kHz frequency for 4 elements

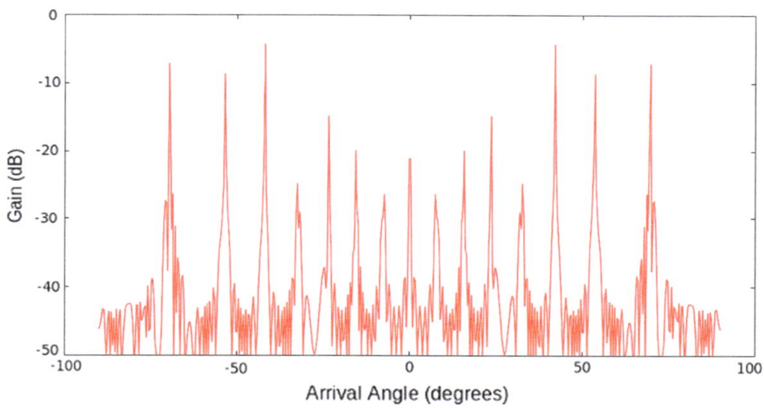

Fig. 13 Plot showing for 12.8 kHz frequency for 144 element array

8 Conclusion and Future Work

The project explored the use of a multicore power architecture processor (QorIQ T4240) with Embedded Linux for use of high-performance computing tasks such as sonar data processing which was traditionally carried out by dedicated digital signal processors. Extensive benchmarking was performed with different tools, and the processor was found adequate for use in sonar data processing applications. The future work involves implementation of adaptive beamforming techniques on the multicore platform and integration with actual sonar.

Acknowledgements I would like to express my special gratitude and thanks to my external guide Mohamed Husain A. K., Sc-E, NPOL (external project guide) whose guidance and immense support encouraged me to complete the project successfully. And I would also like to thank Prakash V (internal project guide) for his suggestions.

References

1. Leppakoski A (2013) Framework for industrial embedded system product development and management. Konecranes Plc, Hyvinkaa, Finland; E. Salminen; T. D. Hamalainen, 23–24 Oct 2013
2. IEEE Std 1275-1994 Standard for Boot (Initialization, Configuration) Firmware, Core Practices and Requirements
3. Farnell, NXP powerpc datasheet
4. Peng D, Zhang H, Weng J, Li H (2010) Research and design of embedded data acquisition and monitoring system based on PowerPC and CAN bus. 8th world congress on intelligent controla and automation (WCICA), 7–9 July 2010
5. Yocto project documentation, Yocto project manual
6. Citron D, Exploiting the AltiVec Unit for Commercial Applications. Hiroshi Inoue, Takao Moriyama, Motohiro Kawahito, Hideaki Komatsu, Toshio Nakatani IBM Tokyo Research Laboratory 1623-14 Shimotsuruma, Yamato-s
7. Fuller S, Motoroals altivec technologys. Networking and computer core technology
8. Chu E (2010) FFT algorithms and their adaptation to parallel processing. Alan George. Department of Computing and Information Science and Department of Mathematics and Statistics, University of Guelph, Guelph. Ont., Canada, International Linear Algebra Society, 2010
9. Quazi A (1981) An overview on the time delay estimate in active and passive systems for target localization. IEEE transactions Accoustic speech and signal processing, June 1981
10. Griffiths L, Jim C (1982) An alternative approach to linearly constrained adaptive beamforming. IEEE Trans Antennas Propag 30(1):27–34
11. Jacek Dmochowski SA, Benesty J (2009) On spatial aliasing in microphone arrays. IEEE Trans Signal Process 57(4):1383–1395
12. Quinn MJ (2004) Parallel programming in C with MPI and OpenMP. McGraw-Hill Higher Education, New York

A Novel Black Box System for Automobiles

S. Sriram and V. Prakash

Abstract According to the World Health Organization (WHO) report, about 1.25 million people die due to road accident injuries every year. And majority of road crashes are caused by human error. The main objective of this research work is to design a dedicated black box system for automobiles which is used to determine the reason for the accident. Data acquisition is an important process that involves gathering information. By analyzing the various data recorded during road accident, the safety interventions and technology can be implemented to avoid such incident in the future. The black box acts as a data acquisition device, and the acquired data are stored in the Raspberry Pi. During the time of collision, the data acquired are transferred to the traffic control server's mail id. Sensors play a vital role in the proposed system to identify and analyze the reason for the accident. Sensors list include accelerometer sensor, carbon monoxide sensor, flame sensor. This paper explores the various reasons for accident and the simultaneous catastrophic effects as the result of the collision. It introduces two concepts: One is to determine the possibility of carbon monoxide leakage inside the vehicle cabin space and the other is to determine whether the vehicle has entered into no entry region; this is done using ultrasonic sensor paired with an Arduino and ZigBee.

Keywords Black box · Data acquisition · Raspberry Pi · Arduino UNO Event data recording · ZigBee

S. Sriram · V. Prakash (✉)
School of Electronics Engineering (SENSE), VIT University, Chennai, Tamil Nadu, India
e-mail: praksh.v@vit.ac.in

S. Sriram
e-mail: 12sriram12@gmail.com

D. Thalmann et al. (eds.), *Intelligent Embedded Systems*, Lecture Notes in Electrical Engineering 492, https://doi.org/10.1007/978-981-10-8575-8_31

1 Introduction

Accident is defined as an undesirable or unfortunate happening that occurs intentionally or unintentionally and usually results in harm, injury, damage or loss, causality, mishap in which the injured person gets compensation. But not all road accidents are considered unintentional. Various factors are involved to determine whether the happening was intentional or unintentional. The major work of accident analysis involves finding out the reason for the accident. Based on the analyzed results, victim can claim for medical insurance and get proper justice. In recent years, people have started using scientific techniques to find out the cause of the accident. That's where black box system comes into action.

Black box is a device which can be viewed in term of input and output, wherein the real internal working is not known. The same concept applies to the proposed black box system for automobiles which uses various sensors which are highly reliable and selected to make the entire system reliable. People who are involved in the accident analysis procedure just collect and analyze data, rather than learning the internal working of the black box. This is the reason why the black box must contain sensors values that are proven to be reliable for the implementation of the proposed idea. The black box data are transferred in the form of mail along with images from the scene. It concludes by briefing the need for black box system and its future enhancements.

The cause of accident can be due to the person or the automobile. If it is due to automobile, the factors must be determined whether it is due to engine malfunction, improper braking system, fuel leakage, worn-out tires, etc. Designing a black box system must take various factors into consideration. Factors such as positioning of the black box, sensors used for analysis and the reliability of each sensor, defining vehicle collision based on the crash sensor data. Classifying these factors to determine the imapct of the collision [1].

Law enforcement is made in such a way that it protects people. The traffic rules are one among those laws that must be followed to protect oneself and the people. Most of the accidents that result in loss of life are due to breaking of these rules. Thus, the black box for automobiles must be designed in such a way that it must find out whether the traffic rules where abided.

2 Related Work

The type of black box system in vehicles is called as event data recorders (EDR). The EDR is manufactured as a product that is present in most of the automobiles such as Hyundai, Ford, GM. EDR system just records various information such as vehicle speed, fuel information, cruise control status. EDR is available in two models; in the first model, the various values are recorded throughout the drive and the sensor values are overwritten after a particular time delay. During the time of crash, the last collected sensor values are stored in the EEPROM memory of the processor.

Second model of the EDR starts recording the events only when the collision occurs. In both models, the sensor data are stored based on the available EEPROM memory of the processor. Traditional EDR system have both notable pros and cons. EDR is manufactured by various automakers such as General Motors, Ford, Fait. Each has their own unique features and notable amount of competition between them to create rivalry and excludability among customers. General Motors have their own home-manufactured event data recorders. The GM EDR is known as sensing and diagnostic module (SDM) [2].

GM's EDR considers various parameters such as vehicle speed, engine throttle position sensor, and brake status, five seconds before the collision. General Motors did not build its EDR overnight; its EDR has evolved over years based on various factors such as the requirement by the customer, rules and regulations posed by the law. GM has constantly updated its EDR based on the needs. Similar to General Motors, Ford also came up with its own event data recorder known as restraint control module (RCM) [3]. Restraint control module focuses much on the passenger restraint system. Restraint control module checks mostly the air bag deployment status. Ford has also included pretensioners in the seat belt.

Pretensioners consist of a piston which is driven when deceleration occurs due to the impact of the accident. This helps in putting the driver back to place to avoid the forward pull toward the dashboard of the car. Air bags include multistage, frontal, and side air bags. Ford has two models of RCM available for download from Vetronix CDR. The former detects both the longitudinal and lateral pules, whereas the latter stores only the longitudinal pulse. Ford's RCM has more resolution than GM's EDR. Since it is known that the typical crash duration exists more than 100 ms, the Ford's EDR is not efficient and capable of storing the entire event.

3 Drawbacks in Existing System

The debate between the necessity of a black box or EDR is still going on. EDR is proved to report accidents accurately, but it takes the cost of passenger privacy. None of the vehicle owners wanted a spying car which spies on their owners. The present EDR technology usually monitors the deployment of air bag system, seat belt status during crash, speed of the vehicle before and after impact termed technically as acceleration and deceleration, brake status.

On an overview, these parameters considered are passenger oriented, and if at all an accident occurs, it might prove the driver has done it. Electronics should not give false information because people believe more in computer than humans. Banking transaction, phone bills, medical record affect consumer's privacy. Deciding the reason for an accident based on the vehicle speed and seat belt has affected the comfort level of the driver.

Security of these recorded information is also a problematic issue for the judiciary and police department. The stored data can be manipulated to falsify the information recorded. In certain EDR, constant update of the software is required to make the system reliable. For example, Ford has two versions of its software. People really do not know what these EDR can do to their vehicle, and all they are aware of is that they are carrying a spy along with them in the vehicle.

The general problem identified is the reading of speed values, wearing of seat belts which most of the passengers feel uncomfortable wearing seat belts. These issues are considered and rectified in the work proposed.

4 Proposed Work

This paper presents a novel black box system for automobiles like in aeroplanes to ensure vehicle safety and recording of various sensor data before and after the collision. The processor used for this purpose is Raspberry Pi. Raspberry Pi is used for collecting various sensor data such as ultrasonic sensor for detecting obstacles and to measure the distance between the vehicle and the object; MEMS vibration sensor is used to determine the brake status, accelerometer, brake oil level sensor, CO sensor, temperature sensor, flame sensor, rain sensor. Raspberry Pi is used for maintaining driver profile using fingerprint module. All these data are stored in the black box or EDR [4].

When the vehicle is about to hit an object, the ultrasonic sensor placed acts like a pre-crash sensor. Braking distance determines the possibilities for the accident [5]. Camera is also used to ensure the recording of incidents after the accident has occurred. The fingerprint module is used to recognize the driver. The owner must decide the person who can drive the vehicle. This ensures the safety of the vehicle and avoids intentional accidents. GPS is included to provide the location of the accident. This helps in multiple ways; first, the information regarding the accident can be reported to the traffic police or call an ambulance automatically for help. Second, if multiple accidents occur frequently, the particular location can be marked as accident-prone area and the necessary details can be sent to the government officials for further actions. The data recorded can be used by police for enquiry and to find out the reason for accident. If the accident has occurred in remote regions, the GPS is more useful in finding the location of the vehicle [6]. Since mail is sent along with pictures of the incident along with various sensor data, the response team can prepare accordingly to react. This saves lot of time.

The reliability of black box is strengthened by using fingerprint module for driver identification. All these data are recorded in the black box in the system. This ensures the identification of the person in the driver seat when he/she meets or causes an accident.

Braking system of a vehicle plays a major role in ensuring the safety of the vehicle and driver. Most of the accidents are caused due to brake failure. The reason for the brake failure is leakage of brake oil, worn-out brake pads, damaged disks in case of disk brakes. The failure of brake lamps is also a reason for accident. Thus, the owner must be aware of all these factors. Level sensor is used to determine the level of the oil; LDR is used to determine the status of the brake lamps. Vibration sensor is used to determine the brake status.

Carbon monoxide leakage is a life-threatening issue in some vehicles [7]. CO can be found in fumes which are produced as a result of burning fuel in vehicle engines, both light and heavy vehicles. If the CO is build inside the cabin, it can poison people. The most common indications of CO poisoning are headache, dizziness, vomiting, and chest pain.

These symptoms can be falsely assumed to be general fever. But constant exposure to carbon monoxide can cause the person to pass out, or it can even kill them [8]. If the person is drunk or sleeping, they can die without any of these symptoms. Carbon monoxide is produced when there is not enough oxygen for the burning process; thus, instead of producing carbon dioxide due to less oxygen, carbon monoxide is produced.

The produced carbon monoxide can enter into the vehicle cabin if there is problem with vehicle exhaust system. If the air conditioner is turned-on, the carbon monoxide may enter the cabin. Since carbon monoxide is colorless, odorless gas, its presence cannot be determined easily [9, 10]. It has been proven that a concentration of carbon monoxide in the range of 30–40 ppm is enough to kill a person for a prolonged exposure [11].

Possibility of the engine setup into fire is less or very rare, but the fuel tank can quickly catch fire when the vehicle is hit near the fuel tank [12]. A flame sensor is used to check whether the fuel tank blew up. Usually, poor quality fuel tanks suffer from this problem. Braking system can be working fine when considering the brake pads, brake oil level, brake lamps. But the time to apply the brake also determines the occurrence of the accident. This is called response time of the driver [13, 14]. Traffic rules are meant to be followed to ensure safety on the road; it includes road signs, danger zones, one way, speed limit zones. If the vehicle enters into the wrong direction in spite of the warning sign, it is proven as the mistake of the driver. For detecting this, a separate setup is used which is located outside the vehicle [15]. This module consists of Arduino, ultrasonic sensor, and ZigBee module.

Fuel leakage from the fuel tank is also possible; fuel leakage in car is very dangerous. If there is a fuel smell is the vehicle, it is a sure sign of problem. It is potentially dangerous and may lead to catastrophic effects such as burning of the vehicle.

The most common reasons for fuel leak are:

1. Injector leak
2. Fuel tank leak
3. Fuel line leak
4. Faulty canister
5. Missing gas cap

The reason for fuel leakage may extend, but it really does not matter what is the reason for fuel smell in vehicle. It must be fixed as soon as possible. If the fuel leakage is present near the engine or exhaust, it can cause catastrophic effects causing potential burning of the vehicle. To find out this problem, a gas sensor (MQ2) is used which is capable of detecting smoke caused as a result of fuel burning. The fuel-level indicator is also included which is of analog. If there is an abrupt decrease in fuel, it might be due to fuel leakage. All these readings along with the photos are mailed to the mail id; in this case, it is the accident response team of the police department. The security and reliability of the sensor data is questionable since it is transferred in the form of mail, and it is known that no system is safe. Here comes the need for encryption. The sensor data are written into a text file which is encrypted and stored in the black box system.

5 Design and Implementation

The design mainly concentrates on the various sensors used, and the processor must be chosen so that it can process the acquired sensor values efficiently. Sensors were handpicked based on the various issues faced by both the vehicle owners and the manufacturers. The sensors were chosen in such a way that it should not cause any privacy issue like the previous event data recorders (EDRs). The design part has taken this into consideration; in fact, the requirement of private transport such as cars, SUVs, trucks is for comfort and above all privacy. EDRs record the speed and the seat belt status prior to collision, and it became easy for the law to end the analysis by blaming on the driver. This is the major reason for which people were afraid to install the EDR in their vehicle. Thus, considering this issue, both the speed sensor and seat belt sensor are carefully eliminated in the design. The list of sensors chosen is to find the various parts in which the manufacturer fails such as faulty exhaust system, fuel tank, brake lamp failure, poor oil storage tanks. All these are the deciding factors for an accident which might end up in fatality. Carbon monoxide has proven to be a constant threat to the passengers. There is possibility of carbon monoxide leakage inside the vehicle cabin. This is caused mainly due to poor exhaust system or damaged air-conditioning system. Camera is included but does not monitor the vehicle the whole time. Camera is activated only when the crash has occurred or sometime prior to crash. The camera is triggered by the ultrasonic sensor. The only restriction for the owners is inclusion of fingerprint which records the person driving the vehicle.

Figure 1 shows the block diagram for the proposed system consisting of Raspberry Pi and various sensors along with fingerprint and ZigBee module. The proposed system requires a vast number of hardware since the system is involved in accident analysis. The Raspberry Pi is chosen as the master processor that records all the sensor data and processes it. Raspberry Pi sends mail to the given mail id; in this case, it is the mail id of traffic police using simple mail transfer protocol (SMTP). Arduino is used for identifying whether the vehicle has entered into a wrong region. If all the vehicles have entered into a wrong region such as no-entry zone, a message reading no entry is transferred to the Raspberry Pi using ZigBee module.

Fingerprint is used to identify the person on the driver seat. Fingerprint-based vehicle entry may cause privacy issues; to avoid this, the Raspberry Pi records the person identity and checks owner or someone else is driving the vehicle. The fingerprint information can be used for investigation and insurance claims.

The hardware consists of Raspberry Pi which is the master processor, Arduino which is a slave that works externally. The slave processor setup is shown in Fig. 2.

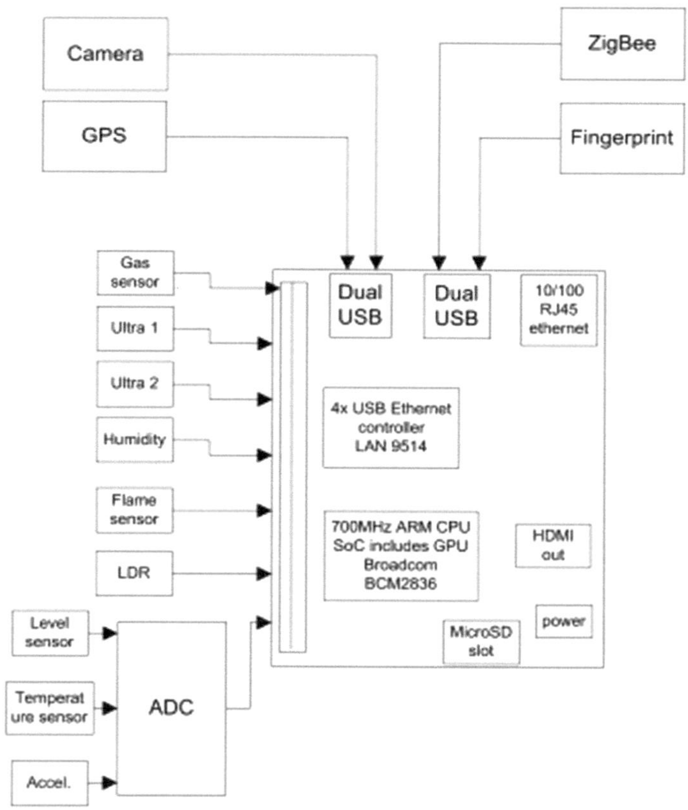

Fig. 1 Overall block diagram for the black box system

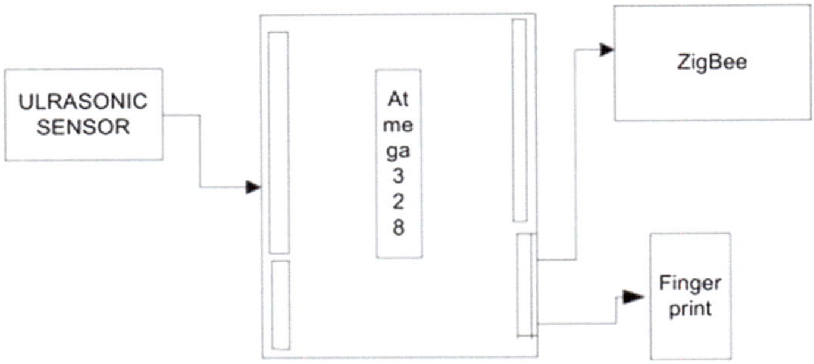

Fig. 2 Arduino and ZigBee setup placed externally

6 Results

The sensors were connected to the Raspberry Pi either directly or through ADC. The ADC used for analog to digital conversion was MCP3008. It is a 10-bit, 8-channel ADC. The fuel level sensor shown in Fig. 3 is analog, it is connected to the ADC, and the type of sensor used for measuring fuel level is contact method which has proven to be more reliable. The average fuel capacity is 50 L for a passenger car. The ratio of fuel decrease entirely depends on the speed of the vehicle. Higher speeds affect the miles per gallon (mpg) value of the vehicle. Figure 7 shows the analog fuel level sensor.

Figure 4 shows various sensors connected to Raspberry Pi to analyze the reason for accident.

Ultrasonic and ZigBee setup to determine whether the vehicle entered into the wrong zone is shown Fig. 5.

The fingerprint sensor is connected to the ZigBee module as shown in Fig. 6 for driver authentication. Fingerprint module determines whether the person driving the vehicle is authorized or not.

Fig. 3 Analog fuel level sensor

Fig. 4 Various sensors attached to Raspberry Pi

Fig. 5 Arduino, ZigBee, and ultrasonic setup

All the connections were made with the system for processing and viewing the results; the overall setup is shown in Fig. 7.

The occurrence of accident is notified to the central server via a mail. The mail contains all the information regarding the accident. The information include sensor data for analysis and image from the accident scenario. The image is shown in Fig. 8.

The SMS alert of the accident is shown in Fig. 9. SMS alert is sent via black box.

The accident location also can be viewed through Google Map app which is shown in Fig. 10.

Fig. 6 Fingerprint sensor module

Fig. 7 Black box setup

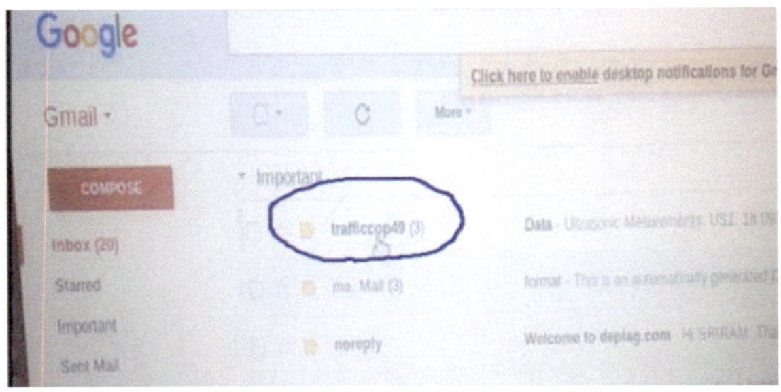

Fig. 8 Mail received indicating the accident and prints the sensor values

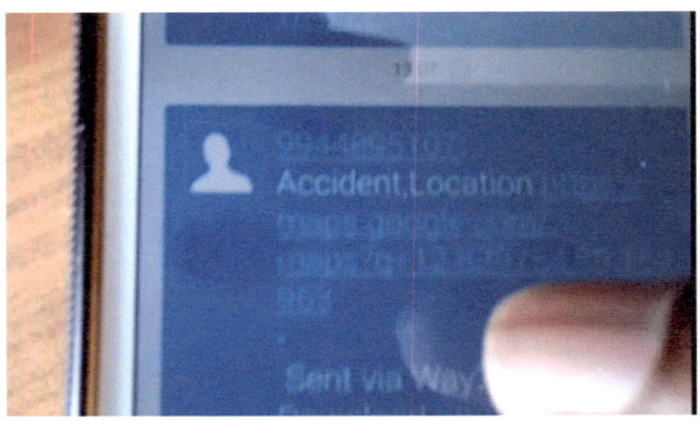

Fig. 9 SMS alert showing the accident location

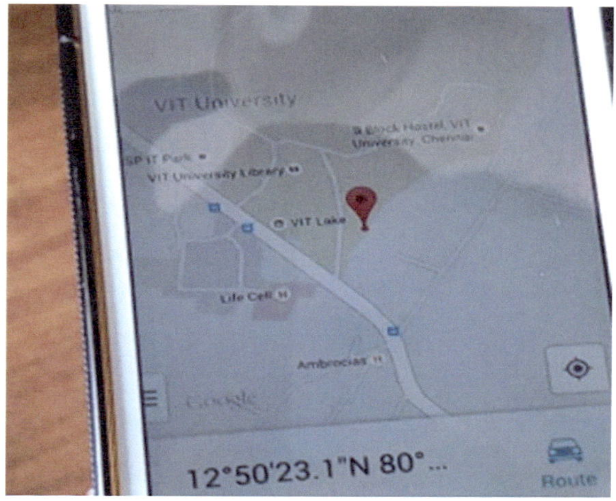

Fig. 10 Accident location shown in Google Maps

7 Conclusion and Future Work

Event data recorders have started to expand across borders; in many countries, more than 90% of vehicles are equipped with the EDR or black box system. Even though the debate for the need for black box is going on, the need of EDR has proven essential for accident analysis. Various factors such as camera, speed sensor, brake position sensor may testify the driver, but they are required for analysis. The black box worked out in this paper addressed the issues other than sensors related to the depletion of air bag which includes seat belt sensor, crash sensor, and pretensioner levels. All the sensors used in a generic EDR are related to air bag system. CO poisoning inside the vehicle is more dangerous than anyone can imagine. The sensors chosen do not affect the privacy in any way. Thus, the goal of the proposed black box was achieved. Since the data are transferred along with the image of the accident scene, data reliability are ensured. GPS is included for locating the region of accident and to respond immediately.

Artificial intelligence can be brought into the proposed system to record the essential data at the required time to save lot of memory. Dedicated servers can be created by the government to maintain the records of the accidents. This will make the analysis even easier.

Acknowledgements Authors would like to thank the anonymous reviewers for their comments in improving the paper, and also, we extend our gratitude to VIT University, Chennai, for their support.

References

1. Kassem A, Jabr R, Salamouni G (2008) Vehicle Black Box System. IEEE Xplore
2. General Motors EDR system, Use of Event Data Recorder for highway crash analysis, National Cooperative Highway Research Program (NCHRP), Transportation Research Board 2008
3. Ford Motors EDR system, Use of Event Data Recorder for highway crash analysis, National Cooperative Highway Research Program (NCHRP), Transportation Research Board 2004
4. A Design approach for intelligent vehicular black box system with intra-vehicular communication using LIN/Flex-ray protocols, IEEE 2008
5. Fambro D, Koppa R, Picha D, Fitzpatrick K (2014) Driver braking performance in stopping sight distance situations. Transp Res Rec: J Transp Res Board
6. Prasad MJ, Arundathi S, Anil N, Harshikha, Kariyappa BS (2014) Automobile black box system for accident analysis. IEEE
7. Ferenczi L, Cadar S, Costiug S, Pitl G, Aciu A, Ghita A (2006) Methane and carbon monoxide gas detection system based on semiconductor sensor. In: Research institute for analytical instrumentation, international conference on automation, quality and testing, robotics. IEEE
8. Nagata S (1994) Carbon monoxide detector unit. Gas Equipment Dev. Div., Yazaki Meter Corp., Shizouka, Japan. IEEE
9. Hampson NB (2009) Cost of accidental carbon monoxide poisoning: a preventable expense. Prev Med Rep
10. Hampson NB (2012) Carbon monoxide poisoning and risk for ischemic stroke. Eur J Intern Med
11. Ramya V, Palaniappan B, Karthik K, Prasad S (2011) Embedded system for vehicle cabin toxic gas detection and alerting. Procedia Eng
12. Where There's Smoke, There's Fire, Walters Forensic Engineering 2010
13. Hosseinlou MH, Ahadi H, Hematian V (2012) A study of the minimum safe stopping distance between vehicles in terms of braking systems, weather and pavement conditions. Indian J Sci Technol
14. Wu D (2010) Test analysis and theoretical calculation on braking distance of automobile with ABS. In: Li J, Shu X, Zha X, Xu B. Affiliated with School of Mechanical and Electronic Engineering, East China Jiao tong University. Springer
15. Landge V, Sharma AK (2013) Identifying Accident prone spot on rural highways. Int J Chem Environ Biol Sci IJCEBS

IOT-Based Automated Aeroponics System

Felin Francis, P. L. Vishnu, Manish Jha and Bharghava Rajaram

Abstract Aeroponics is a soilless method of growing plants in air with the assistance of water vapor or mist environment, using an enriched nutrient solution sprayed as a mist, in order to attain faster plant growth. One of the key ideas behind aeroponics is to keep the growth environment pest- and disease-free, so that the plants can grow faster and healthier. In this paper, we propose an automated aeroponics system in which we use sensors for measuring the temperature, humidity, pH value of water, and the light exposure in the environment where plants are grown. An LED grow light is used as source of light for photosynthesis instead of sunlight. The key irrigation apparatus is spray jet which is used to spray nutrient-mixed mist directly to the dense root system. The temperature sensor will sense the temperature of the environment, and if it exceeds beyond the threshold value, the cooling/ventilation system will compensate for it. Similarly, for humidity, the spray jet and a dehumidifier are used to stabilize the environment. The pH control mechanism will keep the pH value around 6.0. We are incorporating Internet of Things (IoT) for continuous monitoring of data, data analysis, and data logging. Data from each sensor are collected at regular intervals of time and are uploaded to a cloud for monitoring, followed by data analysis, for increased productivity. This data can be used to streamline the automation process, and also for other users to increase crop yield.

Keywords Aeroponics · Precision agriculture · Internet of Things Automation

F. Francis · P. L. Vishnu · M. Jha · B. Rajaram (✉)
School of Electronics Engineering (SENSE), VIT University, Chennai, India
e-mail: bharghava.rajaram@vit.ac.in

D. Thalmann et al. (eds.), *Intelligent Embedded Systems*, Lecture Notes in Electrical Engineering 492, https://doi.org/10.1007/978-981-10-8575-8_32

1 Introduction

Aeroponics is a cultivation system in which plants are grown without soil. Here the roots are hand suspended in air and the nutrient mixture which is required for the plant growth is given to the hanging roots in the form of fine mist [1]. This technique of growing plants was developed in 1920s for the first time by botanists who used this technique to study the structure of root system [2].

Since the aeroponics uses mid-air feeding, it helps the roots to absorb much amount of oxygen. This in turn helps in higher metabolism and growth rate [3]. The experiments show that the growth rate increases up to 10 times than in soil. Also, water loss is very much reduced because of less amount of evaporation [1].

Aeroponics is done with air combined with droplets of nutrient solution. Almost every plant will grow to maturity in short span because of the abundant availability of oxygen nutrients and water [1]. Aeroponics is the best option in areas where the soil is not suited for plant growth like space colonies [2].

Today, aeroponics is gaining more momentum since people are aware about the contaminants and chemicals present in food supply. Aeroponics also provides relief to farmers that they do not have to further worry about the changing climatic conditions. Owing to increased water scarcity in several parts of the world, future commercial food production can be sustained using such soilless growing techniques. NASA is also giving considerable attention to aeroponics, since it is easier to handle mist than a liquid at zero gravity conditions [2].

We exploit the advantages inherent in an aeroponics system in terms of increased controllability and nutrition management. An indoor aeroponics test bed was set up in order to implement and test the automation features. By controlling the root temperature and moisture, ambient light, and nutrient utilization, optimal plant growth can be achieved. In this work, off-the-shelf sensors are used to measure root temperature and moisture, pH (for nutrients), and ambient light. Unlike soil-based temperature and moisture sensors which are more expensive (have to be placed deep in the soil, and thus have to be impervious to corrosion and other contamination), we make use of a cheaper ambient temperature and humidity sensor placed near the roots of the plants. The overall block diagram is shown in Fig. 1. This is further explained in the following sections.

2 Related Work

In this section, we discuss work related to aeroponics with regard to automation and remote monitoring. For example, in the area of hydroponics (growing on water instead of mist) for the controlling and optimization of long period plant growth, a

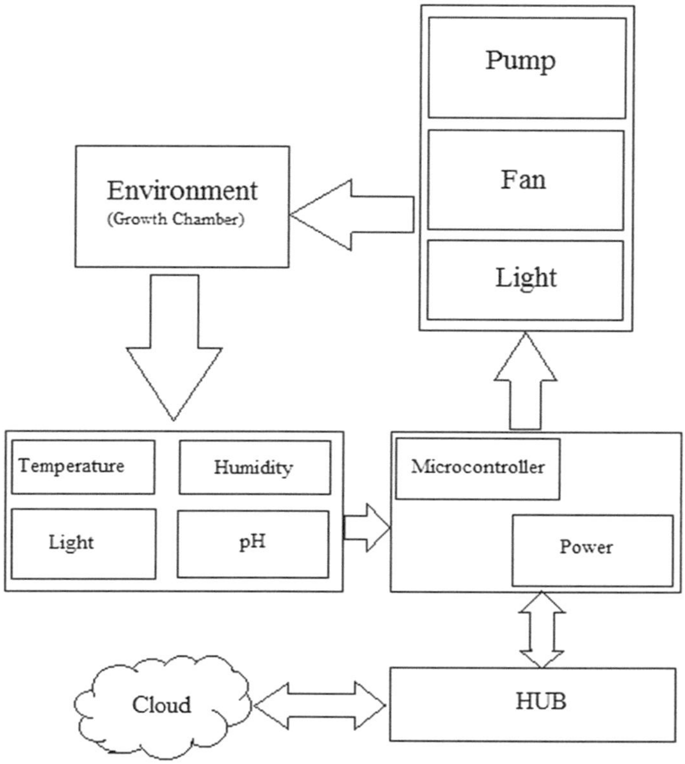

Fig. 1 Overall block diagram. Here, the HUB is an IoT gateway which connects to the cloud using traditional HTTP POST requests. The microcontroller communicates with the HUB over Zigbee (IEEE 802.15.4)

control system with hierarchical intelligence which includes an expert system based on neural networks and genetic algorithms which also includes a hybrid system was done [3]. Depending upon the plant growth, these two control systems were appropriately used. In this paper, the authors propose optimization of plant growth based on the nutrient content in the water used. An expert system was deployed for calculating the appropriate values or set-points of nutrient concentration throughout the whole period of the growth phase of the plant, and the hybrid system is used for calculating the ideal set-points of nutrient concentration which gives the maximum total leaf length (TLL) and maximum stem diameter (SD) during the initial stage of growth also called as seedling stage [3].

One of the biggest threats in a hydroponic system is the vulnerability of spreading diseases so easily [1]. The spreading of disease mainly occurs through the nutrient distribution system. Early disease detection is the only possible solution in a hydroponics system so it is crucial to detect disease at the earliest. JAPIEST is an integral intelligent system which was developed for the control and diagnosis of tomato diseases and pests in hydroponic systems [1]. JAPIEST is capable of diagnosis, prevention and control of diseases that affect tomatoes and is well suited for early detection of diseases, and it also proposes the suitable control treatment [1].

3 Proposed Work

In the proposed model, we are introducing an automated aeroponic system by controlling the humidity, temperature, and light cycles automatically. By automating humidity, wastage of water can be prevented to a great extent. We are also incorporating the concept of Internet of Things for uploading data related to system onto a cloud network. These data will be useful for others who are performing similar cultivations. Also through IoT, the user can monitor the health of plants from anywhere. This section first introduces the hardware components used for our implementation followed with the control algorithm. We construct individual sensor nodes which then communicate to a Central HUB over a Personal Area Network (Zigbee in our case), which then communicates to data to a cloud. Initially, we proceeded with the design of a grow bed for our experimental setup. Multiple LED strip lighting was used to compensate for the ambient light. Aluminum foil was used to enhance the retention of light, temperature, and humidity. This grow bed is shown in Fig. 2.

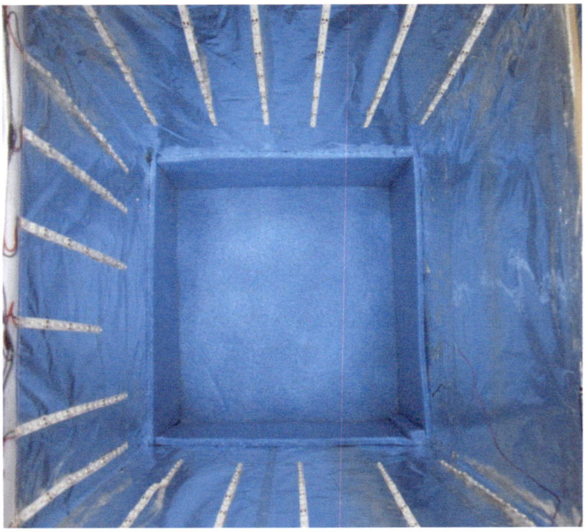

Fig. 2 Grow bed

4 Hardware Components

4.1 Arduino UNO

We make use of an Adruino Uno R3 microcontroller board. The choice of the microcontroller was based on the availability of extensive sensor and interfacing libraries and the presence of an active online support community. The Arduino Uno uses an ATMega328P 8-bit microcontroller with 32 kB flash memory for program and data storage.

4.2 DHT-11 Humidity and Temperature Sensor

Typically, for precision agriculture, sensors like the SHT1x are used. SHT1x is designed to prevent rusting and for increased precision, but at the same time expensive ($\sim$25$). However, in the proposed aeroponics system, we make use of a DHT-11—ambient temperature and humidity sensor. This sensor costs less than 2$, measures only the ambient temperature and humidity and is not specific to agriculture. But since the roots are suspended in air, such a sensor is sufficient in making decisions with regard to root temperature and moisture.

4.3 Real-Time Clock (RTC)

We make use of a Tiny RTC DS1307 serial real-time clock (RTC) to provide the required time pulses needed for automation. An external RTC is required due to the varying time intervals at which the different parameters are controlled. For example, the lighting system operates on a day–night cycle depending on whether the plant is in the growing phase or flowering phase. The RTC used in our paper is a binary coded decimal (BCD), low power, clock/calendar with 56 bytes of NV SRAM. A separate power backup is used in the form of a 3.3 V lithium-ion cell. The RTC is capable of reverting to the backup supply on power interruption. Timekeeping operation continues while the part operates from the backup supply. The RTC is interfaced with the ATMega328P microcontroller using the I2C bus.

4.4 Analog pH sensor

A pH sensor is used to sustain the amount of nutrients in the water used for spraying mist on the roots of the plants. Typical nutrients increase the acidity of the water solution. The ideal pH of the water is 6.0 [4]. We use this as the benchmark pH for the purpose of automation.

4.5 *Zigbee Transceiver Modules*

We use Zigbee as the communication medium between the individual sensor nodes.

4.6 *Raspberry Pi 3*

We use a Raspberry Pi 3 as the IoT gateway which bridges the individual Arduino-based sensor nodes with the cloud (Fig. 3).

5 Implementation

The Arduino Uno board is interfaced with temperature and humidity sensor DHT-11. The data from the sensor are read using the analog pins in Arduino. If the temperature and humidity value falls below the threshold value, the fan and pump will work accordingly. This flow is illustrated in Fig. 4. Similarly, if the sensor values are beyond the threshold, then the fan and pump will be shutdown.

A smart lighting system is implemented in this project. Through this system, the light inside the chamber is adjusted according to the ambient light. The different light cycles depending on the whether the plant is in growth phase or flowering phase are

Fig. 3 Microcontroller and sensor interfacing

Fig. 4 Temperature and humidity control flow

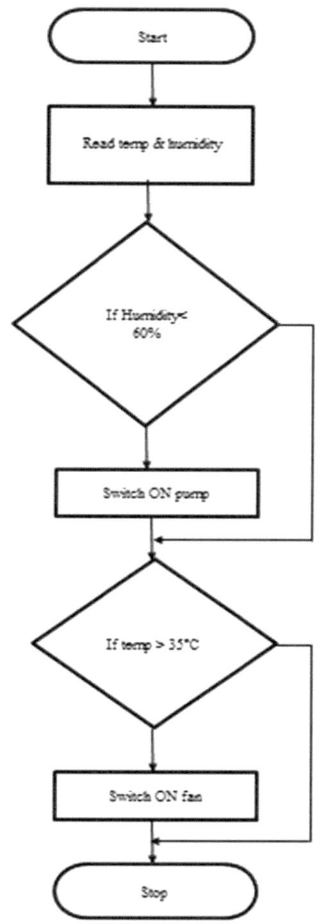

done manually, over the Internet using the cloud service. The time intervals for the different light cycles are implemented with the aid of an RTC module DS1307. The experimental setup used and the automation flow for lighting control is shown in Figs. 5 and 6.

6 Cloud Interfacing

We make use of the ThingSpeak cloud in order to implement data monitoring over the Internet [4]. ThingSpeak is a free to use cloud service tailored to implement IoT applications. HTTP POST and GET requests are used to communicate to the ThingSpeak cloud using the Raspberry Pi HUB. The dashboard of the ThingSpeak cloud service is shown in Fig. 7.

Fig. 5 Experimental setup

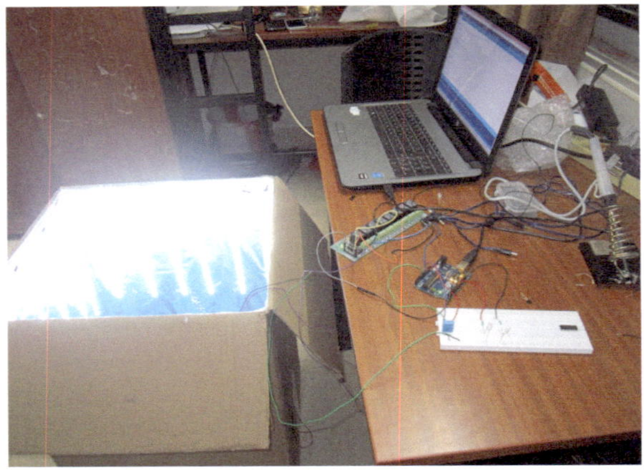

Fig. 6 Light control flow

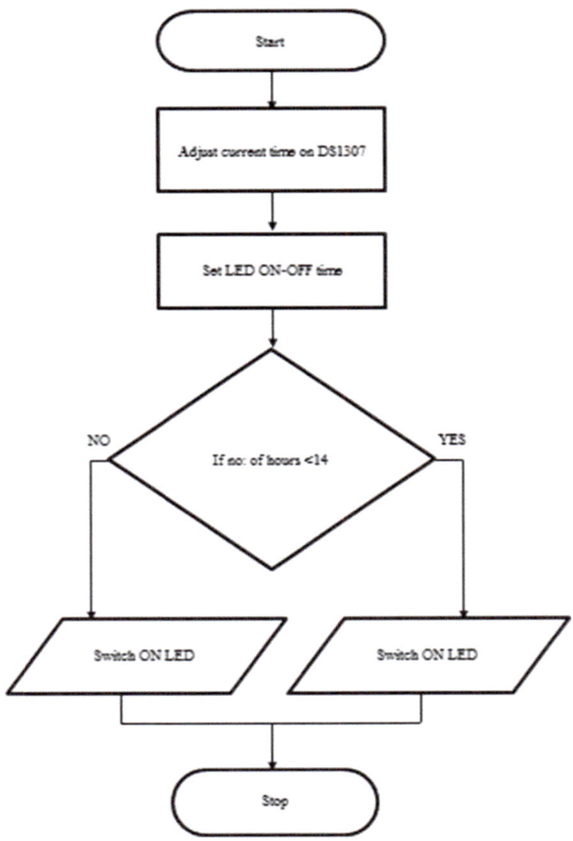

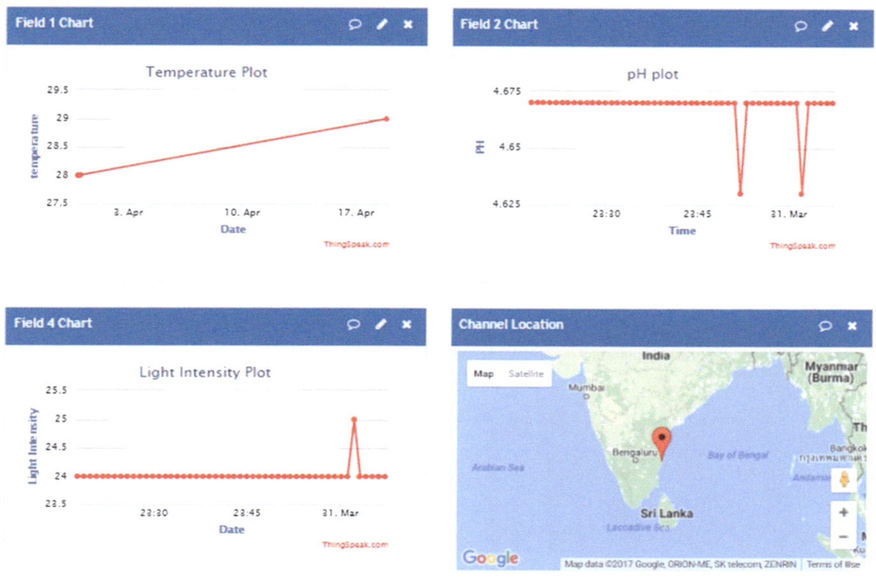

Fig. 7 ThingSpeak cloud interface for the proposed system

7 Conclusion

The monitoring and control system for distribution of water and nutrients in aeroponic, a growing chamber/grow bed, has been successfully designed and implemented. The system features individual data acquisition nodes to measure temperature, moisture, pH, and ambient light in order to control the parameters for optimal plant growth. The threshold values and light cycles can be set manual or over the IoT cloud service set up for the purpose of remote data monitoring.

References

1. Ahonen T, Virrankoski R, Elmusrati M (2008) Greenhouse monitoring with wireless sensor network. In: 2008 IEEE/ASME international conference on embedded systems and mechtronic and applications, Oct 2008. IEEE, pp 403–408
2. NASA Spinoff in 2006, Innovative Partnership Program, Publications and Graphics Department NASA Center for Aerospace Information (CASI), 2006
3. Hou J, Gao Y (2010) Greenhouse wireless sensor network monitoring system design based on solar energy. In: International conference on challenges in environmental science and computer engineering in 2010, vol 2. IEEE, pp 475–479
4. Abdul-Rahman AI, Graves CA (2016) Internet of things application using Tethered MSP430 to Thingspeak cloud. In: 2016 IEEE symposium on service-oriented system engineering (SOSE), Oxford, 2016, pp 352–357